Algorithms and Computation
in Mathematics · Volume 22

Editors

Arjeh M. Cohen   Henri Cohen
David Eisenbud Michael F. Singer   Bernd Sturmfels

Algorithms and Computation
in Mathematics · Volume 22

Editors

J. Rafael Sendra
Franz Winkler
Sonia Pérez-Díaz

Rational Algebraic Curves

A Computer Algebra Approach

With 24 Figures and 2 Tables

Authors

J. Rafael Sendra
Departamento de Matemáticas
Universidad de Alcalá
28871 Alcalá de Henares, Madrid
Spain
E-mail: rafael.sendra@uah.es

Franz Winkler
RISC-Linz
J. Kepler Universität Linz
4040 Linz
Austria
E-mail: franz.winkler@jku.at

Sonia Pérez-Díaz
Departamento de Matemáticas
Universidad de Alcalá
28871 Alcalá de Henares, Madrid
Spain
E-mail: sonia.perez@uah.es

Mathematics Subject Classification (2000): 14H50, 14M20, 14Q05, 68W30

ISSN 1431-1550
ISBN 978-3-642-09291-6   e-ISBN 978-3-540-73725-4

Springer is a part of Springer Science+Business Media
springer.com
© Springer-Verlag Berlin Heidelberg 2008
Softcover reprint of the hardcover 1st edition 2008

Cover design: WMXDesign GmbH, Heidelberg

Preface

Algebraic curves and surfaces are an old topic of geometric and algebraic investigation. They have found applications for instance in ancient and modern architectural designs, in number theoretic problems, in models of biological shapes, in error-correcting codes, and in cryptographic algorithms. Recently they have gained additional practical importance as central objects in computer-aided geometric design. Modern airplanes, cars, and household appliances would be unthinkable without the computational manipulation of algebraic curves and surfaces. Algebraic curves and surfaces combine fascinating mathematical beauty with challenging computational complexity and wide spread practical applicability.

In this book we treat only algebraic curves, although many of the results and methods can be and in fact have been generalized to surfaces. Being the solution loci of algebraic, i.e., polynomial, equations in two variables, plane algebraic curves are well suited for being investigated with symbolic computer algebra methods. This is exactly the approach we take in our book. We apply algorithms from computer algebra to the analysis, and manipulation of algebraic curves. To a large extent this amounts to being able to represent these algebraic curves in different ways, such as implicitly by defining polynomials, parametrically by rational functions, or locally parametrically by power series expansions around a point. All these representations have their individual advantages; an implicit representation lets us decide easily whether a given point actually lies on a given curve, a parametric representation allows us to generate points of a given curve over the desired coordinate fields, and with the help of a power series expansion we can for instance overcome the numerical problems of tracing a curve through a singularity.

The central problem in this book is the determination of rational parametrizability of a curve, and, in case it exists, the computation of a good rational parametrization. This amounts to determining the genus of a curve, i.e., its complete singularity structure, computing regular points of the curve in small coordinate fields, and constructing linear systems of curves with prescribed intersection multiplicities. Various optimality criteria for rational parametrizations of algebraic curves are discussed. We also point to some applications of these techniques in computer aided geometric design. Many of the symbolic algorithmic methods described in our book are implemented in the program system CASA, which is based on the computer algebra system Maple.

Our book is mainly intended for graduate students specializing in constructive algebraic curve geometry. We hope that researchers wanting to get a quick overview of what can be done with algebraic curves in terms of symbolic algebraic computation will also find this book helpful.

This book is the result of several years of research of the authors in the topic, and in consequence some parts of it are based on previous research published in journal papers, surveys, and conference proceedings (see [ReS97a], [Sen02], [Sen04], [SeW91], [SeW97], [SeW99], [SeW01a], [SeW01b]).

We gratefully acknowledge support of our work on this book by FWF (Austria) SFB F013/F1304, ÖAD (Austria) Acc.Int.Proj.Nr.20/2002, (Spain) Acc. Int. HU2001-0002, (Spain) BMF 2002-04402-C02-01, and (Spain) MTM 2005-08690-C02-01.

Alcalá de Henares and Linz,   *J. Rafael Sendra*
June 2007   *Franz Winkler*
                                                              Sonia Pérez-Díaz

Contents

1 Introduction and Motivation 1
  1.1 Intersection of Curves 4
  1.2 Generating Points on a Curve 5
  1.3 Solving Diophantine Equations 6
  1.4 Computing the General Solution
     of First-Order Ordinary Differential Equations 8
  1.5 Applications in CAGD 9

2 Plane Algebraic Curves 15
  2.1 Basic Notions 15
    2.1.1 Affine Plane Curves 16
    2.1.2 Projective Plane Curves 19
  2.2 Polynomial and Rational Functions 24
    2.2.1 Coordinate Rings and Polynomial Functions 24
    2.2.2 Polynomial Mappings 26
    2.2.3 Rational Functions and Local Rings 28
    2.2.4 Degree of a Rational Mapping 32
  2.3 Intersection of Curves 34
  2.4 Linear Systems of Curves 41
  2.5 Local Parametrizations and Puiseux Series 50
    2.5.1 Power Series, Places, and Branches 51
    2.5.2 Puiseux's Theorem and the Newton Polygon Method .. 55
    2.5.3 Rational Newton Polygon Method 61
  Exercises 62

3 The Genus of a Curve 67
  3.1 Divisor Spaces and Genus 67
  3.2 Computation of the Genus 69
  3.3 Symbolic Computation of the Genus 78
  Exercises 85

4   Rational Parametrization 87
    4.1   Rational Curves and Parametrizations 88
    4.2   Proper Parametrizations 95
    4.3   Tracing Index 100
        4.3.1   Computation of the Index of a Parametrization 101
        4.3.2   Tracing Index Under Reparametrizations 104
    4.4   Inversion of Proper Parametrizations 105
    4.5   Implicitization 108
    4.6   Parametrization by Lines 114
        4.6.1   Parametrization of Conics 114
        4.6.2   Parametrization of Curves with a Point of High
                Multiplicity 116
        4.6.3   The Class of Curves Parametrizable by Lines 118
    4.7   Parametrization by Adjoint Curves 119
    4.8   Symbolic Treatment of Parametrization 136
    Exercises 145

5   Algebraically Optimal Parametrization 149
    5.1   Fields of Parametrization 150
    5.2   Rational Points on Conics 154
        5.2.1   The Parabolic Case 155
        5.2.2   The Hyperbolic and the Elliptic Case 156
        5.2.3   Solving the Legendre Equation 157
    5.3   Optimal Parametrization of Rational Curves 169
    Exercises 185

6   Rational Reparametrization 187
    6.1   Making a Parametrization Proper 188
        6.1.1   Lüroth's Theorem and Proper Reparametrizations 188
        6.1.2   Proper Reparametrization Algorithm 190
    6.2   Making a Parametrization Polynomial 194
    6.3   Making a Parametrization Normal 200
    Exercises 207

7   Real Curves 209
    7.1   Parametrization 209
    7.2   Reparametrization 217
        7.2.1   Analytic Polynomial and Analytic Rational Functions .. 217
        7.2.2   Real Reparametrization 220
    7.3   Normal Parametrization 226
    Exercises 235

A   The System CASA 239

B   Algebraic Preliminaries 247
B.1   Basic Ring and Field Theory 247
B.2   Polynomials and Power Series 250
B.3   Polynomial Ideals and Elimination Theory 253
    B.3.1 Gröbner Bases 253
    B.3.2 Resultants 254
B.4   Algebraic Sets 256

References 257

Index 265

Table of Algorithms 269

1

Introduction and Motivation

Summary. In this first chapter, we informally introduce the notion of rational algebraic curves, and we motivate their use by means of some examples of applications. These examples cover the intersection of curves in Section 1.1, the generation of points on curves in Section 1.2, the solution of Diophantine equations in Section 1.3, the solution of certain differential equations in Section 1.4, and applications in computer aided geometric design in Section 1.5.

The theory of algebraic curves has a long and distinguished history, and there is a huge number of excellent books on this topic. In our book we concentrate on the computational aspects of algebraic curves, specially of rational algebraic curves, and we will frequently refer to classical literature. Moreover, our computational approach is not approximative but symbolic and based on computer algebra methods. That means we are dealing with exact mathematical descriptions of geometric objects and both the input and the output of algorithms are exact.

Our book is mainly intended for graduate students specializing in constructive algebraic curve geometry, as well as for researchers wanting to get a quick overview of what can be done with algebraic curves in terms of symbolic algebraic computation. Throughout this book we only consider algebraic curves. So, whenever we speak of a "curve" we mean an "algebraic curve."

In this first chapter, we informally introduce the notion of rational algebraic curves, and we motivate their use by means of some examples of applications.

When speaking about algebraic curves one may distinguish between algebraic plane curves and algebraic space curves. Nevertheless, it is well known (see for instance [Ful89], p. 155) that any space curve can be birationally projected onto a plane curve. This means that there exists a rationally invertible projection (in fact, almost all projections have this property), that maps the space curve onto a plane curve. Using such a projection and its inverse, which can be computed by means of elimination theory techniques, one may reduce the study of algebraic curves in arbitrary dimensional space to the study of

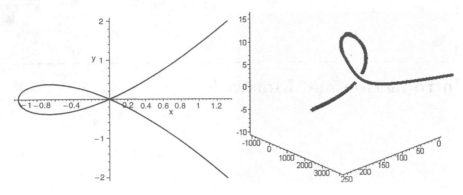

Fig. 1.1. $\pi_z(\mathcal{C}_3) = \mathcal{C}_2$ *(left)*, $\mathcal{C}_3$ *(right)*

plane algebraic curves. In fact, throughout this book we will consider plane algebraic curves, i.e., solution loci of nonconstant bivariate polynomials with coefficients in a field, say $\mathbb{C}$. In general, we will not work specifically over the complex numbers $\mathbb{C}$, but rather over an arbitrary algebraically closed field of characteristic zero.

Let us see an example of a birational projection of a space curve onto a plane curve. We consider in $\mathbb{C}^3$ the space curve $\mathcal{C}_3$ (see Fig. 1.1), defined as the intersection of the surfaces

$$g_1(x, y, z) = y + z - z^3, \quad g_2(x, y, z) = x + 1 - z^2;$$

that is, $\mathcal{C}_3 = \{(x, y, z) \in \mathbb{C}^3 \mid g_1(x, y, z) = g_2(x, y, z) = 0\}$. We consider the projection along the z-axis

$$\pi_z : \mathbb{C}^3 \longrightarrow \mathbb{C}^2; \quad (x, y, z) \mapsto (x, y).$$

$\pi_z(\mathcal{C}_3)$ is the plane curve $\mathcal{C}_2$ (see Fig. 1.1) defined by the polynomial

$$f(x, y) = x^3 + x^2 - y^2$$

(in fact, in this case, f is the resultant of g_1 and g_2 w.r.t. z); i.e. $\mathcal{C}_2 = \{(x, y) \in \mathbb{C}^2 \mid f(x, y) = 0\}$. The restriction of the projection π_z to the curve $\mathcal{C}_3$ is rationally invertible for all but finitely many points on $\mathcal{C}_2$. Indeed, the inverse is

$$\pi_z^{-1} : \mathcal{C}_2 \longrightarrow \mathcal{C}_3; \quad (x, y) \mapsto \left(x, y, \frac{y}{x}\right).$$

Some algebraic plane curves can be represented parametrically by means of rational functions. This means that a pair of rational functions $\chi_1(t), \chi_2(t) \in \mathbb{C}(t)$ generates all (except perhaps finitely many) points on the curve when the parameter t takes values in $\mathbb{C}$. This requirement is equivalent to the condition $f(\chi_1(t), \chi_2(t)) = 0$, assuming that not both rational functions are constant and that $f(x, y) = 0$ is the equation of the curve. Plane curves with this property are called rational curves, and their study is the central topic of this book. Only irreducible curves can be rational. The simplest example of

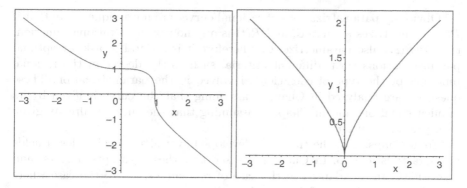

Fig. 1.2. $x^3 + y^3 = 1$ (*left*), $y^3 - x^2 = 0$ (*right*)

a rational curve is a line; the line with equation $ax + by + c = 0$ can be parametrized as $(bt, -at - c/b)$ if b is nonzero, and as $(-c/a, t)$ otherwise. Similarly we see (cf. Sect. 4.6) that all irreducible conics (i.e., plane curves defined by an irreducible polynomial in $\mathbb{C}[x, y]$ of degree 2) are rational. For instance, the circle defined by $x^2 + y^2 = 1$ can be parametrized as

$$\left(\frac{2t}{t^2 + 1}, \frac{t^2 - 1}{t^2 + 1} \right).$$

Therefore all irreducible plane curves of degree 1 or 2 are rational. However, curves of higher degree might or might not be rational. For instance, the cubic curve defined by $x^3 + y^3 = 1$ cannot be parametrized (see Example 4.3), while the cubic defined by $y^3 = x^2$ is parametrized as (t^3, t^2). A criterion for rationality is the genus of the curve (see Chap. 3). Intuitively speaking, the genus of an irreducible plane curve $\mathcal{C}$ of degree d measures the difference between the maximum number of singularities that an arbitrary irreducible curve of degree d may have and the actual number of singularities $\mathcal{C}$ has. For curves of degree 3 the maximum number of singularities is 1. This explains why the cubic defined by $y^3 - x^2 = 0$, having a double point at the origin, is rational, whereas the cubic defined by $x^3 + y^3 = 1$, having no singularity, is not rational (see Fig. 1.2). Of course, for determining the genus of a curve we have to view it in projective space. Note that the cubic defined by $y = x^3$ is smooth in the affine plane but can be parametrized as (t, t^3); it has a double point at infinity.

In this book we are interested in rational curves, and more precisely in the algorithmic treatment of this type of geometric objects. Once the basic notions have been introduced in Chap. 2, we start the study of rational curves with two different computational problems. The first one consists in deciding algorithmically whether a given curve is rational or not, and the second one deals with the question of actually computing rational parametrizations of rational curves. A solution of the rationality problem by means of the genus is described in Chap. 3. In Chap. 4 we give an algorithmic solution of the parametrization problem.

Obviously parametrizations of rational curves are not unique. Note that if $\mathcal{P}(t)$ parametrizes a curve $\mathcal{C}$, and $R(t)$ is any nonconstant rational function, then $\mathcal{P}(R(t))$ also parametrizes $\mathcal{C}$. Therefore it is natural to ask for optimal parametrizations w.r.t. different criteria, such as the degree of the rational functions or the type of coefficients involved in the parametrization. These questions are analyzed in Chap. 5, assuming that the curve is given by its implicit equation, and in Chap. 6, assuming that the curve is already given by a parametrization.

From Chaps. 2–6, the theory is developed over algebraically closed fields of characteristic zero. In Chap. 7, we see how these concepts, results, and algorithms can be adapted to the case of real curves; i.e. to the case where the reference field is the field of real numbers $\mathbb{R}$.

So now that we have an idea what rational curves are and which problems we might have to address when working with them, let us see what can be done with rational curves. The natural question is: if one is given a curve by means of its implicit equation, why do we need to generate a parametric representation of the curve? Of course, for some of us the sheer pleasure of developing the mathematical theory might be justification enough. But beyond this epistemological justification there are other good reasons for the parametrization of a curve. There exist problems in mathematics and its applications where the parametric representation of a curve is much more useful than its implicit representation; and vice versa.

In the following we briefly describe some mathematical problems where the use of parametrizations is helpful. By these examples we intend to convince the reader of the usefulness of rational curves. By no means do we claim to present an exhaustive list of applications. Algebraic curves appear in ancient and modern architectural designs, in number theoretic problems (see [PoV00] and [PoV02]), in biological shapes (see [BLM97]) in error-correcting codes (see [Gop77], [Gop81], [Pre98]), and in cryptographic algorithms (see [BSS99], [Buc01], [Kob98], [Kob02]). Moreover, recently they have gained additional practical importance as central objects in computer aided geometric design (see [Far93], [FHK02], [HoL93], [Sed98]): modern airplanes, cars, and household appliances would be unthinkable without the computational manipulation of algebraic curves and surfaces. Parametrizations also play a role in line integration, plotting, node distribution in polynomial interpolation (see [GMS02]), control theory (see [For92]), etc. The topic of rational algebraic curves and surfaces is an active research area. Recent advances can be found, for instance, in [Baj94], [CoS97b], [HSW97] and [ScS07]. We have chosen a few examples of applications which we want to describe briefly.

1.1 Intersection of Curves

Let us assume that two curves $\mathcal{C}_1$ and $\mathcal{C}_2$ are defined implicitly by the polynomials $f_1(x, y)$ and $f_2(x, y)$; i.e., $\mathcal{C}_i$ consists of the solutions of $f_i(x, y) = 0$. Also, let us assume that $\mathcal{C}_1$ and $\mathcal{C}_2$ do not have common components, i.e.

$\gcd(f_1, f_2) = 1$. In order to determine the intersection points, we compute the roots of $m(x) = \mathrm{res}_y(f_1, f_2)$, and for each root α of the resultant we compute the roots, say $\{\beta_i \,|\, i \in I_\alpha\}$, of $\gcd(f_1(\alpha, y), f_2(\alpha, y))$ (see Sect. 2.3). Finally the (affine) intersection points are

$$\{(\alpha, \beta_i) \,|\, m(\alpha) = 0, i \in I_\alpha\}.$$

This approach is simple and it is not worthy to parametrize the curves in order to solve the problem. However, if one of the curves is rational, and a rational parametrization of it is available, the process can be simplified as follows. Let $\mathcal{P}(t)$ be a parametrization of $\mathcal{C}_2$. Then we simply compute the roots, say $\{\alpha_1, \ldots, \alpha_s\}$, of the numerator of the rational function $f_2(\mathcal{P}(t))$ which are not roots of the denominators of the parametrization. What we get are the (affine) intersection points

$$\{\mathcal{P}(\alpha_i) \,|\, i = 1, \ldots, s\}.$$

In this process the normality of the parametrization (see Sects. 6.3 and 7.3) plays a role. If the parametrization is not normal then we should check whether the critical point is an intersection point.

1.2 Generating Points on a Curve

We start with a typical simple example. Let us consider the problem of generating points on a plane curve $\mathcal{C}$. If the curve is given implicitly by its defining polynomial, say $f(x, y)$, then we can generate points by intersecting $\mathcal{C}$ with a line; i.e., by computing the roots of a univariate polynomial of the form $g(t) = f(at + b, ct + d)$. For each root α of $g(t)$, the point $(a\alpha + b, c\alpha + d)$ is on the curve. In general, these points will have coordinates in an algebraic extension field whose degree is the degree of $\mathcal{C}$. However, if $\mathcal{C}$ is rational, and we have a parametrization $\mathcal{P}(t)$, then by giving values to the parameter we can easily generate points on the curve.

Moreover, let us now assume that the defining polynomial $f(x, y)$ has coefficients in a subfield of $\mathbb{C}$, say for instance $\mathbb{Q}$, and let us assume that the curve is rational. We propose the problem of deciding the existence of points on $\mathcal{C}$ with coordinates in $\mathbb{Q}$. This question can be answered with the algorithms described in Chap. 5. Furthermore, in Chap. 5, we prove that if there exists a simple point on $\mathcal{C}$ with coordinates in $\mathbb{Q}$, then the parametrization algorithm, given in Sect. 5.3, generates a rational parametrization of $\mathcal{C}$ with coefficients in $\mathbb{Q}$, and therefore taking rational parameter values one may generate infinity many points in $\mathcal{C}$ with coordinates in $\mathbb{Q}$. In addition, in the above situation, one may generate points of the curve with coordinates in any algebraic extension $\mathbb{Q}(\alpha)$ of $\mathbb{Q}$. In a similar way, we can generate points over $\mathbb{R}$. Note that these problems are difficult to approach if one only works with the implicit equation of the curve.

On the other hand, if we need to decide whether a point (a, b) is on the curve, we simply has to check whether $f(a, b)$ is zero, while using a parametrization $\mathcal{P}(t) = \left(\dfrac{\chi_{1,1}(t)}{\chi_{1,2}(t)}, \dfrac{\chi_{2,1}(t)}{\chi_{2,2}(t)} \right)$ one would have to check whether

$$\gcd(\chi_{1,2}(t)a - \chi_{1,1}(t), \chi_{2,2}(t)b - \chi_{2,1}(t)) \neq \text{const},$$

where we assume that $\gcd(\chi_{i,1}, \chi_{i,2}) = 1$.

1.3 Solving Diophantine Equations

Now, we show how parametrizations can be used to solve certain types of Diophantine equations. For further details on this application we refer to [PoV00], [PoV02].

We consider a polynomial $f(x, y) \in \mathbb{Z}[x, y]$ of total degree at least 3, such that the curve $\mathcal{C}$ defined by $f(x, y)$ is rational, and such that $\mathcal{C}$ has at least three valuations at infinity (a necessary and sufficient condition for this last requirement can be found in [PoV00]). In [PoV00], the authors present an algorithmic method for the explicit determination of all integer solutions of Diophantine equations of this type. This method is based on the construction of a rational parametrization with coefficients over $\mathbb{Q}$ of the curve $\mathcal{C}$ (see Sect. 5.3 on how to find such a parametrization), and on the practical solution of Thue equations (for solving Thue equations we refer to [TzW89] and [BiH96]).

For simplicity in the explanation, we assume that $(0 : 1 : 0)$ and $(1 : 0 : 0)$ are not points on the projective closure of $\mathcal{C}$. First we decide whether $\mathcal{C}$ can be parametrized over $\mathbb{Q}$ (see Chap. 5). If this is not the case, then the only integer solutions are the integer singular points of the curve. Otherwise, applying the algorithm in Sect. 5.3, we compute a rational proper parametrization of $\mathcal{C}$ over $\mathbb{Q}$ in reduced form,

$$\mathcal{P}(t) = \left(\frac{u(t)}{w_1(t)}, \frac{v(t)}{w_2(t)} \right) \in \mathbb{Q}(t).$$

Afterwards, we homogenize the rational functions of the parametrization, say

$$\mathcal{P}^*(t, s) = \left(\frac{U(t, s)}{W_1(t, s)}, \frac{V(t, s)}{W_2(t, s)} \right).$$

Now, because of our assumptions, either $W_1(t, s)$ or $W_2(t, s)$ have at least three different factors (see [PoV00] for further details). Let us assume w.l.o.g. that W_1 satisfies this property. Then, we compute the resultant $R_1 = \text{res}_t(U(t, 1), W_1(t, 1))$, and the greatest common divisor, δ_1, of the cofactors of the first column of the Sylvester matrix of $U(t, 1), W_1(t, 1)$. A similar strategy is applied to $U(1, s), W_1(1, s)$ to get R_2 and δ_2. Next we determine the integer solutions (t, s) with $\gcd(t, s) = 1$ and $t \geq 0$, of the Thue equations

$$W_1(t, s) = k,$$

where $k \in \mathbb{Z}$ divides $\mathrm{lcm}(R_1/\delta_1, R_2/\delta_2)$. Let us say that $\mathcal{S}$ is the set of integer solutions of these Thue equations. Then, the integer singular points of $\mathcal{C}$ and the points in $\{\mathcal{P}^*(t, s) \,|\, (t, s) \in \mathcal{S}\} \cap \mathbb{Z}^2$ are all the integer solutions of the equation $f(x, y) = 0$.

Let us see an example of this procedure. In fact, this is Example 4.1. in [PoV00]. Let n be a positive integer, and let $\mathcal{C}_n$ be the curve defined by the polynomial

$$f_n(x, y) = x^3 - (n-1)x^2 y - (n+2)xy^2 - y^3 - 2ny(x + y).$$

Applying the algorithms in Chap. 3, we check that all the curves $\mathcal{C}_n$ are rational (in fact, they are irreducible cubics with a double point at the origin). Performing the parametrization algorithms in Chap. 4, we derive the following parametrization of $\mathcal{C}_n$:

$$\mathcal{P}_n(t) = \left(\frac{2nt^2 + 2nt}{t^3 - (n-1)t^2 - (n+2)t - 1}, \frac{2nt + 2n}{t^3 - (n-1)t^2 - (n+2)t - 1} \right).$$

Now, we consider

$$U(n, t, s) = 2nt^2 s + 2nts^2, \quad V(n, t, s) = 2nts^2 + 2ns^3,$$

$$W(n, t, s) = t^3 - (n-1)t^2 s - (n+2)ts^2 - s^3.$$

Note that in this example, $W_1 = W_2 = W(n, t, s)$. Therefore,

$$\mathcal{P}_n^*(t, s) =$$
$$\left(\frac{2nt^2 s + 2nts^2}{t^3 - (n-1)t^2 s - (n+2)ts^2 - s^3}, \frac{2nts^2 + 2ns^3}{t^3 - (n-1)t^2 s - (n+2)ts^2 - s^3} \right).$$

The resultant of $U(n, t, 1), W(n, t, 1)$ is $R_1 = 8n^3$. The greatest common divisor δ_1 of the cofactors of the first column of the Sylvester matrix of $U(n, t, 1), W(n, t, 1)$ is $4n^2$. Thus, $R_1/\delta_1 = -2n$. Reasoning similarly with $U(n, 1, s), W(n, 1, s)$ we get that $R_2/\delta_2 = -2n$, and then $\mathrm{lcm}(R_1/\delta_1, R_2/\delta_2) = 2n$.

Finally, we compute the integer solutions (t, s), with $\gcd(t, s) = 1$ and $t \geq 0$, of the Thue equations $W(n, t, s) = k$, where k divides $2n$. By applying Theorem 3 in [MPL96] we get that $\mathcal{S} =$

$$\{(1, 0), (0, 1), (1, -1), (1, 1), (1, -2), (2, -1), (1, -n-1), (n, 1), (n+1, -n)\}.$$

Moreover,

$$\{\mathcal{P}_n^*(t, s) \,|\, (t, s) \in \mathcal{S}\} \cap \mathbb{Z}^2 = \{(0, 0) = \mathcal{P}_n^*(1, 0), \ (0, -2n) = \mathcal{P}_n^*(0, 1)\}.$$

Since the only singularity of $\mathcal{C}_n$ is $(0, 0)$, one deduces that the integer solutions to $f_n(x, y) = 0$ are $(0, 0)$ and $(0, -2n)$.

1.4 Computing the General Solution
of First-Order Ordinary Differential Equations

Let us show how to deal with the problem of deciding the existence, and actual computation, of rational general solutions of algebraic ordinary differential equations (for further details on this application we refer to [FeG04]).

Let $F(y, y')$ be a first order irreducible differential polynomial with coefficients in $\mathbb{Q}$. If

$$\overline{y} = \frac{\overline{a}_n x^n + \cdots + \overline{a}_0}{x^m + \overline{b}_{m-1} x^{m-1} + \cdots + \overline{b}_0},$$

is a nontrivial solution of $F(y, y') = 0$, where $\overline{a}_i, \overline{b}_j \in \mathbb{Q}$, and $\overline{a}_n \neq 0$, then

$$\hat{y} = \frac{\overline{a}_n (x + c)^n + \cdots + \overline{a}_0}{(x + c)^m + \overline{b}_{m-1}(x + c)^{m-1} + \cdots + \overline{b}_0},$$

is a general solution of $F(y, y') = 0$, where c is an arbitrary constant.

Therefore, the problem of finding a rational general solution is reduced to the problem of finding a nontrivial rational solution. For this purpose, we consider the polynomial $F(y, y_1) \in \mathbb{Q}[y, y_1]$. This polynomial defines an algebraic plane curve $\mathcal{C}$. Now, if $\overline{y} = r(x) \in \mathbb{Q}(t)$ is a nontrivial rational solution of $F(y, y') = 0$, then

$$\mathcal{P}(x) = (r(x), r'(x)) \in \overline{\mathbb{Q}}(x)^2$$

can be regarded as a rational parametrization of $\mathcal{C}$. In fact, one can see that $\mathcal{P}(x)$ is a proper parametrization of $\mathcal{C}$ (see Definition 4.12 for the notion of properness). In [FeG04] it is shown that given a proper rational parametrization $\mathcal{P}(x) = (r(x), s(x)) \in \overline{\mathbb{Q}}(x)^2$ of $\mathcal{C}$, the differential equation $F(y, y')$ has a rational solution if and only if one of the following relations:

$$ar'(x) = s(x) \quad \text{or} \quad a(x - b)^2 r'(x) = s(x), \tag{1.1}$$

is satisfied, where $a, b \in \overline{\mathbb{Q}}$, and $a \neq 0$. Moreover, if one of the above relations holds, replacing x by $a(x + c)$ or by $(ab(x + c) - 1)/(a(x + c))$, respectively, in $y(x) = r(x)$, one obtains a rational general solution of $F(y, y') = 0$, where c is an arbitrary constant. Using the results developed in Sect. 5.3, one may prove that if $F(y, y') = 0$ has a rational general solution, then the coefficients of the rational general solution are in $\mathbb{Q}$. For further developments of this problem see [ACFG05].

Let us see an example of this procedure. We consider the differential equation

$$F(y, y') = 229 - 144y + 16y(y')^2 + 16y^4 - 128y^2 + 4y(y')^3 + 4y^3 - 4y^3(y')^2$$
$$-y^2(y')^2 + 6(y')^2 + (y')^3 + (y')^4 = 0.$$

The curve $\mathcal{C}$ associated to the differential equation is defined by the polynomial $F(y, y_1) =$

$$229 - 144y + 16yy_1^2 + 16y^4 - 128y^2 + 4yy_1^3 + 4y^3 - 4y^3y_1^2 - y^2y_1^2 + 6y_1^2 + y_1^3 + y_1^4.$$

Applying the methods which will be developed in this book (see Chaps. 3–5), we check that $\mathcal{C}$ is rational and we determine the parametrization

$$(r(x), s(x)) := \left(\frac{x^3 + x^4 + 1}{x^2}, \frac{x^3 + 2x^4 - 2}{x} \right)$$

of $\mathcal{C}$. Now, we see that

$$\frac{s}{r'} = x^2.$$

Therefore, the second condition in (1.1) is satisfied with $a = 1, b = 0$. Substituting

$$\frac{ab(x + c) - 1}{a(x + c)} = \frac{-1}{x + c},$$

in $r(x)$ we get the following rational general solution of the differential equation:

$$\hat{y} = \frac{-x - c + 1 + x^4 + 4x^3c + 6x^2c^2 + 4xc^3 + c^4}{(x + c)^2}.$$

1.5 Applications in CAGD

Computer-aided geometric design (CAGD) is a natural environment for practical applications of algebraic curves and surfaces, and in particular of rational curves and rational surfaces. The widely used Bézier curves and surfaces are typical examples of rational curves and surfaces. Offsetting and blending of such geometrical objects lead to interesting problems. The reader may find explanations of these and other problems in the vast literature on CAGD, e.g., [ASS97], [ASS99], [Far93], [FHK02], [FaN90a], [FaN90b], [Har01], [Hof93], [HoL93], [Lü95], [PDS01], [PDS03], [PoW97], [SeS99], [SeS00].

Blending processes appear in the modeling of geometric objects. Usually, one models the object as a collection of surfaces. But, in many cases, one wants this collection to form a composite object whose surface is smooth. This question leads to the blending problem. In fact, a blending surface is a surface that provides a smooth transition between distinct geometric features of an object. Consequently the bending construction basically deals with algebraic surfaces. However, in addition to surfaces, we also encounter certain algebraic curves, called clipping curves, which describe the borders of the geometric features to be blended. When the problem is approached parametrically (see for instance [Har01], [PDS01], [PDS03], [PoW97]), parametrizations of the surfaces and of the clipping curves are required, and in particular polynomial

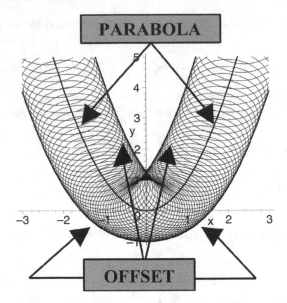

Fig. 1.3. Generation of the offset to the parabola

parametrizations. For this purpose, the results and algorithms developed in Sect. 6.2 can be applied.

The notion of an offset is directly related to the concept of an envelope. More precisely, the offset curve, at distance d, to an irreducible plane curve $\mathcal{C}$ is "essentially" the envelope of the system of circles centered at the points of $\mathcal{C}$ with fixed radius d (see Fig. 1.3). Offsets arise in practical applications such as tolerance analysis, geometric control, robot path-planning and numerical-control machining problems. Typically we may think of describing the curve that a cylindrical tool generates when it moves on a prescribed path.

Frequently offset processes are carried out with rational geometric objects, in particular with rational plane curves. However, in order to guarantee the computability of data structures and algorithms, rational parametrizations of offset curves are required. The main difficulty is that in general the rationality of the original curve is not preserved in the transition to the offset. For instance, while the parabola, the ellipse, and the hyperbola are rational curves (compare Fig. 1.4), the offset of a parabola is rational but the offset of an ellipse or a hyperbola is not rational.

In order to overcome this difficulty one may use different techniques such as Laguerre geometry (see [PeP98a], and [PeP98b]) or parametrization methods (see [ASS97]). Based on some of the algorithms presented in this book, the method described in [ASS97] solves this problem. Essentially, this method works as follows. Let $\mathcal{C}$ be the original rational curve. Let

$$\mathcal{P}(t) = (P_1(t), P_2(t))$$

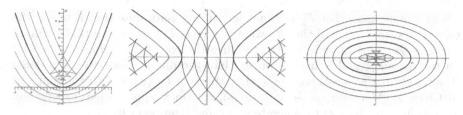

Fig. 1.4. Offset to the parabola (*left*), to the hyperbola (*center*), to the ellipse (*right*)

be a proper rational parametrization of $\mathcal{C}$. In practical applications $\mathcal{C}$ is assumed to be real, and one wants to work with a real parametrization. However, $\mathcal{P}(t)$ might have been produced by some previous process, as for instance an intersection problem, in which case it may happen that the input parametrization is not real. In this situation, the real reparametrization algorithm presented in Sect. 7.2 can be applied. It also may occur that $\mathcal{C}$ is given by means of its implicit equation, in which case the real parametrization algorithm in Section 7.1 can be applied.

Once $\mathcal{P}(t)$ is provided, one computes the normal vector associated to the parametrization $\mathcal{P}(t)$, namely

$$\mathcal{N}(t) := (-P_2'(t), P_1'(t)).$$

Note that the offset at distance d basically consist of the points of the form

$$\mathcal{P}(t) \pm \frac{d}{\sqrt{P_1'(t)^2 + P_2'(t)^2}} \mathcal{N}(t).$$

Now one checks whether this parametrization satisfies the "rational Pythagorean hodograph condition", i.e. whether

$$P_1'(t)^2 + P_2'(t)^2,$$

written in reduced form, is the square of a rational function in t. If the condition holds, then the offset to $\mathcal{C}$ has two components (see Chap. 2 for the precise definition of the concept of curve component), and both components are rational. In fact, these two components are parametrized as

$$\mathcal{P}(t) + \frac{d}{m(t)}\mathcal{N}(t), \quad \text{and} \quad \mathcal{P}(t) - \frac{d}{m(t)}\mathcal{N}(t),$$

where $m(t) = a(t)/b(t)$, assuming that $P_1'(t)^2 + P_2'(t)^2 = a(t)^2/b(t)^2$.

If the rational Pythagorean hodograph condition does not hold, then the offset is irreducible. In this case, in order to analyze whether the offset is rational, one associates to $\mathcal{P}(t)$ an auxiliary plane curve, namely the curve defined by the primitive part w.r.t. x_2 of the numerator of the rational function

$$x_2^2 P_1'(x_1) - P_1'(x_1) - 2x_2 P_2'(x_1),$$

where one assumes that $P_1'(t)$ is not identically zero; note that if this is the case then $\mathcal{C}$ is a line and the situation is trivial. Let us denote this auxiliary curve as $\mathcal{G_P}$. This curve is constructed directly from $\mathcal{P}(t)$, it does not depend on the distance, and is much simpler than the offset. The offset is rational if and only if $\mathcal{G_P}$ is rational (note that this can be check with the material in Chap. 3). Moreover, if $\mathcal{G_P}$ is rational, one may apply a parametrization algorithm (see Chap. 4) to generate a parametrization of it, say

$$\mathcal{R}(t) = (R_1(t), R_2(t)).$$

Then, if $\mathcal{Q}(t) := \mathcal{P}(R(t))$ and $\mathcal{M}(t) := (M_1(t), M_2(t))$ is the normal vector associated to $\mathcal{Q}(t)$, the rational function $M_1(t)^2 + M_2(t)^2$ can be written as the square of a rational function $A(t)/B(t)$, from where one deduces that the offset is parametrized as

$$\mathcal{Q}(t) \pm \frac{dB(t)}{A(t)} \mathcal{M}(t).$$

The study of offsets is an active research area, and many other topics related to algebraic plane curves such as genus, degree, singularities, intersections, implicitization, topology, etc., have been investigated for offset curves (see for instance [ASS07], [AlS07], [ASS97], [ASS99], [FaN90a], [FaN90b], [Hof93], [HSW97], [HoL93], [Lü95], [PeP98b], [PoW97], [Pot95], [SSeS05], [SeS99], [SeS00]).

Let us see an example of the process described above. We consider as initial curve the parabola of equation $y = x^2$, and its proper parametrization $\mathcal{P}(t) = (t, t^2)$. The normal vector associated to $\mathcal{P}(t)$ is $\mathcal{N}(t) = (-2t, 1)$. Now, we check the rational Pythagorean hodograph condition

$$P_1'(t)^2 + P_2'(t)^2 = 4t^2 + 1,$$

and we observe that $4t^2 + 1$ is not the square of a rational function. Therefore, the offset to the parabola is irreducible. We still have to analyze whether the offset is rational. For this purpose, we consider the auxiliary curve $\mathcal{G_P}$ whose implicit equation is

$$x_2^2 - 1 - 4x_2 x_1.$$

We observe that $\mathcal{G_P}$ is rational, and therefore the offset is rational. Moreover, a parametrization of $\mathcal{G_P}$ is

$$\mathcal{R} = \left(\frac{t^2 - 1}{4t}, t \right).$$

Therefore, the reparametrization $\mathcal{Q}(t)$ is

$$\mathcal{Q}(t) := \mathcal{P} \left(\frac{t^2 - 1}{4t} \right),$$

and applying the formula one gets that the offset to the parabola, at a generic distance d, can be parametrized as

$$\left(\frac{(t^2 + 1 - 4dt)(t^2 - 1)}{4t\,(t^2 + 1)},\ \frac{t^6 - t^4 - t^2 + 1 + 32dt^3}{16t^2\,(t^2 + 1)} \right).$$

The implicit equation of the offset to the parabola is

$$-y^2 + 32x^2d^2y^2 - 8x^2yd^2 + d^2 + 20x^2d^2 - 32x^2y^2 + 8d^2y^2 + 2yx^2 - 8yd^2$$
$$+ 48x^4d^2 - 16x^4y^2 - 48x^2d^4 + 40x^4y + 32x^2y^3 - 16d^4y^2 - 32d^4y$$
$$+ 32d^2y^3 - x^4 + 8d^4 + 8y^3 - 16x^6 + 16d^6 - 16y^4 = 0.$$

Of course, one might also first compute the implicit equation of the offset, and then apply directly the genus and parametrization algorithms in the book. Why would not this approach be preferable to the one described above? The answer is clear: first one has to apply elimination techniques for implicitizing, and this might be very time consuming. Second, in general the newly constructed offset curve is much more complicated, and its implicit equation has parametric coefficients in the distance d. However, with the approach we have described above, instead of treating the offset directly, we read the information from the original curve and an auxiliary curve much simpler than the offset. In our example, this should be obvious from a comparison of the equation defining $\mathcal{G}_{\mathcal{P}}$ and the equation of the offset for the parabola.

2

Plane Algebraic Curves

In this chapter we introduce some basic notions on plane algebraic curves, we derive some fundamental properties of algebraic curves, and we outline the general working environment of the book. This chapter consists of five sections. In Sect. 2.1, we present the basic notions on curves distinguishing between affine and projective curves. Section 2.2 is devoted to polynomial and rational functions. The material of this section is presented for the more general case of varieties (i.e., irreducible algebraic sets), and will play an important role in subsequent sections. In Sect. 2.3 we focus again on the case of plane curves. The study of the intersection of curves leads to the notion of multiplicity of intersection and to Bézout's theorem. Section 2.4 is devoted to the study of linear systems of curves. We will see in the following chapters that this notion is crucial for solving the problem of parametrizing a rational curve. The chapter ends with Sect. 2.5 where we show how to locally parametrize a curve around a point of the curve by means of Puiseux series. In addition, based on this fact, we introduce the notion of a place of a curve.

Throughout this chapter let K be an algebraically closed field of characteristic zero, and as usual let the affine plane $\mathbb{A}^2(K)$ be embedded into the projective plane $\mathbb{P}^2(K)$ by identifying the point $(a, b) \in \mathbb{A}^2(K)$ with the point $(a : b : 1) \in \mathbb{P}^2(K)$. Also throughout this book, we assume that the set of natural numbers $\mathbb{N}$ contains 0. Some important algebraic and geometric prerequisites are collected in Appendix B.

2.1 Basic Notions

In this section we introduce the basic notions and results on algebraic plane curves. We first deal with affine plane curves, and afterwards projective plane curves are considered.

2.1.1 Affine Plane Curves

An affine plane algebraic curve C over K is a hypersurface in $\mathbb{A}^2(K)$. Thus, it is an affine algebraic set defined by a nonconstant polynomial f in $K[x, y]$. The squarefree part of f defines the same curve C, so we might as well require the defining polynomial to be squarefree.

Definition 2.1. *An affine plane algebraic curve over K is defined as the set*

$$C = \{(a, b) \in \mathbb{A}^2(K) \mid f(a, b) = 0\}$$

for a nonconstant squarefree polynomial $f(x, y) \in K[x, y]$.

We call f the defining polynomial *of C (of course, a polynomial $g = cf$, for some nonzero $c \in K$, defines the same curve, so f is unique only up to multiplication by nonzero constants).*

We will write f as

$$f(x, y) = f_d(x, y) + f_{d-1}(x, y) + \cdots + f_0(x, y),$$

where $f_k(x, y)$ is a homogeneous polynomial (form) of degree k, and $f_d(x, y)$ is nonzero. The polynomials f_k are called the homogeneous components *of f, and d is called the* degree *of C, denoted by $\deg(C)$. Curves of degree 1 are called* lines, *of degree 2* conics, *of degree 3* cubics, *etc.*

If $f = \prod_{i=1}^{n} f_i$, where f_i are the irreducible factors of f, we say that the affine curve defined by each polynomial f_i is a component *of C. Furthermore, the curve C is said to be* irreducible *if its defining polynomial is irreducible.*

Throughout this book we only consider algebraic curves. So, whenever we speak of a "curve" we mean an "algebraic curve."

Sometimes in subsequent chapters we will need to consider curves with multiple components. This means that the given definition has to be extended to arbitrary polynomials $f = \prod_{i=1}^{n} f_i^{e_i}$, where f_i are the irreducible factors of f, and $e_i \in \mathbb{N}$ are their multiplicities. In this situation, the curve defined by f is the curve defined by its squarefree part, but the component generated by f_i carries multiplicity e_i. Whenever we will use this generalization we will always explicitly say so.

Definition 2.2. *Let C be an affine plane curve over K defined by $f(x, y) \in K[x, y]$, and let $P = (a, b) \in C$. We say that P is of multiplicity r on C if and only if all the derivatives of f up to and including the $(r-1)$th vanish at P but at least one rth derivative does not vanish at P. We denote the multiplicity of P on C by $\mathrm{mult}_P(C)$.*

P is called a simple point *on C iff $\mathrm{mult}_P(C) = 1$. If $\mathrm{mult}_P(C) = r > 1$, then we say that P is a* multiple *or* singular *point (or* singularity*) of multiplicity r on C or an r-fold point; if $r = 2$, then P is called a* double *point, if $r = 3$ a* triple *point, etc. We say that a curve is* nonsingular *if it has no singular points.*

Clearly $P \notin C$ if and only if $\text{mult}_P(C) = 0$. If C is a line, then for every $P \in C$ we have $\text{mult}_P(C) = 1$; i.e., C is nonsingular. The case of conics is investigated in Exercise 2.3. Furthermore, for every point $P \in C$ we have $1 \leq \text{mult}_P(C) \leq \deg(C)$.

The singularities of the curve C defined by f are the points of the affine algebraic set $V(f, \frac{\partial f}{\partial x}, \frac{\partial f}{\partial y})$. Later we will see that this set is 0-dimensional, i.e., every curve has only finitely many singularities.

We leave the proofs of the following two theorems as exercises.

Theorem 2.3. *Let the curve C be defined by f, $P \in C$, and T an invertible linear mapping on $\mathbb{A}^2(K)$ (i.e. a linear change of coordinates) s.t. $T(\tilde{P}) = P$. Let $\tilde{C}$ be defined by $\tilde{f} = f \circ T$. Then the multiplicity of P on C is the same as the multiplicity of $\tilde{P}$ on $\tilde{C}$.*

So the notion of multiplicity is invariant under linear changes of coordinates, cf. Definition 2.28.

Theorem 2.4. *Let C be an affine plane curve defined by $f(x, y)$. The multiplicity of C at the origin of $\mathbb{A}^2(K)$ is the minimum of the degrees of the nonzero homogeneous components of f.*

Hence, taking into account Theorem 2.3, the multiplicity of P can also be determined by moving P to the origin by means of a linear change of coordinates and applying Theorem 2.4.

Let $P = (a, b) \in \mathbb{A}^2(K)$ be an r-fold point ($r \geq 1$) on the curve C defined by the polynomial f. Then the first nonvanishing component in the Taylor expansion of f at P is

$$T_r(x, y) = \sum_{i=0}^{r} \binom{r}{i} \frac{\partial^r f}{\partial x^i \partial y^{r-i}}(P)(x - a)^i (y - b)^{r-i}.$$

By a linear change of coordinates which moves P to the origin the polynomial T_r is transformed to a homogeneous bivariate polynomial of degree r. Hence, since the number of factors of a polynomial is invariant under linear changes of coordinates, we get that all irreducible factors of T_r are linear. They are the tangents to the curve at P.

Definition 2.5. *Let C be an affine plane curve with defining polynomial $f(x, y)$, and $P = (a, b) \in \mathbb{A}^2(K)$ such that $\text{mult}_P(C) = r \geq 1$. Then the tangents to C at P are the irreducible factors of the polynomial*

$$\sum_{i=0}^{r} \binom{r}{i} \frac{\partial^r f}{\partial x^i \partial y^{r-i}}(P)(x - a)^i (y - b)^{r-i}$$

and the multiplicity of a tangent is the multiplicity of the corresponding factor.

For analyzing a singular point P on a curve C we need to know its multiplicity but also the multiplicities of the tangents at P. If all the r tangents at the r-fold point P are different, then this singularity is of well-behaved type. For instance, when we trace the curve through P we can simply follow the tangent and then approximate back onto the curve. This is not possible any more when some of the tangents are the same.

Definition 2.6. *A singular point P of multiplicity r on an affine plane curve C is called* ordinary *iff the r tangents to C at P are distinct, and* nonordinary *otherwise. We also say that the* character *of P is either ordinary of nonordinary.*

Theorem 2.7. *Let the curve C be defined by f, $P \in C$, and T an invertible linear mapping on $\mathbb{A}^2(K)$ (i.e., a linear change of coordinates) s.t. $T(\tilde{P}) = P$. Let $\tilde{C}$ be defined by $\tilde{f} = f \circ T$. Then T defines a 1–1 correspondence, preserving multiplicities, between the tangents to C at P and the tangents to $\tilde{C}$ at $\tilde{P}$.*

We leave the proof of this theorem as an exercise.

Corollary 2.8. *The character of a singular point is invariant under linear changes of coordinates.*

Lemma 2.9. *Let C be an affine plane curve defined by the squarefree polynomial $f = \prod_{i=1}^{n} f_i$, where all the factors f_i are irreducible. Let C_i be the component of C defined by f_i. Let P be a point in $\mathbb{A}^2(K)$. Then the following hold:*

(1) $\text{mult}_P(C) = \sum_{i=1}^{n} \text{mult}_P(C_i)$.
(2) If L is a tangent to C_i at P with multiplicity s_i, then L is a tangent to C at P with multiplicity $\sum_{i=1}^{n} s_i$.

Proof. (1) By Theorem 2.3 we may assume that P is the origin. Let

$$f_i(x, y) = \sum_{j=r_i}^{n_i} g_{i,j}(x, y) \quad \text{for} \quad i = 1, \dots, n,$$

where n_i is the degree of C_i, $r_i = \text{mult}_P(C_i)$, and $g_{i,j}$ is the homogeneous component of f_i of degree j. Then the lowest degree homogeneous component of f is $\prod_{i=1}^{n} g_{i,r_i}$. Hence, (1) follows from Lemma 2.4 (2).

(2) follows directly from Theorem 2.7 and from the expression of the lowest degree homogeneous component of f deduced in the proof of statement (1). $\square$

Theorem 2.10. *An affine plane curve has only finitely many singular points.*

Proof. Let C be an affine plane curve with defining polynomial f, let $f = f_1 \cdots f_r$ be the irreducible factorization of f, and let C_i be the component generated by f_i (note that f is squarefree, so the f_i's are pairwise relatively prime). From Lemma 2.9, we deduce that the singular points of $\mathcal{C}$ are the singular points of each component $\mathcal{C}_i$ and the intersection points of all pairs of components (i.e., the points in the affine algebraic sets $V(f_i, f_j)$, $i \neq j$). Hence the set W of singular points of $\mathcal{C}$ is

$$W = V\left(f, \frac{\partial f}{\partial x}, \frac{\partial f}{\partial y}\right) = \bigcup_{i=1}^{r} V\left(f_i, \frac{\partial f_i}{\partial x}, \frac{\partial f_i}{\partial y}\right) \cup \bigcup_{i \neq j} V(f_i, f_j).$$

Now, observe that $\gcd(f_i, f_j) = 1$ for $i \neq j$ and $\gcd(f_i, \frac{\partial f_i}{\partial x}, \frac{\partial f_i}{\partial y}) = 1$. Therefore (compare Appendix B), we conclude that W is finite. $\qquad\square$

2.1.2 Projective Plane Curves

A projective plane curve is a hypersurface in the projective plane.

Definition 2.11. *A projective plane algebraic curve over K is defined as the set*

$$\mathcal{C} = \{(a : b : c) \in \mathbb{P}^2(K) \mid F(a, b, c) = 0\}$$

for a nonconstant squarefree homogeneous polynomial $F(x, y, z) \in K[x, y, z]$.

We call F the defining polynomial *of $\mathcal{C}$ (of course, a polynomial $G = cF$, for some nonzero $c \in K$ defines the same curve, so F is unique only up to multiplication by nonzero constants).*

The notions of degree, component, and irreducibility (defined in Definition 2.1 for affine curves) can be adapted for projective curves in an obvious way. Also, as in the case of affine curves, we sometimes need to refer to multiple components of a projective plane curve. We introduce this notion by extending the concept of curve to arbitrary forms. We will also always explicitly indicate when we make use of this generalization.

The natural embedding of $\mathbb{A}^2(K)$ into $\mathbb{P}^2(K)$ induces a natural correspondence between affine and projective curves.

Definition 2.12. *The* projective plane curve $\mathcal{C}^*$ corresponding *to an affine plane curve $\mathcal{C}$ over K is the projective closure of $\mathcal{C}$ in $\mathbb{P}^2(K)$.*

If the affine curve $\mathcal{C}$ is defined by the polynomial $f(x, y)$, then (compare Appendix B) we immediately get that the corresponding projective curve $\mathcal{C}^*$ is defined by the homogenization $F(x, y, z)$ of $f(x, y)$. Therefore, if

$$f(x, y) = f_d(x, y) + f_{d-1}(x, y) + \cdots + f_0(x, y)$$

is the decomposition of f into forms, then

$$F(x, y, z) = f_d(x, y) + f_{d-1}(x, y)z + \cdots + f_0(x, y)z^d,$$

and
$$C^* = \{(a : b : c) \in \mathbb{P}^2(K) \mid F(a, b, c) = 0\}.$$

Every point (a, b) on $\mathcal{C}$ corresponds to a point on $(a : b : 1)$ on $\mathcal{C}^*$, and every additional point on $\mathcal{C}^*$ is a point at infinity. In other words, the first two coordinates of the additional points are the nontrivial solutions of $f_d(x, y)$, the third coordinate being 0. Thus, the curve $\mathcal{C}^*$ has only finitely many points at infinity. Of course, a projective curve, not associated to an affine curve, could have $z = 0$ as a component and therefore have infinitely many points at infinity.

On the other hand, associated to every projective curve there are infinitely many affine curves. We may take any line in $\mathbb{P}^2(K)$ as the line at infinity, by a linear change of coordinates move it to $z = 0$, and then dehomogenize. But in practice we mostly use dehomogenizations provided by taking the axes as lines at infinity. More precisely, if $\mathcal{C}$ is the projective curve defined by the form $F(x, y, z)$, we denote by $\mathcal{C}_{*,z}$ the affine plane curve defined by $F(x, y, 1)$. Similarly, we consider $\mathcal{C}_{*,y}$, and $\mathcal{C}_{*,x}$.

So, any point P on a projective curve $\mathcal{C}$ corresponds to a point on a suitable affine version of $\mathcal{C}$. The notions of multiplicity of a point and tangents at a point are local properties. So for determining the multiplicity of P at $\mathcal{C}$ and the tangents to $\mathcal{C}$ at P we choose a suitable affine plane (by dehomogenizing w.r.t. one of the projective variables) containing P, determine the multiplicity and tangents there, and afterwards homogenize the tangents to move them back to the projective plane. This process does not depend on the particular dehomogenizing variable (compare, for instance, [Ful89], Chap. 5). For the case of simple points one has the following explicit expression for the tangent line.

Theorem 2.13. *Let P be a simple point of the projective plane curve $\mathcal{C}$ of defining polynomial $F(x, y, z)$. Then*
$$x\frac{F}{\partial x}(P) + y\frac{F}{\partial y}(P) + z\frac{F}{\partial z}(P)$$

is the defining polynomial of the tangent to $\mathcal{C}$ at P.

Proof. We may assume w.l.o.g. that $P = (a : b : 1)$. The tangent line to $\mathcal{C}_{*,z}$ at (a, b) is given by
$$(x - a)\frac{f}{\partial x}(a, b) + (y - b)\frac{f}{\partial y}(a, b),$$

where $f(x, y) = F(x, y, 1)$. Therefore, the tangent to $\mathcal{C}$ at P is given by
$$(x - az)\frac{F}{\partial x}(P) + (y - bz)\frac{F}{\partial y}(P).$$

Now, the result follows applying Euler's formula to F at P (see Appendix B), i.e.,
$$0 = a\frac{F}{\partial x}(P) + b\frac{F}{\partial y}(P) + 1\frac{F}{\partial z}(P). \qquad \square$$

Projective singularities can be characterized as follows.

Theorem 2.14. $P \in \mathbb{P}^2(K)$ *is a singularity of the projective plane curve* $\mathcal{C}$ *defined by the homogeneous polynomial* $F(x, y, z)$ *if and only if* $\frac{\partial F}{\partial x}(P) = \frac{\partial F}{\partial y}(P) = \frac{\partial F}{\partial z}(P) = 0.$

Proof. Let $d = \deg(F)$. We may assume w.l.o.g. that P is not on the line at infinity $z = 0$, i.e., $P = (a : b : 1)$. Let $\mathcal{C}_{*,z}$ be the affine curve defined by $f(x, y) = F(x, y, 1)$ and let $P_* = (a, b)$ be the image of P in this affine version of the plane. P is a singular point of $\mathcal{C}$ if and only if P_* is a singular point of $\mathcal{C}_{*,z}$, i.e. if and only if

$$f(P_*) = \frac{\partial f}{\partial x}(P_*) = \frac{\partial f}{\partial y}(P_*) = 0.$$

But

$$\frac{\partial f}{\partial x}(P_*) = \frac{\partial F}{\partial x}(P), \quad \frac{\partial f}{\partial y}(P_*) = \frac{\partial F}{\partial y}(P).$$

Furthermore, by Euler's Formula for homogeneous polynomials we have

$$x \cdot \frac{\partial F}{\partial x}(P) + y \cdot \frac{\partial F}{\partial y}(P) + z \cdot \frac{\partial F}{\partial z}(P) = d \cdot F(P).$$

The theorem now follows at once.   □

By an inductive argument this theorem can be extended to higher multiplicities. We leave the proof as an exercise.

Theorem 2.15. $P \in \mathbb{P}^2(K)$ *is a point of multiplicity at least* r *on the projective plane curve* $\mathcal{C}$ *defined by the homogeneous polynomial* $F(x, y, z)$ *of degree* d *(where* $r \leq d$*) if and only if all the* $(r - 1)$*th partial derivatives of* F *vanish at* P.

We finish this section with an example where all these notions are illustrated.

Example 2.16. Let $\mathcal{C}$ be the projective plane curve over $\mathbb{C}$ defined by the homogeneous polynomial

$$\begin{aligned}
F(x, y, z) = {} & x^2 y^3 z^4 - y^6 z^3 - 4 x^2 y^4 z^3 - 4 x^4 y^2 z^3 + 3 y^7 z^2 + 10 x^2 y^5 z^2 \\
& + 9 x^4 y^3 z^2 + 5 x^6 y z^2 - 3 y^8 z - 9 x^2 y^6 z - 11 x^4 y^4 z - 7 x^6 y^2 z \\
& - 2 x^8 z + y^9 + 2 x^2 y^7 + 3 x^4 y^5 + 4 x^6 y^3 + 2 x^8 y .
\end{aligned}$$

The degree of the curve is 9. In Fig. 2.1 the real part of $\mathcal{C}_{*,z}$ is plotted.

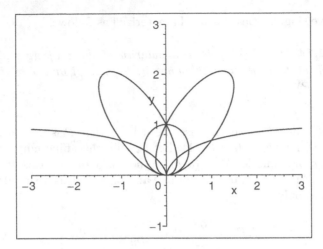

Fig. 2.1. Real part of $\mathcal{C}_{*,z}$

First, we compute the finitely many points at infinity of the curve. We observe that
$$F(x, y, 0) = y(2x^4 + y^4)(y^2 + x^2)^2$$
does not vanish identically, so the line $z = 0$ is not a component of $\mathcal{C}$. In fact, the points at infinity are $(1 : 0 : 0)$, $(1 : \alpha : 0)$ for $\alpha^4 + 2 = 0$, and the cyclic points $(1 : \pm i : 0)$. Hence, the line at infinity intersects $\mathcal{C}$ at seven points (compare to Bézout's Theorem, see Theorem 2.48).

Now, we proceed to determine and analyze the singularities. We apply Theorem 2.14. Solving the system
$$\left\{ \frac{\partial F}{\partial x} = 0, \frac{\partial F}{\partial y} = 0, \frac{\partial F}{\partial z} = 0 \right\}$$
we find that the singular points of $\mathcal{C}$ are
$$(1 : \pm i : 0), \quad (0 : 0 : 1), \quad \left(\pm \frac{1}{3\sqrt{2}} : \frac{1}{3} : 1 \right),$$
$$\left(\pm \frac{1}{2} : \frac{1}{2} : 1 \right), \quad (0 : 1 : 1), \quad \text{and} \quad (\pm 1 : \alpha : 1),$$
where $\alpha^3 + \alpha - 1 = 0$. So $\mathcal{C}$ has 14 singular points.

Now we compute the multiplicity of and the tangents to each singular point. For this purpose, we determine the first nonvanishing term in the

Table 2.1. Singular points of $\mathcal{C}$

Point	Tangents	Multiplicity
P_1^+	$(2y - z - 2ix)(2y + z - 2ix)$	2
P_1^-	$(2y + z + 2ix)(2y - z + 2ix)$	2
P_2	$y^3 x^2$	5
P_3^+	$(8x + \sqrt{2}z - 7\sqrt{2}y)(28x - 5\sqrt{2}z + \sqrt{2}y)$	2
P_3^-	$(8x - \sqrt{2}z + 7\sqrt{2}y)(28x + 5\sqrt{2}z - \sqrt{2}y)$	2
P_4^+	$(2x - z)(2x - 4y + z)$	2
P_4^-	$(2x + z)(2x + 4y - z)$	2
P_5	$(y - z)(3x^2 + 2yz - y^2 - z^2)$	3
P_α^+	$((1 + \frac{1}{2}\alpha + \frac{1}{2}\alpha^2)z + x + (-\frac{5}{2} - \frac{1}{2}\alpha - 2\alpha^2)y)$ $((-\frac{35}{146} - \frac{23}{73}\alpha + \frac{1}{73}\alpha^2)z + x + (-\frac{65}{146} - \frac{1}{73}\alpha - \frac{111}{146}\alpha^2)y)$	2
P_α^-	$((-1 - \frac{1}{2}\alpha - \frac{1}{2}\alpha^2)z + x + (\frac{5}{2} + \frac{1}{2}\alpha + 2\alpha^2)y)$ $((\frac{35}{146} + \frac{23}{73}\alpha - \frac{1}{73}\alpha^2)z + x + (\frac{65}{146} + \frac{1}{73}\alpha + \frac{111}{146}\alpha^2)y)$	2

corresponding Taylor expansion. The result of this computation is shown in Table 2.1. In this table we denote the singularities of $\mathcal{C}$ as follows:

$$P_1^\pm := (1 : \pm i : 0), \ P_2 := (0 : 0 : 1), \ P_3^\pm := \left(\pm\frac{1}{3\sqrt{2}} : \frac{1}{3} : 1\right),$$

$$P_4^\pm := \left(\pm\frac{1}{2} : \frac{1}{2} : 1\right), \ P_5 := (0 : 1 : 1), \ P_\alpha^\pm := (\pm 1 : \alpha : 1)$$

All these singular points are ordinary, except the affine origin P_2. Factoring F over $\mathbb{C}$ we get

$$F(x, y, z) = (x^2 + y^2 - yz)(y^3 + yx^2 - zx^2)(y^4 - 2y^3z + y^2z^2 - 3yzx^2 + 2x^4).$$

Therefore, $\mathcal{C}$ decomposes into a union of a conic, a cubic, and a quartic (see Fig. 2.1). Furthermore, P_2 is a double point on the quartic, a double point on the cubic, and a simple point on the conic. Thus, applying Lemma 2.9(1), the multiplicity of $\mathcal{C}$ at the point P_2 is 5. P_5 is a double point on the quartic and a simple point on the conic. Hence, the multiplicity of $\mathcal{C}$ at P_5 is 3. $P_4^\pm$ are simple points on the conic and the cubic. $P_3^\pm$ are simple points on the quartic and the cubic. Similarly, the points $P_\alpha^\pm$ are also simple points on the quartic and the cubic (two of them are real, and four of them complex). Finally, the cyclic points are simple points on the cubic and the conic. Hence, the singular points $P_1^\pm$, $P_3^\pm$, $P_4^\pm$, $P_\alpha^\pm$, are double points on $\mathcal{C}$.

2.2 Polynomial and Rational Functions

In this section, we do not just consider algebraic curves, but general varieties, i.e., irreducible algebraic sets in $\mathbb{A}^n(K)$, for a fixed n (see Appendix B). This section is devoted to the study of some notations and concepts which are important in subsequent chapters. These results allow to establish the bridge from curves to rational function fields. The reader may skip this section now and return to it when necessary.

All the rings and fields in this section will contain K as a subring. A homomorphism of such rings, $\varphi : R \longrightarrow S$, will always be a ring homomorphism which leaves K fixed.

2.2.1 Coordinate Rings and Polynomial Functions

We consider a variety V in $\mathbb{A}^n(K)$. Note that $I(V)$, the ideal of all polynomials in $K[x_1, \ldots, x_n]$ vanishing on V, is a prime ideal.

Definition 2.17. *Let $V \subseteq \mathbb{A}^n(K)$ be a variety and $I(V)$ its ideal. The integral domain*

$$\Gamma(V) = K[x_1, \ldots, x_n]/I(V)$$

is called the coordinate ring of V. So the elements of $\Gamma(V)$ are of the form $[g]_{I(V)}$, *i.e., equivalence classes modulo $I(V)$.*

Let $\mathcal{J}(V, K)$ be the set of all functions from V to K. The set $\mathcal{J}(V, K)$ becomes a ring if we define

$$(f + g)(x) = f(x) + g(x), \quad (f \cdot g)(x) = f(x) \cdot g(x),$$

for all $f, g \in \mathcal{J}(V, K)$, $x \in V$. The natural homomorphism from K into $\mathcal{J}(V, K)$, which maps a $\lambda \in K$ to the constant function $x \mapsto \lambda$, makes K a subring of $\mathcal{J}(V, K)$.

Definition 2.18. *Let* $V \subseteq \mathbb{A}^n(K)$ *be a variety. A function* $\varphi \in \mathcal{J}(V, K)$ *is called a* polynomial function *on* V, *iff there exists a polynomial* $f \in K[x_1, \ldots, x_n]$ *with*

$$\varphi(a_1, \ldots, a_n) = f(a_1, \ldots, a_n)$$

for all $(a_1, \ldots, a_n) \in V$. *In this case we say that* f represents *the function* φ.

The polynomial functions on a variety V form a subring of $\mathcal{J}(V, K)$ containing K (via the natural homomorphism). Two polynomials f, g represent the same function if and only if $(f - g)(P) = 0$ for all $P \in V$, i.e. $f - g \in I(V)$. So we can identify the polynomial functions on V with the elements of the coordinate ring $\Gamma(V)$.

There is an effective method for computing in $\Gamma(V)$, the coordinate ring of variety V, based on Gröbner bases. Compare Appendix B for the definition and some basic facts about Gröbner bases. Let G be a Gröbner basis for the prime ideal $I(V)$. Then

$$\Gamma(V) \simeq N_G = \{\, f \in K[x_1, \ldots, x_n] \mid f \text{ is in normal form w.r.t. } G \,\}.$$

So, if we have a Gröbner basis G for $I(V)$ w.r.t. any term ordering, then the irreducible terms w.r.t. G are representatives of the elements of $\Gamma(V)$. Addition in $\Gamma(V) = N_G$ is simply addition of the representatives, for multiplication we multiply the representatives and then reduce modulo the Gröbner basis G. If V is a hypersurface, then the ideal $I(V)$ is principal, and the defining polynomial of V is a Gröbner basis for $I(V)$. Hence, arithmetic in the coordinate ring $\Gamma(V)$ can be carried out by means of remainders.

Example 2.19. (a) If $V = \mathbb{A}^n(K)$, then $I(V) = \langle 0 \rangle$ and $\Gamma(V) = K[x_1, \ldots, x_n]$.
(b) Let $V \subseteq \mathbb{A}^n(K)$ be a variety. Then V is a single point if and only if $\Gamma(V) = K$.
(c) Let the hyperbola H be defined by $xy - 1$ in $\mathbb{A}^2(\mathbb{C})$. So

$$\Gamma(H) = \{\, f(x) + g(y) \mid f \in \mathbb{C}[x], \ g \in \mathbb{C}[y] \,\}.$$

(d) Let C be the circle in the space $\mathbb{A}^3(\mathbb{C})$ created by the intersection of the two spheres

$$x^2 + y^2 + z^2 - 9 \quad \text{and} \quad (x - 1)^2 + (y - 1)^2 + (z - 1)^2 - 9.$$

A Gröbner basis for $I(C)$ (w.r.t. to the lexicographic term ordering with $x > y > z$) is

$$G = \{2x + 2y + 2z - 3, \ 8y^2 + 8yz - 12y + 8z^2 - 12z - 27\}.$$

So

$$f = yz + z^2 + 1, \quad g = y + z - 1$$

are two elements of $\Gamma(C)$ in reduced form modulo G. We get the representation of the product modulo G as

$$f \cdot g = yz^2 + \frac{1}{2}yz + y + \frac{1}{2}z^2 + \frac{35}{8}z - 1.$$

There is a 1–1 correspondence between ideals in $\Gamma(V)$ and superideals of $I(V)$ in $K[x_1, \ldots, x_n]$; i.e., ideals of $K[x_1, \ldots, x_n]$ containing $I(V)$ (compare, for instance, Proposition 10 of Chap. 5, pp.223, in [CLO97]). Therefore, the following statement follows from the Noetherianity of $K[x_1, \ldots, x_n]$.

Theorem 2.20. *Let V be a variety. Then $\Gamma(V)$ is a Noetherian ring.*

We do not introduce coordinate rings of projective varieties here. The interested reader is refereed to [Ful89].

2.2.2 Polynomial Mappings

Definition 2.21. *Let $V \subseteq \mathbb{A}^n(K), W \subseteq \mathbb{A}^m(K)$ be varieties. A function $\varphi : V \to W$ is called a* polynomial *or* regular *mapping iff there are polynomials $f_1, \ldots, f_m \in K[x_1, \ldots, x_n]$ such that $\varphi(P) = (f_1(P), \ldots, f_m(P))$ for all $P \in V$.*

Theorem 2.22. *Let $V \subseteq \mathbb{A}^n(K), W \subseteq \mathbb{A}^m(K)$ be varieties. There is a natural 1–1 correspondence between the polynomial mappings $\varphi : V \to W$ and the homomorphisms $\tilde{\varphi} : \Gamma(W) \to \Gamma(V)$.*

Proof. Let $\varphi : V \to W$ be regular. With φ we associate the homomorphism

$$\tilde{\varphi} : \Gamma(W) \to \Gamma(V)$$
$$f \mapsto f \circ \varphi .$$

The map $\tilde{\ } : \varphi \to \tilde{\varphi}$ is 1-1 and onto. Cf. [Ful89], Chap.II.2, for details.   □

Definition 2.23. *A regular mapping $\varphi : V \to W$ is a* regular isomorphism *iff there is a regular mapping $\psi : W \to V$, such that*

$$\varphi \circ \psi = \mathrm{id}_W \quad and \quad \psi \circ \varphi = \mathrm{id}_V.$$

In this case the varieties V and W are regularly isomorphic *(via φ, ψ).*

Theorem 2.24. *V and W are regularly isomorphic via φ if and only if $\tilde{\varphi} : \Gamma(W) \to \Gamma(V)$ is an isomorphism of K-algebras.*

Proof. By Theorem 2.22 $\tilde{\varphi}$ is a homomorphism. Let $\psi : W \to V$ be such that $\varphi \circ \psi = \mathrm{id}_W, \psi \circ \varphi = \mathrm{id}_V$. Then, $\tilde{\psi}$ is a homomorphism from $\Gamma(V)$ to $\Gamma(W)$. $\tilde{\varphi} \circ \tilde{\psi} : \Gamma(V) \to \Gamma(V)$ is the identity on $\Gamma(V)$, since

$$\tilde{\varphi} \circ \tilde{\psi}(f) = \tilde{\varphi}(f \circ \psi) = f \circ \psi \circ \varphi = f.$$

Analogously we get that $\tilde{\psi} \circ \tilde{\varphi} = \mathrm{id}_{\Gamma(W)}$. Thus, $\tilde{\varphi}$ is an isomorphism.

Conversely, if $\tilde{\lambda}$ is an isomorphism from $\Gamma(W)$ to $\Gamma(V)$, then the corresponding λ is an isomorphism from V to W.   □

Example 2.25. (a) Let the (generalized) parabola $V \subset \mathbb{A}^2(\mathbb{C})$ be defined by $y - x^k$. V and $\mathbb{A}^1(\mathbb{C})$ are isomorphic via

$$\varphi : \quad V \quad \to \quad \mathbb{A}^1(\mathbb{C}) \qquad , \qquad \psi : \quad \mathbb{A}^1(\mathbb{C}) \quad \to \quad V$$
$$(x, y) \quad \mapsto \quad x \qquad\qquad\qquad\qquad t \quad \mapsto \quad (t, t^k) \quad .$$

(b) The projection $\varphi(x, y) = x$ of the hyperbola $xy - 1$ to the x-axis $(\mathbb{A}^1(\mathbb{C}))$ is <u>not</u> an isomorphism. There is no point (x, y) on the hyperbola such that $\varphi(x, y) = 0$.

(c) Let $V \subset \mathbb{A}^2(\mathbb{C})$ be defined by $y^2 - x^3$. Then $\Gamma(V) \cong \{p(x) + q(x)y | p, q \in \mathbb{C}[x]\}$. The mapping $\varphi : t \mapsto (t^2, t^3)$ from $\mathbb{A}^1(\mathbb{C})$ to V is 1–1, but not an isomorphism. Otherwise, we would have that $\tilde{\varphi} : \Gamma(V) \to \Gamma(\mathbb{A}^1(\mathbb{C})) = \mathbb{C}[t]$ is an isomorphism. But for arbitrary $p, q \in \mathbb{C}[x]$ we have $\tilde{\varphi}(p(x) + q(x)y) = p(t^2) + q(t^2)t^3 \neq t$.

Theorem 2.26. *Let $V \subseteq \mathbb{A}^n(K), W \subseteq \mathbb{A}^m(K)$ be varieties. Let $\varphi : V \to W$ be a surjective regular mapping, X an algebraic subset of W.*

(a) *$\varphi^{-1}(X)$ is an algebraic subset of V.*
(b) *If $\varphi^{-1}(X)$ is irreducible, then also X is irreducible.*

Proof. (a) Let $\varphi = (\varphi_1, \ldots, \varphi_m)$, $f_1 = \cdots = f_r = 0$ be the defining equations for X, and $g_i = f_i(\varphi_1, \ldots, \varphi_m) \in K[x_1, \ldots, x_n]$ for $1 \leq i \leq r$. Let $P = (a_1, \ldots, a_n)$ be an arbitrary point in V. Then

$$P \in \varphi^{-1}(X) \Longleftrightarrow \varphi(P) \in X \Longleftrightarrow g_1(P) = \cdots = g_r(P) = 0.$$

(b) If $X = X_1 \cup X_2$, then $\varphi^{-1}(X) = \varphi^{-1}(X_1) \cup \varphi^{-1}(X_2)$. If $X_1 \not\subseteq X_2$ then $\varphi^{-1}(X_1) \not\subseteq \varphi^{-1}(X_2)$. So if X is reducible, then so is $\varphi^{-1}(X)$. $\qquad\square$

Example 2.27. We show that $V = V(y - x^2, z - x^3) \subset \mathbb{A}^3(\mathbb{C})$ is a variety. The regular mapping

$$\varphi : \quad \mathbb{A}^1(\mathbb{C}) \quad \to \quad V$$
$$t \quad \mapsto \quad (t, t^2, t^3)$$

is surjective and $\varphi^{-1}(V) = \mathbb{A}^1(\mathbb{C})$ is irreducible. So by Theorem 2.26 also V is irreducible.

There are some kinds of very frequently used and important regular mappings. One such kind of mappings are the *projections*

$$\text{pr} : \quad \mathbb{A}^n(K) \quad \to \quad \mathbb{A}^r(K)$$
$$(a_1, \ldots, a_n) \quad \mapsto \quad (a_1, \ldots, a_r),$$

for $n \geq r$.
Let $V \subseteq \mathbb{A}^n(K)$ be a variety, $f \in \Gamma(V)$. Let

$$G(f) = \{(a_1, \ldots, a_{n+1}) \mid (a_1, \ldots, a_n) \in V, \ a_{n+1} = f(a_1, \ldots, a_n)\} \subseteq \mathbb{A}^{n+1}(K)$$

be the *graph* of f. $G(f)$ is an affine variety, and

$$\varphi : \quad \begin{array}{ccc} V & \rightarrow & G(f) \\ (a_1, \ldots, a_n) & \mapsto & (a_1, \ldots, a_n, f(a_1, \ldots, a_n)) \end{array}$$

is an isomorphism between V and $G(f)$. The projection from $\mathbb{A}^{n+1}(K)$ to $\mathbb{A}^n(K)$ is the inverse of φ.

Another important kind of regular mappings are changes of coordinates.

Definition 2.28. *An* affine change of coordinates *in $\mathbb{A}^n(K)$ is a bijective linear polynomial mapping, i.e. a bijective mapping of the form*

$$T : \quad \begin{array}{ccc} \mathbb{A}^n(K) & \rightarrow & \mathbb{A}^n(K) \\ (a_1, \ldots, a_n) & \mapsto & (T_1(a_1, \ldots, a_n), \ldots, T_n(a_1, \ldots, a_n)), \end{array}$$

where $\deg(T_i) = 1$ for $1 \leq i \leq n$. A projective change of coordinates *in $\mathbb{P}^n$ is a bijective linear mapping of the form*

$$T : \quad \begin{array}{ccc} \mathbb{P}^n(K) & \rightarrow & \mathbb{P}^n(K) \\ (a_1 : \ldots : a_{n+1}) & \mapsto & (T_1(a_1 : \ldots : a_{n+1}) : \ldots : T_n(a_1 : \ldots : a_{n+1})), \end{array}$$

where T_i is a linear form for $1 \leq i \leq n$.

Using column notation for the coordinates of points, every linear polynomial mapping from $\mathbb{A}^n(K)$ into itself can be written as

$$T(\overline{x}) = A \cdot \overline{x} + b$$

for some matrix A and vector b. T is an affine change of coordinates, if and only if A is an invertible matrix.

Affine geometry is the geometry of properties which are invariant under affine changes of coordinates, whereas in *projective geometry* the properties are invariant under projective changes of coordinates.

2.2.3 Rational Functions and Local Rings

The coordinate ring $\Gamma(V)$ of a variety $V \subseteq \mathbb{A}^n(K)$ is an integral domain. So it can be embedded into its quotient field.

Definition 2.29. *The* field of rational functions $K(V)$ *on a variety $V \subseteq \mathbb{A}^n(K)$ is the quotient field of $\Gamma(V)$. So*

$$K(V) \simeq \left\{ \frac{f}{g} \mid f, g \in K[x_1, \ldots, x_n], g \notin I(V) \right\} / \sim ,$$

where $\dfrac{f}{g} \sim \dfrac{f'}{g'} \iff fg' - f'g \in I(V)$.

Definition 2.30. *A* rational function *$\varphi \in K(V)$ is defined or* regular *at $P \in V$ iff φ can be written as $\varphi = f/g$ with $g(P) \neq 0$. In this case $f(P)/g(P)$ is the* value *of φ at P. The set of points in V at which a rational function φ is defined is called the* domain of definition *of φ. A point $P \in V$ at which the function φ is not defined is a* pole *of φ. For $P \in V$ the* local ring *of V at P is defined as $\mathcal{O}_P(V) = \{\varphi \in K(V) \mid \varphi$ regular at $P\}$.*

The notion of value of a rational function at a point on a variety is well defined. The local ring $\mathcal{O}_P(V)$ is indeed a local ring in the sense of having a unique maximal ideal. This maximal ideal is the subset of $\mathcal{O}_P(V)$ containing those rational functions which vanish on P. One easily verifies that $\mathcal{O}_P(V)$ is a subring of $K(V)$ containing $\Gamma(V)$. So we have the following increasing chain of rings:

$$K \subseteq \Gamma(V) \subseteq \mathcal{O}_P(V) \subseteq K(V).$$

Example 2.31. Let V be the unit circle in $\mathbb{A}^2(\mathbb{C})$ defined by $x^2 + y^2 - 1$. The rational function

$$\varphi(x, y) = \frac{1-y}{x}$$

is obviously regular at all points of V except possibly $(0, \pm 1)$. But φ is also regular at $(0, 1)$, which can be seen by the following transformation:

$$\frac{1-y}{x} = \frac{x(1-y)}{x^2} = \frac{x(1-y)}{1-y^2} = \frac{x}{1+y}.$$

Theorem 2.32. *The set of poles of a rational function φ on a variety $V \subseteq \mathbb{A}^n(K)$ is an algebraic set.*

Proof. Consider $J_\varphi = \{g \in K[x_1, \ldots, x_n] \mid g\varphi \in \Gamma(V)\}$. J_φ is an ideal in $K[x_1, \ldots, x_n]$ containing $I(V)$. The points of $V(J_\varphi)$ are exactly the poles of φ: if $P \in V(J_\varphi)$, then for every representation $\varphi = f/g$ we have $g \in J_\varphi$, so $g(P) = 0$, and therefore P is a pole. On the other hand, if $P \notin V(J_\varphi)$, then for some $g \in J_\varphi$ we have $g(P) \neq 0$. So there is an $r \in \Gamma(V)$ such that $\varphi = r/g$ and $g(P) \neq 0$, i.e., P is not a pole. □

Theorem 2.33. *A rational function $\varphi \in K(V)$, which is regular on every point of the variety V, is a regular function on V. So*

$$\Gamma(V) = \bigcap_{P \in V} \mathcal{O}_P(V).$$

Proof. If φ is regular on every point of V, then $V(J_\varphi) = \emptyset$ (proof of Theorem 2.32). So, by Hilbert's Nullstellensatz (see Appendix B) we have $1 \in J_\varphi$, i.e. $1 \cdot \varphi = \varphi \in \Gamma(V)$. □

Example 2.34 (Example 2.31 continued). $\varphi = (1-y)/x$ cannot be regular on the whole circle, because otherwise, by Theorem 2.33, there should be a polynomial $p(x,y) \in \mathbb{C}[x,y]$ such that

$$\varphi(x,y) = \frac{1-y}{x} = p(x,y).$$

This would mean $1 - y - x \cdot p(x,y) \in I(V) = \langle x^2 + y^2 - 1 \rangle$, or, equivalently, $1 - y \in \langle x^2 + y^2 - 1, x \rangle = \langle y^2 - 1, x \rangle$, which is impossible, as can be seen from the theory of Gröbner bases.

As we have extended the notion of a regular function to that of a regular mapping, we will now extend the notion of a rational function on a variety to that of a rational mapping on the variety.

Definition 2.35. *Let* $V \subseteq \mathbb{A}^n(K), W \subseteq \mathbb{A}^m(K)$ *be varieties. An m-tuple* φ *of rational functions* $\varphi_1, \ldots, \varphi_m \in K(V)$ *with the property that for an arbitrary point* $P \in V$, *at which all the* φ_i *are regular, we have* $(\varphi_1(P), \ldots, \varphi_m(P)) \in W$, *is called a* rational mapping *from* V *to* W, $\varphi : V \to W$. φ *is regular at* $P \in V$ *iff all the* φ_i *are regular at* P.

So, a rational mapping is not a mapping of the whole variety V into W, but of a certain nonempty open (in the Zariski topology) subset $U \subseteq V$ into W.

Example 2.36. Let $\mathcal{C}$ be the curve in $\mathbb{A}^2(\mathbb{C})$ defined by (see Fig. 2.2)

$$f(x,y) = y^2 - x^3 - x^2.$$

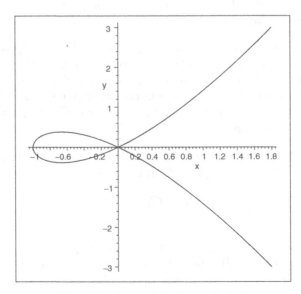

Fig. 2.2. Curve $\mathcal{C}$ in Example 2.36

The tuple of rational (in fact polynomial) functions

$$\varphi_1(t) = t^2 - 1, \quad \varphi_2(t) = t(t^2 - 1)$$

determines a rational mapping φ from $\mathbb{A}^1(\mathbb{C})$ to $\mathcal{C}$. This rational mapping has a rational inverse, i.e., a rational mapping from $\mathcal{C}$ to $\mathbb{A}^1(\mathbb{C})$:

$$\chi(x, y) = y/x.$$

We check that χ really is the inverse of φ:

$$\chi(\varphi_1(t), \varphi_2(t)) = \frac{t(t^2 - 1)}{t^2 - 1} = t,$$

$$\varphi_1(\chi(x, y)) = \frac{y^2}{x^2} - 1 = \frac{y^2 - x^2}{x^2} = \frac{x^3}{x^2} = x,$$

$$\varphi_2(\chi(x, y)) = \frac{y}{x} \cdot \left(\frac{y^2}{x^2} - 1 \right) = \frac{y(y^2 - x^2)}{x^3} = y.$$

So, up to finitely many exceptions, the points in $\mathbb{A}^1(\mathbb{C})$ and $\mathcal{C}$ correspond uniquely to each other.

Definition 2.37. *Let the rational mapping $\varphi : V \to W$ have a rational inverse, i.e., a rational mapping $\psi : W \to V$ such that $\psi \circ \varphi = \mathrm{id}_V$, $\varphi \circ \psi = \mathrm{id}_W$ (wherever the composition of these mappings is defined), and $\varphi(V), \psi(W)$ are dense in W, V, respectively. In this case φ is called a* birational isomorphism *from V to W (and ψ a birational isomorphism from W to V), and V and W are* birationally isomorphic *or* birationally equivalent.

Isomorphism of varieties is reflected in the function fields of these varieties.

Theorem 2.38. *The varieties V and W are birationally isomorphic if and only if the corresponding function fields $K(V)$ and $K(W)$ are isomorphic.*

Proof. Let $V \subseteq \mathbb{A}^n(K)$ and $W \subseteq \mathbb{A}^m(K)$. Let $\varphi : V \to W$ be a birational isomorphism from V to W and let $\psi : W \to V$ be its inverse. Consider the following homomorphisms between the function fields:

$$\tilde{\varphi} : K(W) \longrightarrow K(V) \qquad \tilde{\psi} : K(V) \longrightarrow K(W)$$
$$r \longmapsto r \circ \varphi \qquad\qquad s \longmapsto s \circ \psi$$

(Actually by $r \circ \varphi$ we mean the rational function on $K(W)$ whose restriction to the dense subset $\varphi(V)$ of W is $r \circ \varphi$, and analogously for $s \circ \psi$.) These homomorphisms $\tilde{\varphi}$ and $\tilde{\psi}$ are inverses of each other, so we have an isomorphism of the function fields.

On the other hand, let α be an isomorphism from $K(V)$ to $K(W)$, and β its inverse, i.e.,

$$\alpha : K(V) \to K(W), \qquad \beta : K(W) \to K(V).$$

Let $x_1, \ldots, x_n$ and $y_1, \ldots, y_m$ be the coordinate functions of V and W, respectively. Then

$$\tilde{\beta} = (\beta(y_1)(x_1, \ldots, x_n), \ \ldots, \beta(y_m)(x_1, \ldots, x_n))$$

is a birational isomorphism from V to W and

$$\tilde{\alpha} = (\alpha(x_1)(y_1, \ldots, y_m), \ \ldots, \alpha(x_n)(y_1, \ldots, y_m))$$

is its inverse from W to V. □

2.2.4 Degree of a Rational Mapping

Now let us investigate the degree of rational mappings between varieties. Intuitively speaking, the degree measures how often the mapping traces the image variety. The interested reader is advised to check out [SeW01b] for more details.

Definition 2.39. *Let W_1 and W_2 be varieties over K. Let $\phi : W_1 \to W_2$ be a rational mapping such that $\phi(W_1) \subset W_2$ is dense. Then ϕ is a* dominant *mapping from W_1 to W_2.*

Now, let us assume that $\dim(W_1) = \dim(W_2)$ (cf. Appendix B for the notion of dimension), and we consider the monomorphism induced by a dominant rational mapping ϕ from W_1 to W_2, and the field extensions

$$K \subset \tilde{\phi}(K(W_2)) \subset K(W_1).$$

Then, since the transcendence degree of field extensions is additive, taking into account that $\dim(W_1) = \dim(W_2)$ and that ϕ is dominant, one has that the transcendence degree of $K(W_1)$ over $\tilde{\phi}(K(W_2))$ is zero, and hence the extension is algebraic. Moreover, since $K(W_1)$ can be obtained by adjoining to $\tilde{\phi}(K(W_2))$ the variables of W_1, we see that $[K(W_1) : \tilde{\phi}(K(W_2))]$ is finite.

Definition 2.40. *The* degree *of the dominant rational mapping ϕ from W_1 to W_2, where $\dim(W_1) = \dim(W_2)$, is the degree of the finite algebraic field extension $K(W_1)$ over $\tilde{\phi}(K(W_2))$, that is*

$$\mathrm{degree}(\phi) = [K(W_1) : \tilde{\phi}(K(W_2))].$$

Observe that the notion of degree can be used to characterize the birationality of rational mappings as follows.

Lemma 2.41. *A dominant rational mapping $\phi : W_1 \to W_2$ between varieties of the same dimension is birational if and only if $\mathrm{degree}(\phi) = 1$.*

Taking into account that the degree of algebraic field extensions is multiplicative, one deduces the following lemma.

Lemma 2.42. *Let $\phi_1 : W_1 \to W_2$ and $\phi_2 : W_2 \to W_3$ be dominant rational mappings between varieties of the same dimension. Then*

$$\text{degree}(\phi_2 \circ \phi_1) = \text{degree}(\phi_1) \cdot \text{degree}(\phi_2).$$

One way of computing the degree of a rational mapping is by directly computing the degree of the algebraic field extension. Alternatively, we may use the fact that the degree of the mapping is the cardinality of a generic fibre. Those points where the cardinality of the fibre does not equal the degree of the mapping are called *ramification points* of the rational mapping. More precisely, we may apply the following result (see Proposition 7.16 in [Har95]).

Theorem 2.43. *Let $\phi : W_1 \to W_2$ be a dominant rational mapping between varieties of the same dimension. There exists a nonempty open subset U of W_2 such that for every $P \in U$ the cardinality of the fibre $\phi^{-1}(P)$ is equal to $\text{degree}(\phi)$.*

Thus, a direct application of this result, combined with elimination techniques, provides a method for computing the degree. Let $W_1 \subset A^r(K)$ and $W_2 \subset A^s(K)$ be varieties of the same dimension defined over K by $\{F_1(\bar{x}), \ldots, F_n(\bar{x})\} \subset K[\bar{x}]$, and $\{G_1(\bar{y}), \ldots, G_m(\bar{y})\} \subset K[\bar{y}]$, respectively, where $\bar{x} = (x_1, \ldots, x_r), \bar{y} = (y_1, \ldots, y_s)$. Let

$$\phi = \left(\frac{\phi_1}{\phi_{s+1,1}}, \ldots, \frac{\phi_s}{\phi_{s+1,s}} \right) : W_1 \to W_2$$

be a dominant rational mapping, where $\phi_i, \phi_{s+1,i}$ are polynomials over K, and $\gcd(\phi_i, \phi_{s+1,i}) = 1$. Then, Corollary 2.44 follows from Theorem 2.43.

Corollary 2.44. *Let $\phi : W_1 \to W_2$ be a dominant rational mapping between varieties of the same dimension.*

(a) Let $\bar{b}$ be a generic element of W_2. Then the degree of ϕ is equal to the cardinality of the finite set

$$\left\{ \bar{a} \in W_1 \; \middle| \; \phi(\bar{a}) = \bar{b}, \prod_{i=1}^{s} \phi_{s+1,i}(\bar{a}) \neq 0 \right\}.$$

(b) Let $\bar{a}$ be a generic element of W_1. Then the degree of ϕ is equal to the cardinality of the finite set

$$\left\{ \bar{a}' \in W_1 \; \middle| \; \phi(\bar{a}') = \phi(\bar{a}), \prod_{i=1}^{s} \phi_{s+1,i}(\bar{a}') \neq 0 \right\}.$$

Function fields of projective varieties can be introduced in a similar way. One only has to take care that representations of rational functions should have the same degrees in the numerators and denominators. For details the interested reader is referred to [Ful89], Sect. 4.2.

2.3 Intersection of Curves

In this section, we analyze the intersection of two plane curves. Since curves are algebraic sets, and since the intersection of two algebraic sets is again an algebraic set, we see that the intersection of two plane curves is an algebraic set in the plane consisting of 0-dimensional and 1-dimensional components. The ground field K is algebraically closed, so the intersection of two curves (seen projectively) is nonempty. In fact, the intersection of two curves contains a 1-dimensional component, i.e., a curve, if and only if the gcd of the corresponding defining polynomials is nonconstant, or equivalently if the two curves have a common component. In this case, the 1-dimensional component of the intersection is defined by the gcd.

Therefore, the problem of analyzing the intersection of curves is reduced to the case of two curves without common components. There are two questions which we need to answer. First, we want to compute the finitely many intersection points of the two curves. This means solving a 0-dimensional system of two bivariate polynomials. Second, we also want to analyze the number of intersection points of the curves without actually computing them. This counting of intersections points with proper multiplicities is achieved by Bézout's Theorem (see Theorem 2.48). For this purpose, we introduce the notion of multiplicity of intersection.

We start with the problem of computing the intersection points. Let $\mathcal{C}$ and $\mathcal{D}$ be two projective plane curves defined by $F(x, y, z)$ and $G(x, y, z)$, respectively, such that $\gcd(F, G) = 1$. We want to compute the finitely many points in $V(F, G)$. Since we are working in the plane, the solutions of this system of algebraic equations can be determined by resultants.

First, we observe that if both polynomials F and G are bivariate forms in the same variables, say $F, G \in K[x, y]$ (similarly if $F, G \in K[x, z]$ or $F, G \in K[y, z]$), then each curve is a finite union of lines passing through $(0 : 0 : 1)$. Hence, since the curves do not have common components, they intersect only in $(0 : 0 : 1)$. For instance, the curves defined by the equations $x(x - y)$ and $y(x + y)$ meet only at $(0 : 0 : 1)$.

So now let us assume that at least one of the defining polynomials is not a bivariate form in x and y (or any other pair of variables), say $F \notin K[x, y]$. Then, we consider the resultant $R(x, y)$ of F and G with respect to z. Since $\mathcal{C}$ and $\mathcal{D}$ do not have common components, $R(x, y)$ is not identically zero. Furthermore, since $\deg_z(F) \geq 1$ and G is not constant, the resultant R is a nonconstant bivariate homogeneous polynomial. Hence it factors as

$$R(x, y) = \prod_{i=1}^{s} (b_i x - a_i y)^{r_i}$$

for some $a_i, b_i \in K$, and $r_i \in \mathbb{N}$. Every intersection point of $\mathcal{C}$ and $\mathcal{D}$ must have coordinates $(a : b : c)$ s.t. (a, b) is a root of the resultant R. Therefore, the solutions of R provide the intersection points. Since $(0 : 0 : 0)$ is not a point in $\mathbb{P}^2(K)$, but it might be the formal result of extending the solution $(0,0)$ of R, we check whether $(0:0:1)$ is an intersection point. The remaining intersection points are given by $(a_i : b_i : c_{i,j})$, where $c_{i,j}$ are the roots in K of $\gcd(F(a_i, b_i, z), G(a_i, b_i, z))$.

Example 2.45. Let $\mathcal{C}$ and $\mathcal{D}$ be the projective plane curves defined by $F(x, y, z) = x^2 + y^2 - yz$ and $G(x, y, z) = y^3 + x^2 y - x^2 z$, respectively. Observe that these curves are the conic and cubic of Example 2.16. Since $\gcd(F, G) = 1$, $\mathcal{C}$ and $\mathcal{D}$ do not have common components. Obviously $(0:0:1)$ is an intersection point. For determining the other intersection points, we compute

$$R = \operatorname{res}_z(F, G) = x^4 - y^4 = (x - y)(x + y)(x^2 + y^2).$$

For extending the solutions of R to the third coordinate we compute

$$\gcd(F(1, 1, z), G(1, 1, z)) = z - 2,$$
$$\gcd(F(1, -1, z), G(1, -1, z)) = z + 2,$$
$$\gcd(F(1, \pm i, z), G(1, \pm i, z)) = z.$$

So the intersection points of $\mathcal{C}$ and $\mathcal{D}$ are

$$(0:0:1), \left(\frac{1}{2} : \frac{1}{2} : 1\right), \left(-\frac{1}{2} : \frac{1}{2} : 1\right), (1 : i : 0), (1 : -i : 0)$$

Now we proceed to the problem of analyzing the number of intersections of two projective curves without common components. For this purpose, we first study upper bounds, and then we see how these upper bounds can always be reached by a suitable definition of the notion of intersection multiplicity.

Theorem 2.46. *Let $\mathcal{C}$ and $\mathcal{D}$ be two projective plane curves without common components and degrees n and m, respectively. Then the number of intersection points of $\mathcal{C}$ and $\mathcal{D}$ is at most $n \cdot m$.*

Proof. First observe that the number of intersection points is invariant under linear changes of coordinates. Let k be the number of intersection points of $\mathcal{C}$ and $\mathcal{D}$. W.l.o.g. we assume that (perhaps after a suitable linear change of coordinates) $P = (0 : 0 : 1)$ is not a point on $\mathcal{C}$ or $\mathcal{D}$ (so the leading coefficients of F and G w.r.t. z are constant and every root of the resultant of F and G w.r.t. z is extendable) and also not on a line connecting any pair of intersection points of $\mathcal{C}$ and $\mathcal{D}$. Let $F(x, y, z)$ and $G(x, y, z)$ be the defining polynomials of $\mathcal{C}$ and $\mathcal{D}$, respectively. Note that since $(0 : 0 : 1)$ is not on the curves, we have $F, G \notin K[x, y]$. Let $R(x, y)$ be the resultant of F and G with respect to z. Since the curves do not have common components and the defining polynomials have positive degrees in z, R is a nonzero homogeneous polynomial in $K[x, y]$ of degree $n \cdot m$. Furthermore, as we have already seen, each linear factor of R generates a set of intersection points. Thus, if we can prove that each linear factor generates exactly one intersection point, we have shown that $k \leq \deg(R) = n \cdot m$. Let us assume that $(bx - ay)$ is a linear factor generating at least two different intersection points P_1 and P_2. But this implies that the line $bx - ay$ passes through P_1, P_2 and $(0 : 0 : 1)$, which is impossible. Furthermore, a root (a, b) of R cannot be a common root of the leading coefficients of F and G; so there exists a c s.t. $(a : b : c) \in \mathcal{C} \cap \mathcal{D}$. This shows that every linear factor $(bx - ay)$ generates at least one intersection point.   □

Remarks. Observe that from the proof of Theorem 2.46, we get that every linear factor $(bx - ay)$ of the resultant $R(x, y)$ of F and G with respect to z generates exactly one intersection point of F and G.

This upper bound on the number of intersection points can be turned into an exact number of intersections by a proper counting of multiple intersection points. The definition of intersection multiplicity is motivated by the proof of the previous theorem. First, we present the notion for curves such that the point $(0 : 0 : 1)$ is not on any of the two curves, nor on any line connecting two of their intersection points. Afterwards, we observe that the concept can be extended to the general case by means of linear changes of coordinates.

Definition 2.47. *Let $\mathcal{C}$ and $\mathcal{D}$ be projective plane curves without common components, such that $(0 : 0 : 1)$ is not on $\mathcal{C}$ or $\mathcal{D}$ and also not on any line connecting two intersection points of $\mathcal{C}$ and $\mathcal{D}$. Let $P = (a : b : c) \in \mathcal{C} \cap \mathcal{D}$, and let $F(x, y, z)$ and $G(x, y, z)$ be the defining polynomials of $\mathcal{C}$ and $\mathcal{D}$, respectively. Then, the* multiplicity of intersection *of $\mathcal{C}$ and $\mathcal{D}$ at P, denoted by* $\mathrm{mult}_P(\mathcal{C}, \mathcal{D})$, *is defined as the multiplicity of the corresponding factor $bx - ay$ in the resultant of F and G with respect to z. If $P \notin \mathcal{C} \cap \mathcal{D}$ then we define the multiplicity of intersection at P as 0.*

We observe that the conditions on $(0 : 0 : 1)$, required in Definition 2.47, can be avoided by means of linear changes of coordinates. Moreover, the extension of the definition to the general case does not depend on the particular linear change of coordinates, as remarked in [Wal50], Sect. IV.5. Indeed, since $\mathcal{C}$ and $\mathcal{D}$ do not have common components the number of intersection points is finite. Therefore, there always exist linear changes of coordinates satisfying the conditions required in the definition. Furthermore, as remarked in the last part of the proof of Theorem 2.46, if T is any linear change of coordinates satisfying the conditions of the definition, then each factor of the corresponding resultant is generated by exactly one intersection point. Therefore, the multiplicity of the factors in the resultant is preserved by this type of linear changes of coordinates.

Theorem 2.48 (Bézout's Theorem). *Let $\mathcal{C}$ and $\mathcal{D}$ be two projective plane curves without common components and degrees n and m, respectively. Then*

$$n \cdot m = \sum_{P \in \mathcal{C} \cap \mathcal{D}} \mathrm{mult}_P(\mathcal{C}, \mathcal{D}).$$

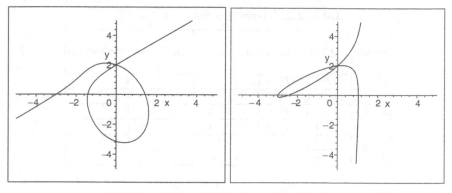

Fig. 2.3. Real part of $\mathcal{C}_{*,z}$ (left),   Real part of $\mathcal{D}_{*,z}$ (right)

Proof. Let us assume w.l.o.g. that $\mathcal{C}$ and $\mathcal{D}$ are such that $(0 : 0 : 1)$ is not on $\mathcal{C}$ or $\mathcal{D}$ nor on any line connecting two of their intersection points. Let $F(x, y, z)$ and $G(x, y, z)$ be the defining polynomials of $\mathcal{C}$ and $\mathcal{D}$, respectively. Then the resultant $R(x, y)$ of F and G with respect to z is a nonconstant homogeneous polynomial of degree $n \cdot m$. Furthermore, if $\{(a_i : b_i : c_i)\}_{i=1,\dots,r}$ are the intersection points of $\mathcal{C}$ and $\mathcal{D}$ (note that $(0 : 0 : 1)$ is not one of them) then we get

$$R(x, y) = \prod_{i=1}^{r}(b_i x - a_i y)^{n_i},$$

where n_i is, by definition, the multiplicity of intersection of $\mathcal{C}$ and $\mathcal{D}$ at $(a_i : b_i : c_i)$.   □

Example 2.49. We consider the two cubics $\mathcal{C}$ and $\mathcal{D}$ of Fig. 2.3 defined by the polynomials $F(x, y, z) =$

$$\frac{516}{85}z^3 - \frac{352}{85}yz^2 - \frac{7}{17}y^2z + \frac{41}{85}y^3 + \frac{172}{85}xz^2 - \frac{88}{85}xyz + \frac{1}{85}y^2x - 3x^2z + x^2y - x^3,$$

and $G(x, y, z) = -132z^3 + 128yz^2 - 29y^2z - y^3 + 28xz^2 - 76xyz + 31y^2x + 75x^2z - 41x^2y + 17x^3$, respectively.

Let us determine the intersection points of these two cubics and their corresponding multiplicities of intersection. For this purpose, we first compute the resultant

$$R(x, y) = \mathrm{res}_z(F, G) = \frac{5474304}{25}x^4\,y\,(3x + y)\,(x + 2y)\,(x + y)\,(x - y).$$

For each factor $(bx - ay)$ of the resultant $R(x, y)$ we obtain the polynomial $D(z) = \gcd(F(a, b, z), G(a, b, z))$ in order to find the intersection points generated by this factor. Table 2.2 shows the results of this computation (compare Fig. 2.4):

Table 2.2. Intersection points of $\mathcal{C}$ and $\mathcal{D}$

Factor	$D(z)$	Intersection point	Multipl. of intersection
x^4	$(2z-1)^2$	$P_1 = (0:2:1)$	4
y	$z+1/3$	$P_2 = (-3:0:1)$	1
$3x+y$	$z-1$	$P_3 = (1:-3:1)$	1
$x+2y$	$-z+1$	$P_4 = (-2:1:1)$	1
$x+y$	$-z+1$	$P_5 = (-1:1:1)$	1
$x-y$	$z-1$	$P_6 = (1:1:1)$	1

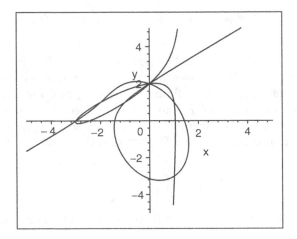

Fig. 2.4. Joint picture of the real parts of $\mathcal{C}_{\ast,z}$ and $\mathcal{D}_{\ast,z}$

Furthermore, since $(0:0:1)$ is not on the cubics nor on any line connecting their intersection points, the multiplicity of intersection is 4 for P_1, and 1 for the other points. It is also interesting to observe that P_1 is a double point on each cubic (compare Theorem 2.50(6)).

As we have seen above, the computational determination of the multiplicity of intersection requires a linear change of coordinates and the absolute factorization of the resultant of the defining polynomials. However, if the defining polynomials F are G are given over a computable subfield $\mathbb{K}$ of K, then we can proceed as follows to factor the resultant. We factor the resultant R of F and G over $\mathbb{K}$. Let $M(x,y)$ be an irreducible factor of R. Then for every root α of $M(x,1)$ we may compute $\gcd(F(\alpha,1,z),G(\alpha,1,z))$ over $\mathbb{K}(\alpha)$ to determine the corresponding intersection point $(\alpha:1:\beta_\alpha) \in \mathbb{P}^2(\mathbb{K}(\alpha))$. By the same arguments as in the previous proofs we see that this gcd is linear in z. Note that for all the conjugate roots of $M(x,1)$ the corresponding multiplicity of intersection is the same. Furthermore, if α_1, α_2 are two conjugate roots of $M(x,1)$ and φ is the automorphism on the splitting field of $M(x,1)$ that maps α_1 to α_2 then $(\alpha_2 : 1 : \beta_{\alpha_2}) = (\varphi(\alpha_1) : 1 : \varphi(\beta_{\alpha_1}))$. In the sequel we will frequently use this idea, and we will then speak about *families of conjugate points* (see Definition 3.15).

Some authors introduce the notion of multiplicity of intersection axiomatically (see, for instance, [Ful89], Sect. 3.3). The following theorem shows that these axioms are satisfied for our definition.

Theorem 2.50. *Let C and D be two projective plane curves without common components, defined by the polynomials F and G, respectively, and let $P \in \mathbb{P}^2(K)$. Then the following statements hold:*

(1) $\mathrm{mult}_P(C, D) \in \mathbb{N}$.
(2) $\mathrm{mult}_P(C, D) = 0$ *if and only if* $P \notin C \cap D$.
(3) If T is a linear change of coordinates, and C', D' are the imagines of C and D under T, respectively, then $\mathrm{mult}_P(C, D) = \mathrm{mult}_{T(P)}(C', D')$.
(4) $\mathrm{mult}_P(C, D) = \mathrm{mult}_P(D, C)$
(5) $\mathrm{mult}_P(C, D) \geq \mathrm{mult}_P(C) \cdot \mathrm{mult}_P(D)$.
(6) $\mathrm{mult}_P(C, D) = \mathrm{mult}_P(C) \cdot \mathrm{mult}_P(D)$ *if and only C and D intersect transversally at P (i.e. if the curves have no common tangents at P).*
(7) Let $C_1, \ldots, C_r$ and $D_1, \ldots, D_s$ be the irreducible components of C and D respectively. Then

$$\mathrm{mult}_P(C, D) = \sum_{i=1}^{r} \sum_{j=1}^{s} \mathrm{mult}_P(C_i, D_j).$$

(8) $\mathrm{mult}_P(C, D) = \mathrm{mult}_P(C, D_H)$, *where D_H is the curve defined by $G + HF$ for an arbitrary form $H \in K[x, y, z]$; i.e., the intersection multiplicity does not depend on the particular representative G in the coordinate ring of C.*

Proof. We have already remarked above that the intersection multiplicity is independent of a particular linear change of coordinates. Statements (1), (2), and (4) can be easily deduced from the definition of multiplicity of intersection, and we leave them to the reader.

A proof of (5) and (6) can be found for instance in [Wal50], Chap. IV.5, Theorem 5.10.

(7) Let us assume without loss of generality that C and D satisfy the requirements of Definition 2.47. That is, $(0 : 0 : 1)$ is not on the curves nor on any line connecting their intersection points. Let $F_i, i = 1, \ldots, r$, and $G_j, j = 1 \ldots, s$, be the defining polynomials of C_i and D_j, respectively. Then we use the following fact: if $A, B, C \in D[x]$, where D is an integral domain, then $\mathrm{res}_x(A, B \cdot C) = \mathrm{res}_x(A, B) \cdot \mathrm{res}_x(A, C)$ (see, for instance, [BCL83] Theorem 3, p. 178). Hence, (7) follows immediately from

$$\mathrm{res}_z\left(\prod_{i=1}^{r} F_i, \prod_{j=1}^{s} G_j\right) = \prod_{i=1}^{r} \mathrm{res}_z\left(F_i, \prod_{j=1}^{s} G_j\right) = \prod_{i=1}^{r} \prod_{j=1}^{s} \mathrm{res}_z(F_i, G_j).$$

(8) Let $H \in K[x, y, z]$ be a form. Obviously $P \in \mathcal{C} \cap \mathcal{D}$ if and only if $P \in \mathcal{C} \cap \mathcal{D}_H$. W.l.o.g. we assume that the conditions of Definition 2.47 are satisfied by $\mathcal{C}$ and $\mathcal{D}$, and also by $\mathcal{C}$ and $\mathcal{D}_H$. Then, $\text{mult}_P(\mathcal{C}, \mathcal{D})$ and $\text{mult}_P(\mathcal{C}, \mathcal{D}_H)$ are given by the multiplicities of the corresponding factors in $\text{res}_z(F, G)$ and $\text{res}_z(F, G + HF)$, respectively. Now, we use the following property of resultants: if $A, B, C \in D[x]$, where D is an integral domain, and a is the leading coefficient of A, then

$$\text{res}_x(A, B) = a^{\deg_x(B) - \deg_x(AC+B)} \, \text{res}_x(A, AC + B)$$

(see, for instance, Theorem 4 on p.178 of [BCL83]). Now (8) follows directly from this fact, since the leading coefficient of F in z is a nonzero constant (note that $(0 : 0 : 1)$ is not on $\mathcal{C}$).   □

From Theorem 2.50, and the proof of uniqueness and existence of the notion of intersection multiplicity given in [Ful89], one can extract an alternative algorithm for computing the intersection multiplicity as it is illustrated in the next example. We leave this as an exercise.

Example 2.51 ([Ful89]). We determine the intersection multiplicity at the origin $O = (0, 0)$ of the affine curves $\mathcal{E}, \mathcal{F}$ defined by

$$\mathcal{E}: \; e(x, y) = (x^2 + y^2)^2 + 3x^2 y - y^3, \quad \mathcal{F}: \; f(x, y) = (x^2 + y^2)^3 - 4x^2 y^2.$$

For ease of notation we do not distinguish between the curves and their defining polynomials. We replace f by the following curve g:

$$f(x, y) - (x^2 + y^2)e(x, y) = y \cdot ((x^2 + y^2)(y^2 - 3x^2) - 4x^2 y) = y \cdot g(x, y).$$

Now we have

$$\text{mult}_O(e, f) = \text{mult}_O(e, y) + \text{mult}_O(e, g).$$

We replace g by h:

$$g + 3e = y \cdot (5x^2 - 3y^2 + 4y^3 + 4x^2 y) = y \cdot h(x, y).$$

So

$$\text{mult}_O(e, f) = 2 \cdot \text{mult}_O(e, y) + \text{mult}_O(e, h).$$

By relations (4), (7) in Theorem 2.50, $\text{mult}_O(e, y)$ is equal to $\text{mult}_O(x^4, y)$, which in turn is equal to 4 by relations (6), (5). By relation (5), $\text{mult}_O(e, h)$ is equal to $\text{mult}_O(e) \cdot \text{mult}_O(h)$, which is 6. Thus, $\text{mult}_O(\mathcal{E}, \mathcal{F}) = \text{mult}_O(e, f) = 14$.

2.4 Linear Systems of Curves

Linear systems of curves are an indispensably tool in algebraic geometry. In this section we derive some basic properties of linear systems of curves, based on the exposition in [Mir99]. The idea of linear systems of curves is to work with sets of curves of fixed degree related by means of some linear conditions; for instance, sets of curves of fixed degree passing through some specific points with at least some fixed multiplicities. Such conditions are captured by the notion of a divisor.

Definition 2.52. *A* divisor *is a formal expression of the type*

$$\sum_{i=1}^{m} r_i P_i,$$

where $r_i \in \mathbb{Z}$, and the P_i are different points in $\mathbb{P}^2(K)$. If all integers r_i are non-negative we say that the divisor is effective *or* positive.

We will identify projective algebraic curves with forms in $K[x, y, z]$. Thus, throughout this section, we allow curves to have multiple components; i.e., we consider curves as defined by arbitrary nonconstant, not necessarily square-free, forms. This means that set theoretically the curves are defined by the squarefree parts of homogeneous polynomials, but the components generated by their irreducible factors carry the corresponding multiplicity.

Let $T_1, \ldots, T_n$ be a fixed ordering of the set of monomials in x, y, z of degree d. It is clear that

$$n = \frac{1}{2}(d+1)(d+2) = \frac{1}{2}d(d+3) + 1.$$

Then, for every curve $\mathcal{C}$ of degree d there exists $(a_1 : \cdots : a_n) \in \mathbb{P}^{n-1}(K)$, such that $F = a_1 T_1 + \cdots + a_n T_n$ defines $\mathcal{C}$, and vice versa. Observe that F is defined only up to multiplication by nonzero constants. Thus, one may identify the set of all projective curves of degree d with $\mathbb{P}^{n-1}(K)$.

Definition 2.53. *A* linear system of curves *of degree d and dimension r is a linear subvariety of dimension r of $\mathbb{P}^{\frac{d(d+3)}{2}}(K)$. If the dimension is one, the linear system is also called a* pencil *of curves.*

An interesting type of linear systems arises when we require the curves to pass through given points with given multiplicities. This motivates the following definition.

Definition 2.54. *$P \in \mathbb{P}^2(K)$ is a* base point *of multiplicity $r \in \mathbb{N}$ of a linear system $\mathcal{H}$ of curves of fixed degree, if every curve $\mathcal{C}$ in $\mathcal{H}$ satisfies* $\mathrm{mult}_P(\mathcal{C}) \geq r$.

Definition 2.55. *We define the* linear system *of curves of degree d generated by the effective divisor* $D = r_1 P_1 + \cdots + r_m P_m$ *as the set of all curves* $\mathcal{C}$ *of degree d such that* $\mathrm{mult}_{P_i}(\mathcal{C}) \geq r_i$, *for* $i = 1, \ldots, m$, *and we denote it by* $\mathcal{H}(d, D)$.

Example 2.56. We compute the linear system of quintics generated by the effective divisor $D = 3P_1 + 2\,P_2 + P_3$, where $P_1 = (0 : 0 : 1)$, $P_2 = (0 : 1 : 1)$, and $P_3 = (1 : 1 : 1)$. For this purpose, we consider the generic form of degree 5:

$$
\begin{aligned}
H(x, y, z) = {} & a_0\, z^5 + a_1\, yz^4 + a_2\, y^2 z^3 + a_3\, y^3 z^2 + a_4\, y^4 z + a_5\, y^5 + a_6\, xz^4 \\
& + a_7\, xyz^3 + a_8\, xy^2 z^2 + a_9\, xy^3 z + a_{10}\, xy^4 + a_{11}\, x^2 z^3 \\
& + a_{12}\, x^2 yz^2 + a_{13}\, x^2 y^2 z + a_{14}\, x^2 y^3 + a_{15}\, x^3 z^2 + a_{16}\, x^3 yz \\
& + a_{17}\, x^3 y^2 + a_{18}\, x^4 z + a_{19}\, x^4 y + a_{20}\, x^5.
\end{aligned}
$$

The linear conditions that we have to impose are (see Theorem 2.15):

$$
\frac{\partial^{k+i+j} H}{\partial x^k \partial y^i \partial z^j}(P_1) = 0, \ \ i + j + k \leq 2, \qquad \frac{\partial^{k+i+j} H}{\partial x^k \partial y^i \partial z^j}(P_2) = 0, \ \ i + j + k \leq 1,
$$

$$
H(P_3) = 0.
$$

Solving them one gets that the linear system is defined by $H(x, y, z) =$

$$
\begin{aligned}
& a_3\, y^3 z^2 - 2\, a_3\, y^4 z + a_3\, y^5 + (-a_9 - a_{10})\, xy^2 z^2 + a_9\, xy^3 z + a_{10}\, xy^4 + \\
& (-a_{13} - a_{14} - a_{15} - a_{16} - a_{17} - a_{18} - a_{19} - a_{20})\, x^2 yz^2 + a_{13}\, x^2 y^2 z + \\
& a_{14}\, x^2 y^3 + a_{15}\, x^3 z^2 + a_{16}\, x^3 yz + a_{17}\, x^3 y^2 + a_{18}\, x^4 z + a_{19}\, x^4 y + a_{20}\, x^5
\end{aligned}
$$

Hence, the dimension of the system is 10 (compare Theorem 2.59). Finally, we take two particular curves in the system, $\mathcal{C}_1$ and $\mathcal{C}_2$, defined by the polynomials

$$
\begin{aligned}
H_1(x, y, z) = {} & 3\, y^3 z^2 - 6\, y^4 z + 3\, y^5 - x\, y^3 z + x\, y^4 - 5\, x^2 y z^2 + 2\, x^2 y^2 z \\
& + x^3 y^2 + x^4 z + y\, x^4,
\end{aligned}
$$

$$
\begin{aligned}
H_2(x, y, z) = {} & y^3 z^2 - 2\, y^4 z + y^5 - \tfrac{8}{3} z^2 x\, y^2 + 3\, x\, y^3 z - \tfrac{1}{3} x\, y^4 - 8\, x^2 y z^2 \\
& + 2\, x^2 y^2 z + y^3 x^2 + 2\, y\, x^3 z + x^3 y^2 - x^4 z + 2\, y\, x^4 + x^5,
\end{aligned}
$$

respectively.

In Fig. 2.5 the real parts of the affine curves $\mathcal{C}_{1_{*,z}}$ and $\mathcal{C}_{2_{*,z}}$ are plotted.

Clearly, a linear system $\mathcal{H}(d, D)$ is the solution variety of a system of linear equations. This system is directly derived from the conditions imposed by the effective divisor D. The conditions imposed by the divisor D might be linearly dependent, i.e., the codimension of $\mathcal{H}(d, D)$ might be less than the number of conditions. In the following we analyze the behavior of the dimension of linear systems generated by effective divisors.

Theorem 2.57. *Let* $P \in \mathbb{P}^2(K)$, *and* $r \in \mathbb{N}$. *Then, for every* $d \in \mathbb{N}$

$$
\dim(\mathcal{H}(d, rP)) = \begin{cases} \frac{d(d+3)}{2} - \frac{r(r+1)}{2} & \text{if } d \geq r, \\ -1 & \text{if } d < r. \end{cases}
$$

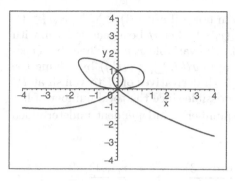 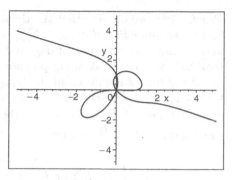

Fig. 2.5. real part of $\mathcal{C}_{1_{*,z}}$ (left),   real part of $\mathcal{C}_{2_{*,z}}$ (right)

Proof. First we observe that the inequality $d(d+3) - r(r+1) \geq 0$ is equivalent to $d \geq r$. Let H be a generic form, with undetermined coefficients, of degree d in the variables x, y, z. Then, by Theorem 2.15, we get the defining polynomial of $\mathcal{H}(d, rP)$ by solving the linear conditions requiring all the $(r-1)$th derivatives of H to vanish at P; i.e., it is obtained by requiring all the terms in the Taylor expansion of H at P up to degree r to vanish. Since Taylor expansion is unique, we see that these $\frac{r(r+1)}{2}$ conditions are independent. Hence, if $d < r$, then H is identically zero, and thus $\dim(\mathcal{H}(d, rP)) = -1$. On the other hand, if $d \geq r$,

$$\dim(\mathcal{H}(d, rP)) = \frac{d(d+3)}{2} - \frac{r(r+1)}{2}. \qquad \square$$

Corollary 2.58. *Let $P \in \mathbb{P}^2(K)$ be a fixed point. Then the linear system of curves of degree d generated by $1 \cdot P$ has codimension 1; i.e., it is a hyperplane in $\mathbb{P}^{\frac{d(d+3)}{2}}(K)$.*

The situation is much more complicated if the generating effective divisor consists of more than one point. One might expect that for each new constraint the dimension drops by one. That is, one might expect the dimension of $\mathcal{H}(d, r_1 P_1 + \cdots + r_m P_m)$ to be

$$\mu = \max\left\{-1, \frac{d(d+3)}{2} - \sum_{i=1}^{m} \frac{r_i(r_i+1)}{2}\right\}. \qquad (2.1)$$

However, this number is only a lower bound, since these constraints may be dependent.

Theorem 2.59. *Let $P_1, \ldots, P_m \in \mathbb{P}^2(K)$, and $r_1, \ldots, r_m \in \mathbb{N}$. Then, for every $d \in \mathbb{N}$,*

$$\dim(\mathcal{H}(d, \sum_{i=1}^{m} r_i P_i)) \geq \frac{d(d+3)}{2} - \sum_{i=1}^{m} \frac{r_i(r_i+1)}{2}.$$

Proof. First, we observe that the theorem holds if $d(d+3) < \sum_{i=1}^{m} r_i(r_i+1)$. So let us assume that $d(d+3) \geq \sum_{i=1}^{m} r_i(r_i+1)$. Let H be a generic form, with undetermined coefficients, of degree d in the variable x, y, z. Then, by Theorem 2.15, we get the defining polynomial of $\mathcal{H}(d, \sum_{i=1}^{m} r_i P_i))$ by solving the linear conditions requiring all the $(r_i - 1)$th derivatives of H to vanish at P_i, for $i = 1, \ldots, m$. Clearly, this is always possible since the total number of linear equations is $\sum_{i=1}^{m} \frac{r_i(r_i+1)}{2}$, and the number of independent undetermined coefficients in H is $\frac{d(d+3)}{2}$. Hence,

$$\dim\left(\mathcal{H}\left(d, \sum_{i=1}^{m} r_i P_i\right)\right) \geq \frac{d(d+3)}{2} - \sum_{i=1}^{m} \frac{r_i(r_i+1)}{2}. \qquad \square$$

From Bézout's Theorem (Theorem 2.48) and the previous theorem one may derive bounds for the number of singularities of an irreducible curve.

Theorem 2.60. *Let C be an irreducible projective plane curve of degree d. Then*

$$\sum_{P \in C} \mathrm{mult}_P(C)(\mathrm{mult}_P(C) - 1) \leq (d-1)(d-2).$$

Proof. Obviously the statement holds if C is a line. So let us assume that $d > 1$. Let F be the defining polynomial of C. By Theorem 2.10, C has only finitely many singular points. Let us denote these singular points by $P_1, \ldots, P_m$, and let us assume that $r_i = \mathrm{mult}_{P_i}(C)$ for $i = 1, \ldots, m$. By Euler's formula for homogeneous polynomials (see Appendix B), it is clear that not all the first derivatives of F are identically zero. Let us assume that $\frac{\partial F}{\partial x}$ is not identically zero, and let $\mathcal{D}$ be the curve defined by this derivative. The degree of $\mathcal{D}$ is $d - 1 > 0$. Furthermore, $\mathcal{D}$ has no components in common with C, since F is irreducible and $\deg(C) > \deg(\mathcal{D})$. Moreover, $\{P_1, \ldots, P_m\} \subset C \cap \mathcal{D}$. In fact, $\mathrm{mult}_{P_i}(\mathcal{D}) = r_i - 1$. Therefore, by Bézout's Theorem (Theorem 2.48), we get

$$d(d-1) = \sum_{P \in C \cap \mathcal{D}} \mathrm{mult}_P(C, \mathcal{D}) \geq \sum_{i=1}^{m} \mathrm{mult}_{P_i}(C, \mathcal{D}),$$

and by Theorem 2.50(5) we get

$$d(d-1) \geq \sum_{i=1}^{m} \mathrm{mult}_{P_i}(C) \cdot \mathrm{mult}_P(\mathcal{D}) \geq \sum_{i=1}^{m} r_i(r_i - 1).$$

Consequently, since $d > 1$, we get

$$\frac{(d-1)(d+2)}{2} - \sum_{i=1}^{m} \frac{r_i(r_i-1)}{2} > \frac{d(d-1)}{2} - \sum_{i=1}^{m} \frac{r_i(r_i-1)}{2} \geq 0.$$

Now, we consider the linear system $\mathcal{H}$ of curves of degree $(d-1)$ generated by the effective divisor $((r_1-1)P_1 + \cdots + (r_m-1)P_m)$. Using Theorem 2.59, we deduce that

$$\dim(\mathcal{H}) \geq \frac{(d-1)(d+2)}{2} - \sum_{i=1}^{m} \frac{r_i(r_i-1)}{2} > 0.$$

We choose $\ell = \frac{(d-1)(d+2)}{2} - \sum_{i=1}^{m} \frac{r_i(r_i-1)}{2} > 0$ simple points $Q_1, \ldots, Q_\ell$ on $\mathcal{C}$, and we consider the linear subsystem $\mathcal{H}'$ of $\mathcal{H}$ generated by the effective divisor $(1 \cdot Q_1 + \cdots + 1 \cdot Q_\ell)$. Clearly

$$\dim(\mathcal{H}') \geq \dim(\mathcal{H}) - \ell \geq 0.$$

Hence $\mathcal{H}'$ is not empty. Let $\mathcal{C}'$ be a curve in $\mathcal{H}'$. Then from Bézout's Theorem (note that $\mathcal{C}$ is irreducible and $\deg(\mathcal{C}) > \deg(\mathcal{C}')$) and Theorem 2.50(5) we get

$$d(d-1) = \sum_{P \in \mathcal{C} \cap \mathcal{C}'} \mathrm{mult}_P(\mathcal{C}, \mathcal{C}')$$

$$\geq \sum_{i=1}^{m} \mathrm{mult}_{P_i}(\mathcal{C}, \mathcal{C}') + \sum_{j=1}^{\ell} \mathrm{mult}_{Q_j}(\mathcal{C}, \mathcal{C}')$$

$$\geq \sum_{i=1}^{m} \mathrm{mult}_{P_i}(\mathcal{C}) \cdot \mathrm{mult}_{P_i}(\mathcal{C}') + \sum_{j=1}^{\ell} \mathrm{mult}_{Q_j}(\mathcal{C}) \cdot \mathrm{mult}_{Q_j}(\mathcal{C}')$$

$$\geq \sum_{i=1}^{m} r_i(r_i-1) + \ell$$

$$= \sum_{i=1}^{n} \frac{r_i(r_i-1)}{2} + \frac{(d-1)(d+2)}{2}.$$

Hence,

$$\sum_{P \in \mathcal{C}} \frac{\mathrm{mult}_P(\mathcal{C})(\mathrm{mult}_P(\mathcal{C}) - 1)}{2} \leq \frac{(d-1)(d-2)}{2}. \qquad \square$$

As we have seen in Theorem 2.59, the expected dimension μ (see (2.1)) is a lower bound for the actual dimension. For instance, if P and Q are two different points, the expected dimension μ of the linear system $\mathcal{H}(2, 2P+2Q)$ is -1. Nevertheless a double line passing through these two points is in the linear system, and hence it is not empty.

Even for the case of linear systems generated by divisors of the form $P_1 + \cdots + P_m$ the actual dimension may be different from μ. For example, take $m = d^2$ and consider two curves $\mathcal{C}_1, \mathcal{C}_2$ of degree d, without common components, and such that all their intersections are transversal and occur at regular points. Then, by Bézout's Theorem the number of intersection points is m; denote them by $P_1, \ldots, P_m$. In this situation, $\mathcal{C}_1, \mathcal{C}_2 \in \mathcal{H}(d, P_1 + \cdots + P_m)$, and therefore $\mathcal{H}(d, P_1 + \cdots + P_m)$ has positive dimension. But, if $d \geq 3$, then

$$\frac{d(d+3)}{2} - d^2 \leq 0.$$

So the actual dimension and the expected dimension μ do not agree. We illustrate this reasoning by a specific example.

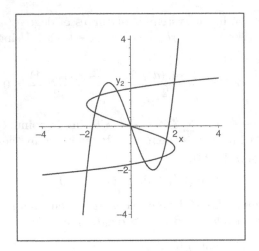

Fig. 2.6. Joint picture of the real parts of C_1 and C_2

Example 2.61. Let us consider the projective cubics C_1 and C_2 defined by the polynomials

$$z^2 x - y^3 + 3yz^2, \quad \text{and} \quad z^2 y - x^3 + 3xz^2,$$

respectively (see Fig. 2.6). The intersection points of the two cubics are the real points

$$P_1 = (0 : 0 : 1), \qquad\qquad P_2 = \left(\frac{-\sqrt{5}-1}{2} : \frac{\sqrt{5}-1}{2} : 1\right),$$

$$P_3 = \left(\frac{\sqrt{5}-1}{2} : \frac{-\sqrt{5}-1}{2} : 1\right), \qquad P_4 = \left(\frac{-\sqrt{5}+1}{2} : \frac{\sqrt{5}+1}{2} : 1\right),$$

$$P_5 = \left(\frac{\sqrt{5}+1}{2} : \frac{-\sqrt{5}+1}{2} : 1\right), \qquad P_6 = (2 : 2 : 1),$$

$$P_7 = (-2 : -2 : 1), \qquad\qquad P_8 = (-\sqrt{2} : \sqrt{2} : 1),$$

$$P_9 = (\sqrt{2} : -\sqrt{2} : 1).$$

Therefore, $C_1, C_2 \in \mathcal{H}(3, P_1 + \cdots + P_9)$, but the expected multiplicity is 0. In fact, $\dim(\mathcal{H}(3, P_1 + \cdots + P_9)) = 1$.

Another way to generate examples of linear systems whose actual dimension is higher than the formally expected dimension is to force linear dependencies by taking all base points on a line. More precisely, let us consider the linear system $\mathcal{H}(d, P_1 + \cdots + P_m)$, where $d < m$ and all the points P_i are on a line $\mathcal{L}$. Then, every curve $C \in \mathcal{H}(d, P_1 + \cdots + P_m)$ intersects $\mathcal{L}$ at more than d

points. Hence, by Bézout's Theorem, $\mathcal{L}$ must be a component of $\mathcal{C}$. Therefore, if we take any collection of more than d different points on $\mathcal{L}$ then the corresponding linear system of curves of degree d is in fact $\mathcal{H}(d, P_1 + \cdots + P_m)$, while the expected dimension decreases.

Now we analyze the behavior of the dimension of the linear system if the points are taken in *general* position. One could expect that if the points are selected *generally* enough, the dimension of the linear system agrees with the lower bound μ (see (2.1)). But, as we have seen in the example $\mathcal{H}(2, 2P+2Q)$, no matter how *generally* we choose P and Q, the expected dimension μ will always be less that the actual dimension. Thus, this lower bound μ is not optimal in general. So, the idea is to define the notion of *general position* such that one reaches the optimal lower bound; that is, such that the dimension of the linear system is minimal.

Therefore, one has to require the solution space of the system of linear equations generated by the divisor to have minimal dimension. This condition can be stated by means of the rank of the matrix of the system. To be more precise, let us fix a degree d, and let us also fix a tuple of multiplicities $(r_1, \ldots, r_m)$. Now let $H(\Lambda, x, y, z)$ be a generic form with undetermined coefficients Λ, and let $P_i = (x_i : y_i : z_i)$, $i = 1 \ldots, m$, be arbitrary points in $\mathbb{P}^2(K)$. Let $\mathcal{S}(\Lambda, x_1, y_1, z_1, \ldots, x_m, y_m, z_m)$ be the system of linear equations, in the undetermined coefficients Λ, such that $\mathcal{H}(d, r_1 P_1 + \cdots + r_m P_m)$ is the solution space of $\mathcal{S}(\Lambda, x_1, y_1, z_1, \ldots, x_m, y_m, z_m)$. Now, let $A(x_1, y_1, z_1, \ldots, x_m, y_m, z_m)$ be the matrix of the system $\mathcal{S}$; note that the entries of A are polynomials in x_i, y_i, z_i. Then, for every non-negative integer j we consider the set

$$\mathcal{R}_j(d, r_1, \ldots, r_m)$$

$$= \{((\tilde{x}_1 : \tilde{y}_1 : \tilde{z}_1), \ldots, (\tilde{x}_m : \tilde{y}_m : \tilde{z}_m)) \in (\mathbb{P}^2(K))^m \mid \mathrm{rank}(A(\tilde{x}_1, \ldots, \tilde{z}_m)) \leq j\}.$$

Clearly $\mathcal{R}_j(d, r_1, \ldots, r_m)$ is a projective algebraic set, since the above rank condition is achieved by means of the vanishing of certain minors of the matrix. On the other hand, it is also clear that

$$\mathcal{R}_0(d, r_1, \ldots, r_m) \subseteq \cdots \subseteq \mathcal{R}_k(d, r_1, \ldots, r_m) \subsetneq \mathcal{R}_{k+1}(d, r_1, \ldots, r_m) = (\mathbb{P}^2(K))^m$$

for some minimal k. Note that $\mathcal{R}_0(d, r_1, \ldots, r_m) \neq (\mathbb{P}^2(K))^m$, and that $\mathcal{R}_s(d, r_1, \ldots, r_m) = (\mathbb{P}^2(K))^m$ for $s \in \mathbb{N}$ larger than the size of the matrix.

Then we say that a particular divisor $D = r_1 Q_1 + \cdots + r_m Q_m$ is in *d-general position* if $Q_1 \times \cdots \times Q_m \in (\mathbb{P}^2(K))^m \setminus \mathcal{R}_k(d, r_1, \ldots, r_m)$, and therefore the specialization of the system $\mathcal{S}$ at $Q_1, \ldots, Q_m$ has minimal dimension. In other words, the divisor $D = r_1 Q_1 + \cdots + r_m Q_m$ is in d-general position if for any other divisor $\tilde{D} = r_1 \tilde{Q}_1 + \cdots + r_m \tilde{Q}_m$ we have

$$k + 1 = \mathrm{rank}(A(Q_1, \ldots, Q_m)) \geq \mathrm{rank}(A(\tilde{Q}_1, \ldots, \tilde{Q}_m)).$$

Or equivalently

$$\frac{d(d+3)}{2} - (k+1) = \dim(\mathcal{H}(d, D)) \le \dim(\mathcal{H}(d, \tilde{D})).$$

We illustrate these ideas by the next example.

Example 2.62. (a) We consider the linear system $\mathcal{H}(1, 2P)$. Clearly, for every $P \; dim(\mathcal{H}(1, 2P)) = -1$. Let us see what general position means here. Let $P = (x_1 : y_1 : z_1)$ be an arbitrary point in $\mathbb{P}^2(K)$. Then $A(x_1, y_1, z_1)$ is the 3×3 identity matrix. Thus,

$$\emptyset = \mathcal{R}_2(1, 2) \subsetneq \mathcal{R}_3(1, 2) = \mathbb{P}^2(K).$$

Therefore, for every $P \in \mathbb{P}^2(K) \setminus \mathcal{R}_2(1, 2) = \mathbb{P}^2(K)$, the divisor $2P$ is in 2-general position, and $dim(\mathcal{H}(1, 2P)) = -1$.

(b) We consider the situation analyzed in Theorem 2.57. Let $P \in \mathbb{P}^2(K)$ and let $r \le d$. Because of the uniqueness of the Taylor expansion of a polynomial we have that the rank of the matrix of the system $\mathcal{S}$ is precisely $\frac{r(r+1)}{2}$. Hence,

$$\emptyset = \mathcal{R}_{\frac{r(r+1)}{2}-1}(d, r) \subset \mathcal{R}_{\frac{r(r+1)}{2}}(d, r) = \mathbb{P}^2(K).$$

Thus,

$$\mathbb{P}^2(K) \setminus \mathcal{R}_{\frac{r(r+1)}{2}-1}(d, r) = \mathbb{P}^2(K).$$

Therefore, every divisor of the form $D = rP$, with $0 < r \le d$ is in d-general position. Furthermore,

$$\dim(\mathcal{H}(d, D)) = \frac{d(d+3)}{2} - \frac{r(r+1)}{2},$$

in accordance with Theorem 2.57.

(c) Now we study the special example of the linear system of conics $\mathcal{H}(2, 2P + 2Q)$. Let $P = (x_1 : y_1 : z_1)$ and $Q = (x_2 : y_2 : z_2)$ be arbitrary points in $\mathbb{P}^2(K)$. Then,

$$A(x_1, y_1, z_1, x_2, y_2, z_2) = \begin{pmatrix} 0 & 0 & 0 & z_1 & y_1 & 2x_1 \\ 0 & z_1 & 2y_1 & 0 & x_1 & 0 \\ 2z_1 & y_1 & 0 & x_1 & 0 & 0 \\ 0 & 0 & 0 & z_2 & y_2 & 2x_2 \\ 0 & z_2 & 2y_2 & 0 & x_2 & 0 \\ 2z_2 & y_2 & 0 & x_2 & 0 & 0 \end{pmatrix}.$$

Since $\det(A(x_1, y_1, z_1, x_2, y_2, z_2)) = 0$, we have $\mathcal{R}_5(2, 2, 2) = \mathcal{R}_6(2, 2, 2) = (\mathbb{P}^2(K))^2$. However, taking random values for $x_1, y_1, z_1, x_2, y_2, z_2$, we deduce that $\mathcal{R}_4(2, 2, 2) \subsetneq \mathcal{R}_5(2, 2, 2)$. Furthermore, the algebraic conditions defining

$\mathcal{R}_4(2,2,2)$ are $\{x_1y_2 = y_1x_2, x_1z_2 = z_1x_2, y_2z_1 = y_1z_2\}$. Thus $2P + 2Q$ is in 2-general position if $P \neq Q$. Now, observe that, because of Theorem 2.60, if a conic $\mathcal{C}$ belongs to $\mathcal{H}(2, 2P + 2Q)$ with $P \neq Q$, then it must be reducible; i.e. a pair of lines. Moreover, since it must have two different double points, namely P and Q, then $\mathcal{C}$ must be the double line passing through P and Q. So, $\dim(\mathcal{H}(2, 2P + 2Q)) = 0$.

We finish this study by proving that for d-general divisors of the form $D = P_1 + \cdots + P_m$, the dimension of the linear system, generated by D, is the expected one (see (2.1)). For further details and considerations on this topic we refer to [Mir99].

Lemma 2.63. *Let $D = r_1P_1 + \cdots + r_mP_m$ be an effective divisor in d-general position. Then for every $1 \leq i \leq m$ the divisor $D_i = r_1P_1 + \cdots + r_iP_i$ is also in d-general position.*

Proof. This follows from the fact that the set of linear equations generated by D_i is contained in the linear equations generated by D. □

Lemma 2.64. *Let $D = r_1P_1 + \cdots + r_mP_m$ be an effective divisor in d-general position, and let Ω be a nonempty open subset of $(\mathbb{P}^2(K))^m$. Then*

$$\max\{\operatorname{rank}(A(Q_1, \ldots, Q_m)) \mid Q_1 \times \cdots \times Q_m \in \Omega\} = \operatorname{rank}(A(P_1, \ldots, P_m)).$$

Proof. Since D is in d-general position, there exists a nonempty open subset $\tilde{\Omega} \subset (\mathbb{P}^2(K))^m$ such that, for all $Q_1 \times \cdots \times Q_m \in \tilde{\Omega}$, $\operatorname{rank}(A(Q_1, \ldots, Q_m))$ is the same and its value is maximal. Therefore, since $(\mathbb{P}^2(K))^m$ is irreducible, one gets that $\Omega \cap \tilde{\Omega} \neq \emptyset$. Thus, the proof is finished by taking a point in $\Omega \cap \tilde{\Omega}$. □

Theorem 2.65. *Let $D = P_1 + \cdots + P_m$ be in d-general position. Then $\mathcal{H}(d, D)$ has the expected dimension, i.e.:*

$$\dim(\mathcal{H}(d, D)) = \max\left\{ -1, \frac{d(d+3)}{2} - m \right\}.$$

Proof. We prove this by induction on m. For $m = 1$, the statement follows from Theorem 2.57. Let us assume that the result holds for $P_1 + \cdots + P_{i-1}, 1 < i \leq m$. Then,

$$\dim(\mathcal{H}(d, P_1 + \cdots + P_{i-1})) = \max\left\{ -1, \frac{d(d+3)}{2} - (i-1) \right\}.$$

If $\dim(\mathcal{H}(d, P_1 + \cdots + P_{i-1})) = -1$, then $\dim(\mathcal{H}(d, P_1 + \cdots + P_i)) = -1$ because $\mathcal{H}(d, P_1 + \cdots + P_i) \subset \mathcal{H}(d, P_1 + \cdots + P_{i-1})$, and therefore the statement holds for the divisor $P_1 + \cdots + P_i$. Otherwise, $\frac{1}{2}(d(d+3) - (i-1)) \geq 0$; or equivalently $\mathcal{H}(d, P_1 + \cdots + P_{i-1}) \neq \emptyset$. Now, from Lemma 2.63 we get that $P_1 + \cdots + P_i$ is in d-general position. Furthermore, taking into account Lemma 2.64, in order to prove the result for $P_1 + \cdots + P_i$ we just have to show that there exists a point Q such that linear equation introduced by Q is linearly independent from the linear equations generated by the divisor $P_1 + \cdots + P_{i-1}$. If this is the case, then taking $\Omega = (\mathbb{P}^2(K))^i$ in Lemma 2.64, we get

$$\mathrm{rank}(A(P_1, \ldots, P_i)) = \max\{\mathrm{rank}(A(Q_1, \ldots, Q_i)) \mid Q_1 \times \cdots \times Q_i \in (\mathbb{P}^2(K))^i\}$$
$$= \mathrm{rank}(A(P_1, \ldots, P_{i-1}, Q)) = i.$$

Now, take $\mathcal{C} \in \mathcal{H}(d, P_1 + \cdots + P_{i-1})$ and $Q \in \mathbb{P}^2(K) \backslash \mathcal{C}$; observe that $\mathcal{H}(d, P_1 + \cdots + P_{i-1}) \neq \emptyset$. This point satisfies all our requirements. $\square$

2.5 Local Parametrizations and Puiseux Series

Although the main topic of this book is global rational parametrization of algebraic curves, local parametrization by Puiseux series is also an important tool in the theory of algebraic curves. In this section we recall some basic definitions and facts relating to local parametrizations.

Let us start out with an example of what we want to do in this section.

Example 2.66. Consider the plane algebraic curve $\mathcal{C} \subset \mathbb{A}^2(\mathbb{C})$ defined by the polynomial

$$f(x, y) = y^5 - 4y^4 + 4y^3 + 2x^2y^2 - xy^2 + 2x^2y + 2xy + x^4 + x^3$$

(see Fig. 2.7). Note that the affine point $(0, 2)$ is an isolated singularity of $\mathcal{C}$. Around the origin, $\mathcal{C}$ is parametrized by two different pairs of analytic

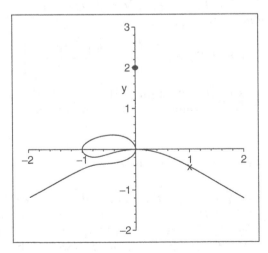

Fig. 2.7. Real part of $\mathcal{C}$

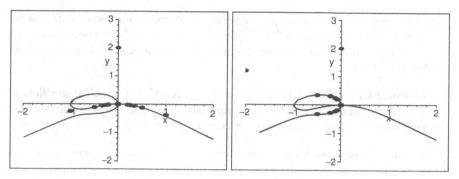

Fig. 2.8. Real part of C and some points generated by $(A_1(t), B_1(t))$ (*left*), real part of C and some points generated by $(A_2(t), B_2(t))$ (*right*)

functions (x_1, y_1) and (x_2, y_2) which have the following power series expansions:

$$(A_1(t), B_1(t)) = \left(t, \; -\frac{1}{2}t^2 + \frac{1}{8}t^4 - \frac{1}{8}t^5 + \frac{1}{16}t^6 + \frac{1}{16}t^7 + \cdots \right),$$

$$(A_2(t), B_2(t)) = \left(-2t^2, \; t + \frac{1}{4}t^2 - \frac{27}{32}t^3 - \frac{7}{8}t^4 - \frac{4057}{2048}t^5 + \cdots \right).$$

In a neighborhood around the origin these power series actually converge to points of the curve C. In fact, these two power series correspond to what we want to call the two branches of C through the origin. In Fig. 2.8 we exhibit how $(A_i(t), B_i(t))$ approaches the curve C in a neighborhood of the origin. We will be interested in determining such power series.

2.5.1 Power Series, Places, and Branches

We denote by $K[[t]]$ the domain of *formal power series* in the indeterminate t with coefficients in the field K, i.e., the set of all sums of the form $\sum_{i=0}^{\infty} a_i t^i$, where $a_i \in K$. The quotient field of $K[[t]]$ is called the field of *formal Laurent series* and is denoted by $K((t))$. As is well known, every nonzero formal Laurent series $A(t) \in K((t))$ can be written in the form

$$A(t) = t^k \cdot (a_0 + a_1 t + a_2 t^2 + \cdots), \quad \text{where } a_0 \neq 0 \text{ and } k \in \mathbb{Z}.$$

The exponent k, i.e., the exponent of the first nonvanishing term of A, is called the *order* of A. We denote it by $\operatorname{ord}(A)$. We let the order of 0 be ∞.

The units in $K[[t]]$ are exactly the power series of order 0, i.e., those having a nonzero constant term. If $\operatorname{ord}(A) = 0$, then A^{-1} can be computed by an obvious recursive process, in which linear equations over K have to be solved. It is easy to check whether a power series is a multiple of another one: $A \mid B$ if and only if $\operatorname{ord}(A) \leq \operatorname{ord}(B)$.

In the sequel we will need power series with fractional exponents. So we will consider Laurent series $K((t^{1/n}))$ in $t^{1/n}$, $n \in \mathbb{N}$. In fact, the union of all these fields of Laurent series with denominator n, for $n \in \mathbb{N}$, is again a field (see Exercise 2.24).

Definition 2.67. *The field $K \ll t \gg := \bigcup_{n=1}^{\infty} K((t^{1/n}))$ is called the field of* formal Puiseux series. *The* order *of a nonzero Puiseux series A is the smallest exponent of a term with nonvanishing coefficient in A. The* order *of 0 is ∞.*

Note that Puiseux series are power series with fractional exponents. In addition, every Puiseux series has a bound n for the denominators of exponents with nonvanishing coefficients.

The substitution of constants for the indeterminate x in a formal power series is usually meaningless. This operation only makes sense for convergent power series in a certain neighborhood of the origin. But we can always substitute 0 for the variable in a power series $A = a_0 + a_1 t + a_2 t^2 + \cdots$, getting the constant coefficient a_0.

It is useful to define the substitution of a power series into another. Let $A, B \in K[[t]]$, $A = a_0 + a_1 t + a_2 t^2 + \cdots$, $B = b_1 t + b_2 t^2 + \cdots$, i.e. $\mathrm{ord}(B) \geq 1$. Then the substitution $A(B)$ is defined as

$$A(B) = a_0 + a_1 B + a_2 B^2 + a_3 B^3 + \cdots =$$
$$= a_0 + a_1 b_1 t + (a_1 b_2 + a_2 b_1^2) t^2 + (a_1 b_3 + 2 a_2 b_1 b_2 + a_3 b_1^3) t^3 + \cdots .$$

In order to avoid the problem of substitution of constants we have to request that $\mathrm{ord}(B) \geq 1$. The following properties of the substitution operation can be easily proved.

Lemma 2.68. *Let $A, B, C \in K[[x]]$, $\mathrm{ord}(B), \mathrm{ord}(C) \geq 1$.*

(a) $(A(B))(C) = A(B(C))$.
(b) *If $\mathrm{ord}(B) = 1$ then there exists a power series B' of order 1, such that* $A = (A(B))(B')$.
(c) *The mapping $A \longrightarrow A(B)$ is an endomorphism on $K[[x]]$.*
(d) *If $\mathrm{ord}(B) = 1$ then the mapping $A \longrightarrow A(B)$ is an automorphism of $K[[x]]$ over K which preserves the order of the elements.*

A curve defined over the field K can be considered to have points over the bigger field $K((t))$ of Laurent series, i.e. in $\mathbb{P}^2(K((t)))$. Such a point, not being constant, is called a local parametrization of the curve. $\mathbb{P}^2(K)$ is naturally embedded in $\mathbb{P}^2(K((t)))$. $\mathbb{P}^2(K)$ corresponds to those points $(x : y : z) \in \mathbb{P}^2(K((t)))$, such that $u \cdot (x, y, z) \in K^3$ for some $u \in K((t))^*$. These considerations lead to the following definition.

Definition 2.69. *Let $\mathcal{C} \subset \mathbb{P}^2(K)$ be a curve defined by the homogeneous polynomial $F(x, y, z) \in K[x, y, z]$. Let $A(t), B(t), C(t)$ be in $K((t))$ such that*

(i) $F(A, B, C) = 0$, and

(ii) there is no $D(t) \in K((t))^*$ such that $D \cdot (A, B, C) \in K^3$.

Then the point $\mathcal{P}(t) = (A : B : C) \in \mathbb{P}^2(K((t)))$ is called a (projective) local parametrization *of C.*

So, obviously, A, B, C are just one possible set of projective coordinates for the local parametrization $\mathcal{P}(t) = (A : B : C)$. For every $D \in K((t))^*$, $(DA : DB : DC)$ is another set of projective coordinates for $\mathcal{P}(t)$.

Lemma 2.70. *Every local parametrization of a projective curve C defined over K has coordinates $(A_1 : A_2 : A_3)$ with $A_i \in K[[t]]$ for $i = 1, 2, 3$, and the minimal order of the nonzero components A_i is 0.*

Proof. Let $(\tilde{A}_1 : \tilde{A}_2 : \tilde{A}_3)$ be a local parametrization of C. Let $\tilde{h}$ be the minimal order of the nonzero components $\tilde{A}_i$. Let $h := -\tilde{h}$. We set $A_i := t^h \cdot \tilde{A}_i$. Then $(A_1 : A_2 : A_3)$ satisfies the conditions of the lemma. $\qquad\square$

Definition 2.71. *Let $\mathcal{P} = (A : B : C)$ be a local parametrization of C with $\min\{\operatorname{ord}(A), \operatorname{ord}(B), \operatorname{ord}(C)\} = 0$. Let a, b, c be the constant coefficients of A, B, C, respectively. Then the point $(a : b : c) \in \mathbb{P}^2(K)$ is called the* center *of the local parametrization $\mathcal{P}$.*

Since local parametrizations are just points in the projective space over a bigger field, we can also introduce the notion of affine local parametrization in the obvious way. Namely, let C be an affine curve, and C^* the corresponding projective curve. Let $(A^* : B^* : C^*)$ be a projective local parametrization of C^*. Setting $A := A^*/C^*$ and $B := B^*/C^*$ we get

(i) $f(A, B) = F(A, B, 1) = 0$, and
(ii) not both A and B are in K,

where f defines C and F defines C^*. Then the pair of Laurent series (A, B) is called an *affine local parametrization* of the affine curve C.

Let $(A^*(t) : B^*(t) : C^*(t))$ be a (projective) local parametrization of the projective curve C^* corresponding to the affine curve C, such that $A^*, B^*, C^* \in K[[t]]$ and $\min\{\operatorname{ord}(A^*), \operatorname{ord}(B^*), \operatorname{ord}(C^*)\} = 0$ (cf. Lemma 2.70). If $\operatorname{ord}(C^*) = 0$, then $\operatorname{ord}(C^{*-1}) = 0$, so if we set $A := A^*/C^*, B := B^*/C^*$, then (A, B) is an affine local parametrization of C with $A, B \in K[[t]]$, i.e., with center at a finite affine point. Conversely, every affine local parametrization with center at a finite affine point has coordinates in $K[[t]]$.

Substituting a nonzero power series of positive order into the coordinates of a local parametrization yields a parametrization with the same center.

Definition 2.72. *Two (affine or projective) local parametrizations $\mathcal{P}_1(t)$, $\mathcal{P}_2(t)$ of an algebraic curve C are called* equivalent *iff there exists $A \in K[[t]]$ with $\operatorname{ord}(A) = 1$ such that $\mathcal{P}_1 = \mathcal{P}_2(A)$.*

By Lemma 2.68 we see that this equivalence of local parametrizations is actually an equivalence relation.

Theorem 2.73. *In a suitable affine coordinate system any given local parametrization is equivalent to one of the type*

$$(\, t^n, \ a_1 t^{n_1} + a_2 t^{n_2} + a_3 t^{n_3} + \cdots \,),$$

where $0 < n$, *and* $0 < n_1 < n_2 < n_3 < \cdots$.

Proof. We choose the origin of the affine coordinate system to be the center of the parametrization. This means the parametrization will have the form (A, B), with

$$A(t) = t^n(a_0 + a_1 t + a_2 t^2 + \cdots), \ n > 0,$$
$$B(t) = t^m(b_0 + b_1 t + b_2 t^2 + \cdots), \ m > 0.$$

At least one of a_0, b_0 is not 0; w.l.o.g. (perhaps after interchanging the axes) we may assume $a_0 \neq 0$. So now we have to find a power series $C(t)$ of order 1 such that $A(C) = t^n$. This can be done by making an undetermined ansatz for $C(t)$, and solving the linear equations derived from $A(C) = t^n$. The condition $a_0 \neq 0$ guarantees that these linear equations are solvable. $\qquad \square$

Definition 2.74. *If a local parametrization* $\mathcal{P}(t)$, *or one equivalent to it, has coordinates in* $K((t^n))$, *for some natural number* $n > 1$, *i.e.,* $\mathcal{P}(t) = \mathcal{P}'(t^n)$ *for some parametrization* $\mathcal{P}'(t)$, *then* $\mathcal{P}(t)$ *is said to be* reducible. *Otherwise,* $\mathcal{P}(t)$ *is said to be* irreducible.

The following criterion for irreducibility is proved in [Wal50].

Theorem 2.75. *The local parametrization* $(\, t^n, \ a_1 t^{n_1} + a_2 t^{n_2} + a_3 t^{n_3} + \cdots \,)$, *where* $0 < n$, $0 < n_1 < n_2 < n_3 < \cdots$ *and* $a_i \neq 0$, *is reducible if and only if the integers* $n, n_1, n_2, n_3, \ldots$ *have a common factor greater than* 1.

Now, we are ready to introduce the concept of a place.

Definition 2.76. *An equivalence class of irreducible local parametrizations of the algebraic curve* $\mathcal{C}$ *is called a* place *of* $\mathcal{C}$. *The common center of the local parametrizations is the* center *of the place.*

By abuse of notation, we will denote places of $\mathcal{C}$ by any irreducible local parametrization representative of the equivalence class.

This notion of a place on a curve $\mathcal{C}$ can be motivated by looking at the case $K = \mathbb{C}$. Let us assume that $\mathcal{C}$ is defined by $f \in \mathbb{C}[x, y]$ and the origin O of the affine coordinate system is a point on $\mathcal{C}$. We want to study the local parametrizations of $\mathcal{C}$ around O. If O is a regular point, we may assume w.l.o.g. that $\frac{\partial f}{\partial y}(0,0) \neq 0$. Then, by the Implicit Function Theorem (see Appendix B), there exists a function $y(x)$, analytic in some neighborhood of $x = 0$, such that

- $y(0) = 0$,
- $f(x, y(x)) = 0$, and
- for all (x_0, y_0) in some neighborhood of $(0,0)$ we have $y_0 = y(x_0)$.

This means that the pair of analytic functions $(x, y(x))$ parametrizes $\mathcal{C}$ around the origin. The analytic function $y(x)$ defined by $f(x, y(x)) = 0$ can be expanded into a Taylor series $\sum_{i=0}^{\infty} c_i t^i$ for some $c_i \in \mathbb{C}$, convergent in a certain neighborhood of the origin. If we set $X(t) = t, Y(t) = \sum_{i=0}^{\infty} c_i t^i$, then $f(X(t), Y(t)) = 0$ and X and Y are convergent around $t = 0$. Hence, $(X(t), Y(t))$ is a local parametrization of $\mathcal{C}$ with center at the origin.

If the origin is a singular point of $\mathcal{C}$ then, there exist finitely many pairs of functions $(x(t), y(t))$, analytic in some neighborhood of $t = 0$, such that

- $x(0) = 0, y(0) = 0$,
- $f(x(t), y(t)) = 0$, and
- for every point $(x_0, y_0) \neq (0, 0)$ on $\mathcal{C}$ in a suitable neighborhood of $(0, 0)$ there is exactly one of the pairs of functions $(x(t), y(t))$ for which there exists a unique t_0 such that $x(t_0) = x_0$ and $y(t_0) = y_0$.

Again the pairs of analytic functions can be expanded into power series $(X(t), Y(t))$, resulting in local parametrizations of $\mathcal{C}$. These parametrizations are irreducible because of the claim of uniqueness of t_0.

It is important to note that the pairs of analytic functions parametrizing $\mathcal{C}$ are not unique. However, any such collection of parametrizations gives the same set of points in a suitable neighborhood.

Let $(x'(t), y'(t))$ be a pair of analytic functions different from $(x(t), y(t))$ but giving the same set of points in a suitable neighborhood of $t = 0$. Then there exists a nonconstant analytic function $v(t)$ with $v(0) = 0$, such that $(x(t), y(t)) = (x'(v(t)), y'(v(t)))$. So the two parametrizations are equivalent.

All parametrizations in an equivalence class determine the same set of points as t varies in a certain neighborhood of 0. So all these parametrizations determine a *branch* of $\mathcal{C}$, a branch being a set of all points $(x(t), y(t))$ obtained by allowing t to vary within some neighborhood of 0 within which $x(t)$ and $y(t)$ are analytic. A place on $\mathcal{C}$ is an algebraic counterpart of a branch of $\mathcal{C}$. Places and branches can also be interpreted in terms of valuations rings. For further details in this direction see [Orz81].

It is not hard to see that the center of a parametrization of $\mathcal{C}$ is a point on $\mathcal{C}$, and the proof is left to the reader. The converse, namely that every point on $\mathcal{C}$ is the center of a least one place of $\mathcal{C}$, follows from the fundamental theorem of Puiseux about the algebraic closure of the field of Puiseux series.

2.5.2 Puiseux's Theorem and the Newton Polygon Method

Let us view $f \in K[x, y]$ as a polynomial in y with coefficients in the field of formal Puiseux series $K \ll x \gg$. Computing a power series expansion for y can be seen as solving a polynomial equation in one variable over the field of Puiseux series. Puiseux's Theorem states that a root always exists. In fact, the proof is constructive and provides a method, the so-called Newton polygon method, for actually constructing solutions.

Theorem 2.77 (Puiseux's Theorem). *The field $K \ll x \gg$ is algebraically closed.*

A proof of Puiseux's Theorem can be given constructively by the Newton polygon method. We describe the Newton polygon method here, and point out how it solves the construction of solutions of univariate polynomial equations over $K \ll x \gg$.

We are given a polynomial $f \in K \ll x \gg [y]$ of degree $n > 0$, i.e.

$$f(x, y) = A_0(x) + A_1(x)y + \cdots + A_n(x)y^n, \quad \text{with } A_n \neq 0.$$

If $A_0 = 0$, then obviously $y = 0$ is a solution. So now let us assume that $A_0 \neq 0$. Let $\alpha_i := \mathrm{ord}(A_i)$ and a_i the coefficient of x^{α_i} in A_i, i.e.,

$$A_i(x) = a_i x^{\alpha_i} + \text{terms of higher order.}$$

We will recursively construct a solution $Y(x)$, a Puiseux series in x, of the equation $f(x, y) = 0$. $Y(x)$ must have the form

$$Y(x) = c_1 x^{\gamma_1} + \underbrace{c_2 x^{\gamma_2} + c_3 x^{\gamma_3} + \cdots}_{Y_1(x)},$$

with $c_j \neq 0, \gamma_j \in \mathbb{Q}, \gamma_j < \gamma_{j+1}$ for all j. In order to get necessary conditions for c_1 and γ_1, we substitute the ansatz $Y(x) = c_1 x^{\gamma_1} + Y_1(x)$ for y in $f(x, y)$, getting

$$f(x, Y(x)) = A_0(x) + A_1(x) \cdot (c_1 x^{\gamma_1} + Y_1(x)) + \cdots + A_n(x) \cdot (c_1 x^{\gamma_1} + Y_1(x))^n = 0.$$

The terms of lowest order must cancel. Therefore there must exist at least two indices j, k with $j \neq k$ and $0 \leq j, k \leq n$ such that

$$c_1^j A_j(x) x^{j\gamma_1} = c_1^j a_j x^{\alpha_j + j\gamma_1} + \cdots \quad \text{and} \quad c_1^k A_k(x) x^{k\gamma_1} = c_1^k a_k x^{\alpha_k + k\gamma_1} + \cdots$$

have the same order and this order is minimal. So if we think of the pairs (i, α_i), for $0 \leq i \leq n$, as points in the affine plane over $\mathbb{Q}$ (if $A_i(x) = 0$ then $\alpha_i = \infty$ and this point is not contained in the affine plane) then this condition means that all the points (i, α_i) are on or above the line L connecting (j, α_j) and (k, α_k). If we set $\beta_1 := \alpha_j + j\gamma_1$, then the points (u, v) on this line L satisfy $v = -\beta_1 - u\gamma_1$, i.e., γ_1 is the negative slope of L.

A convenient way of determining the possible values for γ_1 is to consider the so-called Newton polytope of f. This is the smallest convex polytope in the affine plane over $\mathbb{Q}$, which contains all the points (i, α_i). Those faces of the Newton polygon, s.t. all the P_i's lie on or above the corresponding line, have possible values for γ_1 as their negative slopes.

There can be at most n possible values for γ_1. Having determined a value for γ_1, we now take all the points (i, α_i) on the line L. They correspond to the terms of lowest order in $f(x, Y(x))$. So we have to determine a c_1 such that

$$\sum_{\alpha_i + i\gamma_1 = \beta_1} a_i c_1^i = 0.$$

Since K is algebraically closed, this equation will always have nonzero solutions in K. The possible values for c_1 are the nonzero roots of this equation.

So after γ_1 and c_1 have been determined, the same process is performed on $Y_1(x)$, which must be a root of the equation

$$f_1(x, y_1) = f(x, c_1 x^{\gamma_1} + y_1) = 0.$$

Again the Newton polygon may be used to derive necessary conditions on c_2 and γ_2. However, this time only those lines are considered whose corresponding negative slope γ_2 is greater than γ_1.

This recursive process in the Newton polygon method can be iterated until the desired number of terms is computed, or no further splitting of solutions is possible.

A detailed proof of the fact, that the Newton polygon method can be performed on any polynomial f and that it actually yields Puiseux series (with bounded denominators of exponents) is given in [Wal50].

Now we are ready to see that every point P on an affine curve C has a corresponding place with center at P. For a proof of the following theorem we refer to [Wal50], Theorem 4.1 in Chap. 4.

Theorem 2.78. *Let $f(x, y)$ be a polynomial in $K[x, y]$, and let C be the curve defined by f. To each root $Y(x) \in K \ll x \gg$ of $f(x, y) = 0$ with $\text{ord}(Y) > 0$ there corresponds a unique place of C with center at the origin. Conversely, to each place $(X(t), Y(t))$ of C with center at the origin there correspond $\text{ord}(X)$ roots of $f(x, y) = 0$, each of order greater than zero.*

If $Y(x)$ is a Puiseux series solving $f(x, y) = 0$, $\text{ord}(Y) > 0$, and n is the least integer for which $Y(x) \in K((x^{\frac{1}{n}}))$, then we put $x^{\frac{1}{n}} = t$, and (t^n, Y) is a local parametrization with center at the origin. The solutions of $f(x, y)$ of order 0 correspond to places with center on the y-axis (but different from the origin), and the solutions of negative order correspond to places at infinity.

Example 2.79. We consider the curve of Example 2.66. So the defining polynomial for C is

$$f(x, y) = y^5 - 4y^4 + 4y^3 + 2x^2 y^2 - xy^2 + 2x^2 y + 2xy + x^4 + x^3.$$

Figure 2.9 shows the Newton polygon of f. There are three segments on the lower left boundary of the Newton polygon of f. These three segments give three possible choices for the first exponent γ_1 in the Puiseux series expansion of a solution, namely

$$\gamma_1 \in \left\{ 2, \frac{1}{2}, 0 \right\}.$$

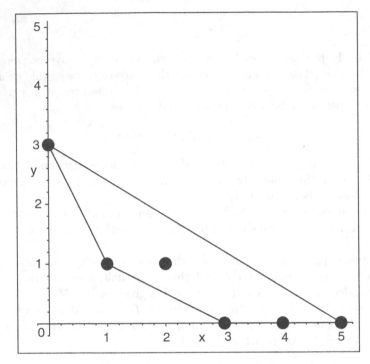

Fig. 2.9. Newton polygon of f

In all three cases the corresponding equation has nonzero roots. In the case of $\gamma_1 = 2$, there are two points on the segment of the Newton polygon, and the corresponding equation

$$1 + 2c_1 = 0$$

has the solution $c_1 = -\frac{1}{2}$. For $\gamma_1 = \frac{1}{2}$ the equation is $4c_1^3 + 2c_1 = 0$, the nonzero solutions are $\pm\frac{1}{\sqrt{-2}}$. Finally, for $\gamma_1 = 0$ the equation is $c_1^5 - 4c_1^4 + 4c_1^3 = c_1^3(c_1 - 2)^2 = 0$, the nonzero solution is 2.

So we get four possible smallest terms of Puiseux series solving $f(x, y) = 0$. Since the field of Puiseux series is algebraically closed, and f is a squarefree polynomial of degree 5, there must be 2 solutions starting with the same term. This is the case for the series starting with the term $2x^0$. Continuing the process with this series, we would see that in the next step it splits into the two different solutions

$$2 + \frac{1 + \sqrt{-95}}{8}x + \cdots \quad and \quad 2 + \frac{1 - \sqrt{-95}}{8}x + \cdots.$$

We continue to expand the series starting with $-\frac{1}{2}x^2$. For determining the next highest exponent γ_2 and nonzero coefficient c_2, we make the ansatz

$Y(x) = -\frac{1}{2}x^2 + Y_1(x)$. Now $Y_1(x)$ must solve the modified equation

$$f_1(x, y_1) = f\left(x, -\frac{1}{2}x^2 + y_1\right)$$

$$= y_1^5 - \left(\frac{5}{2}x^2 + 4\right)y_1^4 + \left(\frac{5}{2}x^4 + 8x^2 + 4\right)y_1^3$$

$$- \left(\frac{5}{4}x^6 + 6x^4 + 4x^2 + x\right)y_1^2$$

$$+ \left(\frac{5}{16}x^8 + 2x^6 + x^4 + x^3 + 2x^2 + 2x\right)y_1 - \frac{1}{32}x^{10} - \frac{1}{4}x^8 - \frac{1}{4}x^5.$$

The Newton polygon of f_1 has only one segment with negative slope greater than $\gamma_1 = 2$. So we get $\gamma_2 = 4$ and $c_2 = \frac{1}{8}$.

Repeating this process, we finally get the following series expansions for the solutions of $f(x, y) = 0$:

$$Y_1(x) = -\frac{1}{2}x^2 + \frac{1}{8}x^4 - \frac{1}{8}x^5 + \frac{1}{16}x^6 + \frac{1}{16}x^7 + \cdots ,$$

$$Y_2(x) = \frac{\sqrt{-2}}{2}x^{\frac{1}{2}} - \frac{1}{8}x + \frac{27\sqrt{-2}}{128}x^{\frac{3}{2}} - \frac{7}{32}x^2 - \frac{4057\sqrt{-2}}{16384}x^{\frac{5}{2}} + \cdots ,$$

$$Y_3(x) = -\frac{\sqrt{-2}}{2}x^{\frac{1}{2}} - \frac{1}{8}x - \frac{27\sqrt{-2}}{128}x^{\frac{3}{2}} - \frac{7}{32}x^2 + \frac{4057\sqrt{-2}}{16384}x^{\frac{5}{2}} + \cdots ,$$

$$Y_4(x) = 2 + \frac{1 + \sqrt{-95}}{8}x + \frac{1425 - 47\sqrt{-95}}{3040}x^2 + \cdots ,$$

$$Y_5(x) = 2 + \frac{1 - \sqrt{-95}}{8}x + \frac{1425 + 47\sqrt{-95}}{3040}x^2 + \cdots .$$

Y_1, Y_2, Y_3 have order greater than 0, so they correspond to places of $\mathcal{C}$ centered at the origin. Y_1 corresponds to the local parametrization

$$(A_1(t), B_1(t)) = \left(t, -\frac{1}{2}t^2 + \frac{1}{8}t^4 - \frac{1}{8}t^5 + \frac{1}{16}t^6 + \frac{1}{16}t^7 + \cdots\right),$$

and Y_2, Y_3 both correspond to the local parametrization

$$(A_2(t), B_2(t)) = \left(-2t^2, \ t + \frac{1}{4}t^2 - \frac{27}{32}t^3 - \frac{7}{8}t^4 - \frac{4057}{2048}t^5 + \cdots\right).$$

Y_4, Y_5 have order 0, and they correspond to parametrizations centered at $(0, 2)$. Not all the branches corresponding to these parametrizations can be seen in Fig. 2.7, since they are complex, except for the point $(0, 2)$.

In Definition 4.1, we will define the notion of rational parametrization. In fact, such rational parametrizations might be called global parametrizations versus the local ones. In the following example we determine a global polynomial parametrization by the Newton polygon method.

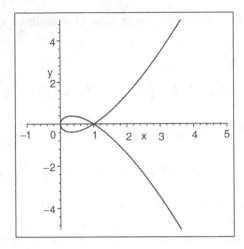

Fig. 2.10. Real part of $\mathcal{C}$

Example 2.80. Let us consider the curve $\mathcal{C}$ in $\mathbb{A}^2(\mathbb{C})$ defined by

$$f(x, y) = y^2 - x^3 + 2x^2 - x.$$

A plot of $\mathcal{C}$ around the origin is given in Fig. 2.10. Let us determine the local parametrizations of $\mathcal{C}$ centered at the origin. $\mathcal{C}$ has only one branch at $(0,0)$, so there should be exactly one place at the origin. We have

$$A_0(x) = -x + 2x^2 - x^3, \quad \alpha_0 = 1, \quad a_0 = -1,$$
$$A_1(x) = 0, \quad\quad\quad\quad\quad \alpha_1 = \infty,$$
$$A_2(x) = 1, \quad\quad\quad\quad\quad \alpha_2 = 0, \quad a_2 = 1.$$

The Newton polygon of f is given in Fig. 2.11 (left).

So $\gamma_1 = \frac{1}{2}$, and c_1 is the solution of the equation $-1 + c_1^2 = 0$, i.e. $c_1 = \pm 1$. We get two different Puiseux series solutions of $f(x, y) = 0$, starting with

$$Y_1(x) = x^{\frac{1}{2}} + \cdots, \quad \text{and} \quad Y_2(x) = -x^{\frac{1}{2}} + \cdots.$$

We continue to expand Y_1. For determining the next term in Y_1, we get the equation

$$f_1(x, y_1) = f(x, x^{\frac{1}{2}} + y_1) = y_1^2 + 2x^{\frac{1}{2}}y_1 + 2x^2 - x^3.$$

The Newton polygon of f_1 is given in Fig. 2.11 (right). There is only one segment with negative slope greater than $\frac{1}{2}$, namely $\gamma_2 = \frac{3}{2}$. The corresponding equation $2 + 2c_2 = 0$ yields $c_2 = -1$. So now we have

$$Y_1(x) = x^{\frac{1}{2}} - x^{\frac{3}{2}} + \cdots .$$

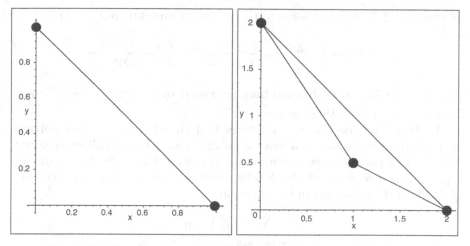

Fig. 2.11. Newton polygon of f (*left*), Newton polygon of f_1 (*right*)

For determining the next term, we consider the equation

$$f_2(x, y_2) = f(x, x^{\frac{1}{2}} - x^{\frac{3}{2}} + y_2) = y_2(y_2 + 2x^{\frac{1}{2}} - 2x^{\frac{3}{2}}).$$

Since y_2 divides $f_2(x, y_2)$, $y_2 = 0$ is a solution. Thus,

$$Y_1(x) = x^{\frac{1}{2}} - x^{\frac{3}{2}}.$$

In the same way we could expand Y_2 further, and we would get

$$Y_2(x) = -x^{\frac{1}{2}} + x^{\frac{3}{2}}.$$

So, by setting $x^{\frac{1}{2}} = \pm t$ in Y_1, Y_2, respectively, we get the local parametrization

$$\mathcal{P}(t) = (t^2, t - t^3).$$

The series in this local parametrization converge for every t, and in fact $\mathcal{P}(t)$ is a global parametrization of $\mathcal{C}$ (compare Chap. 4).

2.5.3 Rational Newton Polygon Method

As we have seen above, several different Puiseux series may correspond to the same parametrization. So we have the problem of identifying equivalent Puiseux series. On the other hand, if $\mathbb{K}$ is a subfield of K and $f \in \mathbb{K}[x, y]$, these series may have coefficients in a more complicated algebraic extension field of $\mathbb{K}$ than the corresponding parametrizations. Compare, for instance, the Puiseux series

$$Y_2(x) = \frac{\sqrt{-2}}{2}x^{\frac{1}{2}} - \frac{1}{8}x + \frac{27\sqrt{-2}}{128}x^{\frac{3}{2}} - \frac{7}{32}x^2 - \frac{4057\sqrt{-2}}{16384}x^{\frac{5}{2}} + \cdots,$$

$$Y_3(x) = -\frac{\sqrt{-2}}{2}x^{\frac{1}{2}} - \frac{1}{8}x - \frac{27\sqrt{-2}}{128}x^{\frac{3}{2}} - \frac{7}{32}x^2 + \frac{4057\sqrt{-2}}{16384}x^{\frac{5}{2}} + \cdots,$$

of Example 2.79, corresponding to the parametrization (here $\mathbb{K} = \mathbb{Q}$)

$$(A_2(t), B_2(t)) = \left(-2t^2,\ t + \frac{1}{4}t^2 - \frac{27}{32}t^3 - \frac{7}{8}t^4 - \frac{4057}{2048}t^5 + \cdots\right).$$

The parametrization is obtained from $(x, Y_2(x))$ by setting $t = \frac{\sqrt{-2}}{2}x^{1/2}$ and from $(x, Y_3(x))$ by setting $t = -\frac{\sqrt{-2}}{2}x^{1/2}$.

D. Duval has developed a refinement of the classical Newton polygon method, the so-called rational Newton polygon method, which allows to detect this kind of parameter substitutions at the stage where the corresponding equation of a segment of the Newton polygon has to be solved [Duv87], [Duv89]. If the exponents in the equation

$$\sum_{\alpha_i + i\gamma_1 = \beta_1} a_i c_1^i = 0$$

have a greatest common divisor $q > 1$, then

$$\sum_{\alpha_i + i\gamma_1 = \beta_1} a_i c_1^i = \sum_{\alpha_i + i\gamma_1 = \beta_1} a_i \tilde{c}_1^{i/q} \quad \text{for } \tilde{c}_1 = c_1{}^q.$$

It suffices to solve for $\tilde{c}_1$. In terms of parametrizations, this amounts to setting $X(t) = t^{mq}$, if $X(t) = t^m$ has been computed so far, starting with $X(t) = t$. Additionally it is sometimes possible to avoid algebraic extension for the coefficients if one allows $X(t) = \lambda t^m$.

Applying this idea to the computation in Example 2.79, we see that the segment corresponding to $\gamma = \frac{1}{2}$ has the equation $4c^2 + 2 = 0$, which can be simplified to $4\tilde{c} + 2 = 0$. This equation has the root $\tilde{c} = -\frac{1}{2}$. For determining the corresponding parametrization, it is sufficient to set $X(t) = -2t^2$ and $Y(t) = t$. In this way the algebraic extension of the coefficient field $\mathbb{Q}$ by $\sqrt{-2}$ can be avoided.

For a detailed complexity analysis of Puiseux series expansion and local parametrization we refer to [Sta00].

Exercises

2.1. Extend the notion of multiplicity at a point and of tangent to curves with multiple components. Generalize Theorem 2.4 to curves with multiple components. Find a curve with multiple components having infinitely many double points.

2.2. Prove that every reducible curve has singularities.

2.3. Let A be a symmetric matrix of order 3 over K, and let C be the conic defined by A; i.e. the conic defined by the polynomial

$$F(x, y, z) = (x, y, z) \cdot A \cdot \begin{pmatrix} x \\ y \\ z \end{pmatrix}.$$

(i) Prove that C is irreducible if and only if $\det(A) \neq 0$.

(ii) Prove that if C is irreducible then C is nonsingular. What are the singularities of C if it is reducible?

(iii) Now, let $K = \mathbb{C}$, and let A be real and regular. Let λ_1, λ_2 be the eigenvalues of the 2×2 principal submatrix of A. Prove that

    a. if $\lambda_1 \cdot \lambda_2 > 0$ then C is a circle iff $\lambda_1 = \lambda_2$; otherwise it is a ellipse.

    b. if $\lambda_1 \cdot \lambda_2 < 0$ then C is a hyperbola.

    c. if $\lambda_1 \cdot \lambda_2 = 0$ then C is a parabola.

2.4. Compute the singular points of the *three-leafed rose*; i.e. of the projective plane curve defined over $\mathbb{C}$ by $F(x, y, z) = (x^2 + y^2)^2 + rx(3y^2 - x^2)z$, where $r \in \mathbb{C}, r \neq 0$. Determine the tangents to the curve at each singularity.

2.5. Let C be the projective plane curve defined by the irreducible form $F \in K[x, y, z]$, and let $P = (a : b : c) \in \mathbb{P}^2(K)$ be such that the polynomial

$$G(x, y, z) = a\frac{\partial F}{\partial x} + b\frac{\partial F}{\partial y} + c\frac{\partial F}{\partial z}$$

is not identically zero.

    a. If D is the projective curve defined by G, prove that $C \cap D$ is the set of singular points of C and those points Q on C at which the line passing through P and Q is tangent to C at Q. The curve D is called the *polar curve* of P with respect to C.

    b. Check that, for conics, the notion of polar curve coincides with the elementary geometric concept of polar line to a conic.

2.6. Let C be the *Folium of Descartes*, i.e. C is the projective curve defined over $\mathbb{C}$ by $x^3 + y^3 - 3axyz$, where $a \in \mathbb{R}, a \neq 0$. Obtain the real regular points on C (i.e. regular points with coordinates over $\mathbb{R}$) at which the tangent to C passes through the point $(4a : 4a : 1)$.

2.7. Let C be the irreducible projective plane curve defined by the irreducible form $F \in K[x, y, z]$, and suppose that C is not a line. We consider the ring homomorphism ψ between $K[x, y, z]$ and the coordinate ring $\Gamma(C)$ such that

$$\psi(x) = \frac{\partial F}{\partial x} \bmod I(C), \quad \psi(y) = \frac{\partial F}{\partial y} \bmod I(C), \quad \psi(z) = \frac{\partial F}{\partial z} \bmod I(C).$$

Prove that $\ker(\psi)$ is a homogeneous prime ideal.

2.8. With the notation of Exercise 2.7:

(i) Prove that $V(\ker(\psi))$ is an irreducible curve. This curve is called the *dual curve of* C.

(ii) Prove that the dual curve of C is the algebraic closure of the set

$$\left\{ \left(\frac{\partial F(P)}{\partial x} : \frac{\partial F(P)}{\partial y} : \frac{\partial F(P)}{\partial z} \right) \middle/ P \in C \text{ is simple} \right\}.$$

(iii) Let A be a symmetric regular matrix of order 3 over K, and let C be the irreducible conic defined by A (see Exercise 2.3). Prove that the dual curve of C is the conic defined by A^{-1}.

(iv) Compute the dual curve of the cubic $x^3 + y^3 - z^3$.

2.9. Let C be the *tacnode*; i.e., C is the curve defined by the polynomial $F(x, y, z) = 2x^4 - 3x^2yz + y^2z^2 - 2y^3z + y^4$, and let $\mathcal{D}$ be the cubic defined by $G(x, y, z) = -1850\,xyz + 1850\,xy^2 - 90\,x^2z - 3114\,x^2y + 2617\,x^3$.

(i) Compute the singularities, their multiplicity, tangents, and character of C and $\mathcal{D}$.

(ii) Compute the multiplicity of intersection at the intersection points of C and $\mathcal{D}$. Compare to Bézout's theorem.

2.10. Generalize Lemma 2.9 to curves with multiple components.

2.11. Consider the tacnode curve C defined by $y^4 - 2y^3 + y^2 - 3x^2y + 2x^4$. Let $f = xy^5 + xy^3$ and $g = 2xy^4 + 3x^3y^2 - 2x^5y$. Decide whether $f = g$ as polynomial functions on C.

2.12. Consider the variety V in $\mathbb{A}^3(\mathbb{C})$ defined by $x^2 + y^2 - 4 = (x - 2)^2 + y^2 + z^2 - 16 = 0$. Let $f = 4x + 15z^2$ and $g = 16y^2 + z^4 - 8$. Decide whether $f = g$ as polynomial functions on V.

2.13. Consider the circle C in $\mathbb{A}^2(\mathbb{C})$ defined by $x^2 + y^2 - 1$ and the polynomial mapping $\varphi(x, y) = (x + y, x - y)$ from the plane to itself. What is the image of C under φ?

2.14. Consider the circle C in $\mathbb{A}^2(\mathbb{C})$ defined by $x^2 + y^2 - 1$ and the polynomial mapping $\varphi(x, y) = (x + y, x + 2y, x + 3y)$ from $\mathbb{A}^2(\mathbb{C})$ to $\mathbb{A}^3(\mathbb{C})$. What is the image of C under φ? Give defining equations for $\varphi(C)$. Is $\varphi(C)$ irreducible?

2.15. Is the algebraic curve C defined by $x^6 + 3x^4y^2 - x^2y^2 + 3x^2y^4 + y^6$ in $\mathbb{A}^2(\mathbb{C})$ irreducible?

2.16. Let C be the parabola defined in $\mathbb{A}^2(\mathbb{C})$ by $y - 2x^2$. Determine the poles of the rational function $\varphi(x, y) = \frac{2x}{y - 2x}$ on C, if any.

2.17. Extract an algorithm for the computation of intersection multiplicity from Theorem 2.50.

2.18. Extend the notion of multiplicity of intersection to curves with multiple components.

2.19. Generalize Theorem 2.50(7) to curves with multiple components.

2.20. Compute the intersection multiplicity of the curves C and D defined by the polynomials

$$F(x, y, z) = x^3 + y^3 - 2xyz,$$

$$G(x, y, z) = 2x^3 - 4x^2y + 3xy^2 + y^3 - 2y^2z$$

at the intersection points.

2.21. Compute a linear system of quintics generated by six double points; i.e. generated by $2P_1 + 2P_2 + 2P_3 + 2P_4 + 2P_5 + 2P_6$, for some $P_i \in \mathbb{P}^2(\mathbb{C})$.

2.22. Determine an irreducible quintic such that all its singularities are double points, and the equality in Theorem 2.60 holds.

2.23. Let $P = (1 : 1 : 1)$ and $Q = (1 : 0 : 1)$. Prove that $D = 2P + Q$ is in 3-general position. Compute $\dim(3, D)$.

2.24. Prove that the union of all fields of Laurent series with denominator n, for $n \in \mathbb{N}$, is a field.

2.25. Prove that the center of a local parametrization of a plane curve C is a point on C.

2.26. Consider the local parametrizations of the curve C of Examples 2.66 and 2.79 centered at $(0, 2)$. Find a suitable coordinate system, such that the parametrization is equivalent to the form in Theorem 2.73.

2.27. Carry out the Newton polygon method for determining the exponents and coefficients of the first three terms of the solutions Y_2, Y_3 in Example 2.79.

2.28. Suppose the Newton polygon of the polynomial $f(x, y)$ has a segment of positive slope such that all the points (i, α_i) are on or above this segment. What does this mean for the corresponding Puiseux series solutions?

3

The Genus of a Curve

The genus of a curve is a birational invariant which plays an important role in the parametrization of algebraic curves (and in the geometry of algebraic curves in general). In fact, only curves of genus 0 can be rationally parametrized. So in the process of parametrization we will first compute the genus of the curve $\mathcal{C}$. This will involve an analysis of the singularities of $\mathcal{C}$, and we will determine the genus as the deficiency between a bound on the number of singularities and the actual number of singularities of $\mathcal{C}$. But in order to arrive at a definition of the genus, we first need to consider divisors on $\mathcal{C}$ and their associated linear spaces. We do not want to repeat this classical development here (for further details see for instance [Ful89] or [Wal50]), but we give a kind of road map for getting to the definition of the genus and from there to a method for computing it.

The chapter consists of three sections. In Sect. 3.1 we present the formal definition of genus. In Sect. 3.2 we see how the genus can be computed by blowing up the singularities, and in Sect. 3.3 we study the problem of carrying out symbolically the algorithmic methods in Sect. 3.2

3.1 Divisor Spaces and Genus

In this section we review some standard notions and properties of divisors of curves. We state the facts without proofs. The reader may consult [LiV00] or [Ful89] for details. We will also make use of Sect. 2.4.

Let $\mathcal{C}$ be a nonsingular irreducible algebraic curve in $\mathbb{P}^2(K)$, defined by the homogeneous polynomial F. Consider an element $[G] \in \Gamma(\mathcal{C})$, the coordinate ring of $\mathcal{C}$, and let $\mathcal{D}$ be the curve defined by G. Let P be a point on $\mathcal{C}$. We define the *order of* $[G]$ *at* P *w.r.t.* $\mathcal{C}$ to be

$$\mathrm{ord}_{P,\mathcal{C}}([G]) = \mathrm{mult}_P(\mathcal{C}, \mathcal{D}).$$

Because of the properties of the intersection multiplicity (compare Theorem 2.50), this definition is independent of the representative G.

Now consider a nonzero rational function $\varphi = G/H$ on $\mathcal{C}$, G and H having no common factors. Since φ is in the function field of the projective curve $\mathcal{C}$, the degrees of G and H have to be the same. Let $\mathcal{E}$ be the curve defined by H. By Bézout's Theorem (see Theorem 2.48), $\mathcal{D}$ and $\mathcal{E}$ have the same number of intersections with $\mathcal{C}$. So if we define the *order of* φ *at* P *w.r.t.* $\mathcal{C}$ to be

$$\operatorname{ord}_{P,\mathcal{C}}(\varphi) = \operatorname{ord}_{P,\mathcal{C}}([G]) - \operatorname{ord}_{P,\mathcal{C}}([H]),$$

then

$$\sum_{p \in \mathcal{C}} \operatorname{ord}_{P,\mathcal{C}}(\varphi) = 0,$$

i.e., φ has as many zeros as poles on $\mathcal{C}$ (see Exercise 3.2).

In Sect. 2.4 we have introduced the notion of a divisor. Here we focus on divisors of curves.

Definition 3.1. *Let* $\operatorname{Div}(\mathcal{C})$ *be the free abelian group generated by the points of* $\mathcal{C}$; *i.e. an element of* $\operatorname{Div}(\mathcal{C})$ *is of the form*

$$D = \sum_{P \in \mathcal{C}} n_P P,$$

where $n_P \in \mathbb{Z}$ *and almost all the* n_P *'s are zero. The elements of* $\operatorname{Div}(\mathcal{C})$ *are called* divisors *of* $\mathcal{C}$. *Furthermore, the* support *of* D *is the set of those points, whose coefficients are nonzero; i.e.*

$$\operatorname{Supp}(D) = \{P \in \mathcal{C} \mid n_P \neq 0\}.$$

As we have defined in Section 2.4, a divisor D *of* $\mathcal{C}$ *is* positive *or* effective *iff* $n_P > 0$ *for all* $P \in \operatorname{Supp}(D)$. *In this case we write* $D \succ 0$. *Furthermore, we say* $D \succ D'$ *iff* $D - D' \succ 0$. *The degree of the divisor* D *is* $\deg(D) = \sum n_P$.

Note that the 0-divisor, $\sum 0 \cdot P$, is effective.

Let $\varphi = G/H$ be a nonzero rational function on $\mathcal{C}$. Then the following divisor $\div(\varphi)$ is generated in a natural way by φ:

$$\div(\varphi) := \sum_{P \in \mathcal{C}} \operatorname{ord}_{P,\mathcal{C}}(\varphi) P.$$

If we let

$$\div_0(\varphi) := \sum_{\operatorname{ord}_{P,\mathcal{C}}(\varphi) > 0} \operatorname{ord}_{P,\mathcal{C}}(\varphi) P, \qquad \text{the } divisor \text{ } of \text{ } zeros \text{ } of \text{ } \varphi,$$

$$\div_\infty(\varphi) := - \sum_{\operatorname{ord}_{P,\mathcal{C}}(\varphi) < 0} \operatorname{ord}_{P,\mathcal{C}}(\varphi) P, \qquad \text{the } divisor \text{ } of \text{ } poles \text{ } of \text{ } \varphi,$$

then $\div(\varphi) = \div_0(\varphi) - \div_\infty(\varphi)$. The 0-divisor is generated by constant rational functions.

Now we associate a linear space of rational functions with a divisor D.

Definition 3.2. *Let D be a divisor of $\mathcal{C}$. The* linear space *of D over K is defined as*

$$\mathcal{L}(D) := \{\varphi \in K(\mathcal{C}) \mid D + \div(\varphi) \succ 0\}.$$

By $\ell(D)$ we denote the dimension of $\mathcal{L}(D)$.

If $D = D_+ - D_-$ for effective divisors D_+, D_- of disjoint support, then the condition $D + \div(\varphi) \succ 0$ means that φ has zeros of order larger or equal to D_- and poles of order smaller or equal to D_+, i.e. $\div_0(\varphi) - D_- \succ 0$ and $D_+ - \div_\infty(\varphi) \succ 0$.

Riemann's Theorem connects the degree of divisors to the dimension of their linear spaces. It will allow us to define the notion of genus of a curve.

Theorem 3.3 (Riemann's Theorem). *There is a constant g ($\in \mathbb{N}$) depending on $\mathcal{C}$ such that $\ell(D) \geq \deg(D) + 1 - g$ for all divisors D of $\mathcal{C}$.*

Definition 3.4. *The* genus *of a nonsingular irreducible curve $\mathcal{C}$ is the least possible value of the constant g in Riemann's Theorem.*

In [Ful89] this theory of divisors and also Riemann's Theorem are developed in an analogous way for nonsingular not necessarily plane curves, e.g., nonsingular models of plane curves. Roughly speaking, a nonsingular model of an irreducible plane curve is a nonsingular not necessarily plane curve which is birationally equivalent to $\mathcal{C}$ (see [Ful89] Sect. 7.5 for details). The whole theory of divisors depends only on the function field of $\mathcal{C}$, so it is invariant under birational transformations. Therefore, also the genus is a birational invariant. This leads to an extension of Definition 3.4 to arbitrary irreducible plane curves, because every such curve has a nonsingular model.

Definition 3.5. *Let $\mathcal{C}$ be an irreducible plane curve, and let $\mathcal{X}$ be its nonsingular model. The* genus *of $\mathcal{C}$ is the genus of $\mathcal{X}$. We denote it by* genus($\mathcal{C}$).

The genus of a curve $\mathcal{C}$ can also be introduced in different ways. So, for instance, we might view a complex curve as a surface in real 4-space (the *Riemann surface* of $\mathcal{C}$), and define the genus of $\mathcal{C}$ as the number of topological handles of this surface. Compare Sect. 9.2 in [BrK86]. In this way we get an equivalent definition of the notion of genus.

3.2 Computation of the Genus

For computing the genus of an irreducible plane curve, we will apply quadratic transformations (blow-ups) for birationally transforming the curve into a curve with only ordinary singularities (see Definition 2.6). For such a curve the genus can be readily determined by proper counting the singularities. Alternatively one could apply, for instance, local parametrizations by Puiseux series (see Sect. 2.5 and [Sta00]).

Throughout this section and also in subsequent chapters we will denote by $\mathrm{Sing}(\mathcal{C})$ the *singular locus* of a projective curve $\mathcal{C}$, i.e., if $F(x, y, z)$ is the homogeneous defining polynomial of $\mathcal{C}$, then

$$\mathrm{Sing}(\mathcal{C}) \;=\; V\left(\frac{\partial F}{\partial x}, \frac{\partial F}{\partial y}, \frac{\partial F}{\partial z}\right)$$

(compare Theorem 2.14). In Theorem 2.60 we have proved that for an irreducible projective curve $\mathcal{C}$ of degree d we have the bound

$$\sum_{P \in \mathrm{Sing}(\mathcal{C})} \mathrm{mult}_P(\mathcal{C})(\mathrm{mult}_P(\mathcal{C}) - 1) \;\leq\; (d-1)(d-2)$$

on the multiplicities of singular points.

This bound is actually sharp. Every irreducible conic achieves this bound, and also every irreducible cubic having a double point. As we will see later, the curves achieving this bound can be rationally parametrized; i.e., they are birationally equivalent to a line. Such curves are of central importance in the field of computer aided geometric design (CAGD). Curves achieving this bound are, in fact, curves of genus 0. Moreover, if a curve $\mathcal{C}$ has only ordinary singularities, then the difference between this bound and the actual number of singularities on $\mathcal{C}$ is basically the genus of $\mathcal{C}$ (actually $2 \cdot \mathrm{genus}(\mathcal{C})$), as proved for instance in [Ful89] Sect. 8.3. This gives us a good computational method for determining the genus of a curve.

Theorem 3.6. *Let $\mathcal{C}$ be a curve with only ordinary singularities, and let d be the degree of $\mathcal{C}$. Then*

$$\mathrm{genus}(\mathcal{C}) \;=\; \frac{1}{2}\left[(d-1)(d-2) \;-\; \sum_{P \in \mathrm{Sing}(\mathcal{C})} \mathrm{mult}_P(\mathcal{C})(\mathrm{mult}_P(\mathcal{C}) - 1)\right].$$

Nonordinary singularities have to be treated specially in the genus formula. In some sense, which will be made precise below, a nonordinary singularity might have other singularities in its "neighborhood," which have to be counted properly in the genus formula. The analysis of such neighborhoods is the topic of the field of resolution of singularities (cf. [Abh66], [Zar39]). Here we treat only a specific subproblem in the resolution of curve singularities, namely the determination of the so-called neighboring singularities and their multiplicities.

More precisely, the plan is the following: since the genus of a curve is invariant under birational transformation (compare the remark after Definition 3.4), we show that there is a sequence of birational transformations $\mathcal{Q} = (\mathcal{Q}_1, \ldots, \mathcal{Q}_n)$ such that

$$\mathcal{C} = \mathcal{C}_0 \xrightarrow{\;\mathcal{Q}_1\;} \mathcal{C}_1 \xrightarrow{\;\mathcal{Q}_2\;} \cdots \xrightarrow{\;\mathcal{Q}_n\;} \mathcal{C}_n,$$

where C_n has only ordinary singularities. So, no matter which sequence Q we take, we get genus(C) = genus(C_n).

The problem with nonordinary singularities is that they have coinciding tangents, i.e., multiple tangents. We will resolve these multiple tangents by "blowing up" the singularity into a line. Then the tangents at the singularity correspond to points on this line. A multiple tangent will correspond to a multiple point on the blow-up. This point will also have to be properly counted in the genus formula. Now this multiple point, a "neighboring singularity," can be investigated further. If it is ordinary, then the process stops, and we have "resolved" the nonordinariness of this singularity. Otherwise the process is continued with the next transformation. It can be shown that after finitely many such blow-ups every nonordinary singularity can be resolved.

We can achieve the blow-ups by quadratic transformations of the plane. These quadratic transformations are special birational maps of the projective plane onto itself, so-called *Cremona transformations*.

Definition 3.7. *The transformation Q of the projective plane $\mathbb{P}^2(K)$ defined by $x' = yz$, $y' = xz$, $z' = xy$, is called the* standard quadratic transformation *or* standard Cremona transformation. *For any change of coordinates T we call $Q \circ T$ a* quadratic transformation.

For the special points $(1 : 0 : 0)$, $(0 : 1 : 0)$ and $(0 : 0 : 1)$ the quadratic transformation is not defined. These points are called the *fundamental points* of the transformation. Every point lying on one of the lines $x = 0$, $y = 0$ or $z = 0$ is sent to the point $(1 : 0 : 0)$, $(0 : 1 : 0)$ or $(0 : 0 : 1)$, respectively. These lines are called the *irregular lines* of the transformation. One can easily prove that this transformation defines a one to one correspondence between points of $\mathbb{P}^2(K)$ not on irregular lines. So the quadratic transformation is a birational map between $\mathbb{P}^2(K)$ and itself. In fact, Q is its own inverse, as we can see from

$$Q(Q(x : y : z)) = (xzxy : yzxy : yzxz) = (x : y : z).$$

Now we study the action of quadratic transformations on an irreducible projective curve.

Definition 3.8. *Let the projective curve C be defined by the homogeneous polynomial $F(x, y, z)$. Then the polynomial $G(x, y, z) = F(yz, xz, xy)$ is called the* algebraic transform *of F. If $G(x, y, z) = H(x, y, z)F'(x, y, z)$, where $H(x, y, z)$ is a product of powers of x, y, z, and F' is not divisible by any x, y, z, we say that F' is the* quadratic transform *of F. We will also say that the curve C' defined by F' is the* quadratic transform *of C.*

The following theorem, taken from Sect. 7.4 of [Wal50], gives a collection of effects of a quadratic transformation on the singularities of a curve. For this purpose, in the sequel, when we speak of nonfundamental intersections of a curve with an irregular line, we mean intersection points different from the fundamental points.

Theorem 3.9. *Let C be a projective curve of degree d defined by F and having the fundamental points $(1:0:0)$, $(0:1:0)$ and $(0:0:1)$ as points of multiplicity r_1, r_2 and r_3, respectively. Let F' be the quadratic transform of F and C' the curve defined by F'. Then if no tangent at any of the fundamental points is an irregular line, the following holds:*

(1) The degree of F' is $2d - r_1 - r_2 - r_3$ and $F'(x, y, z) = F(yz, xz, xy)/x^{r_1} y^{r_2} z^{r_3}$. Furthermore, if $F(x, y, z) = f_d(x, y) + \cdots + f_{r_3}(x, y) z^{d-r_3}$, then

$$F' = x^{d-r_3-r_1} y^{d-r_3-r_2} f_{r_3} + \cdots + z^{d-r_3-1} x^{1-r_1} y^{1-r_2} f_{d-1}$$
$$+ z^{d-r_3} x^{-r_1} y^{-r_2} f_d.$$

(2) *There is a one to one correspondence, preserving multiplicities, between the tangents to C at $(1:0:0), (0:1:0)$ and $(0:0:1)$ and the nonfundamental intersections of C' with the irregular lines $x = 0$, $y = 0$, and $z = 0$, respectively.*

(3) *An r–fold point of C not on an irregular line is transformed into an r–fold point on C', and the tangents at these two points correspond in multiplicity. In particular, the character of the r–fold point is preserved.*

(4) *C' has multiplicity $d - r_2 - r_3, d - r_1 - r_3, d - r_1 - r_2$ at $(1:0:0), (0:1:0)$, $(0:0:1)$, respectively, the tangents being distinct from the irregular lines and corresponding to the nonfundamental intersections of C with $x = 0$, $y = 0$, $z = 0$, respectively.*

Let us clarify Statement (2) in this theorem. Let P be one of the fundamental points, say $P = (0:0:1)$, then $F(x, y, z) = f_d(x, y) + \cdots + f_{r_3}(x, y) z^{d-r_3}$. Let the form f_{r_3} factor as

$$f_{r_3} = (a_1 x - b_1 y)^{\ell_1} \cdots (a_s x - b_s y)^{\ell_s}.$$

Then the nonfundamental intersections of C' with the irregular line $z = 0$ are $\{P_i = (b_i : a_i : 0)\}_{i=1,\ldots,s}$ (i.e., these intersections correspond to the tangents of C at P). To prove this, note that, by Statement (1), the quadratic transform of F satisfies

$$F'(x, y, 0) = x^{d-r_1-r_3} y^{d-r_2-r_3} f_{r_3}(y, x),$$

and therefore the nonfundamental intersections are given by the factors of $f_{r_3}(x, y)$.

Observe that the singularities introduced at the fundamental points (compare Theorem 3.9 (4)) are ordinary, if the nonfundamental intersections of C with the irregular lines are simple points.

Now we could proceed in the following way for obtaining this sequence of quadratic transformations resolving the singularities of a given irreducible curve C. The method consists in recursively "blowing up" C at the nonordinary singularities, i.e.:

(1) Choose a nonordinary singularity P of $\mathcal{C}$. Choose a coordinate system (i.e., apply a linear change of coordinates) such that the singularity P is moved to $(0 : 0 : 1)$, none of its tangents is an irregular line, and no other point on an irregular line is a singular point on $\mathcal{C}$.

(2) Apply the standard quadratic transformation to $\mathcal{C}$, getting the transform curve $\mathcal{C}'$. Outside of the irregular lines, this transform preserves the multiplicity of points and their tangents. New ordinary singularities might be created at the fundamental points. The new curve $\mathcal{C}'$ might have singularities, also nonordinary ones, on the irregular line $z = 0$.

(3) Apply steps (1) and (2) recursively to $\mathcal{C}'$, until no nonordinary singularity is left.

This method selects a coordinate system, and also the order in which the nonordinary singularities of the curve are moved to the fundamental points. One can prove (compare [Ful89], Sect. 7.4) that independently of these selections, the method always achieves an irreducible curve having only ordinary singularities in a finite number of steps. In the sequel, when we will speak about finite sequences of quadratic transformations reducing a given curve, we will assume that these sequences are obtained by the preceding method.

For the purpose of describing this blowing-up process in more detail, we introduce the concept of neighboring points. Let $\mathcal{C}$ be the irreducible curve of degree d defined by $F(x, y, z)$, and $\mathcal{Q} = (\mathcal{Q}_1, \ldots, \mathcal{Q}_n)$ a finite sequence of quadratic transformations constructed as it has been described above and reducing $\mathcal{C}$ to a curve which has only ordinary singularities. We adopt the convention that $\mathcal{Q}_i$ represents the composition of the quadratic transformation with a suitable change of the coordinate system that moves one of the singularities to a fundamental point. Let us also assume that $\mathcal{Q}$ generates the sequence of irreducible curves

$$\mathcal{C} = \mathcal{C}_0 \xrightarrow{\mathcal{Q}_1} \mathcal{C}_1 \xrightarrow{\mathcal{Q}_2} \cdots \xrightarrow{\mathcal{Q}_n} \mathcal{C}_n,$$

where $\mathcal{C}_{i+1}$ is the quadratic transform obtained from $\mathcal{C}_i$ by $\mathcal{Q}_{i+1}$, for $0 \leq i \leq n-1$. Given an r-fold point P on $\mathcal{C}$, suppose that during the process described by $\mathcal{Q}$ the point P has not been translated to a fundamental point till the action of the ith quadratic transformation. Then the *first neighborhood* of P with respect to $\mathcal{Q}$ is defined as the set of all the nonfundamental intersections of the curve $\mathcal{C}_{i+1}$ with the irregular line $z = 0$, assuming that P was moved to $(0 : 0 : 1)$ by the according change of coordinates. Similarly, we take the nonfundamental intersections of $\mathcal{C}_{i+1}$ with $x = 0$ or $y = 0$ if P was translated to $(1 : 0 : 0)$ or $(0 : 1 : 0)$, respectively. The points in the first neighborhood of P with respect to $\mathcal{Q}$ are called the *neighboring points of P at its first neighborhood*. Using the fact that every neighboring point P' of P at its first neighborhood is a point on $\mathcal{C}_{i+1}$, we define the multiplicity and the character of P' as the multiplicity and character of P' as a point on $\mathcal{C}_{i+1}$. Similarly, if $\{P'_1, \ldots, P'_s\}$ is the first neighborhood of P with respect to $\mathcal{Q}$, we get the second neighborhood of P with respect to $\mathcal{Q}$ as the union of the first neighborhoods of P'_k, $k = 1, \ldots, s$. The points in the second neighborhood of P with respect to Q are called the *neighboring points of P at its second neighborhood*. The multiplicity and character of points at the second neighborhood are defined in a way analogous to the one for points in the first neighborhood. But, one must realize that now it may happen that not all the neighboring points are lying on the same curve. These notions are easily extended to neighborhoods of arbitrarily high order. In general, we will call any point in one of the neighborhoods of P a *neighboring point* of P. The neighboring points of P with multiplicity higher than 1 will be called the *singular neighboring points* of P.

Now let us define the neighborhood graph of an irreducible curve.

Definition 3.10. *Let C be an irreducible projective plane curve, $P \in \mathrm{Sing}(C)$. If P is ordinary, then the neighborhood tree at P consists of the single node P. If P is nonordinary, then the neighborhood tree at P has P as its root and the neighborhood trees of the singular neighboring points of P at its first neighborhood as subtrees.*

The neighborhood graph of C, denoted by $\mathrm{Ngr}(C)$, is the collection of the neighborhood trees of all the singular points of C.

The neighboring points of simple points are always simple points, and if P is an ordinary r-fold point its first neighborhood contains exactly r simple points. Therefore, whenever a neighborhood tree contains an ordinary singular point P, then the associated branch of the tree terminates in P. So the neighborhood graph of any curve is finite.

Let us continue using the notation introduced above. That is, C is an irreducible projective plane curve, $Q = (Q_1, \ldots, Q_n)$ is a sequence of quadratic transformations reducing C to a curve with only ordinary singularities, and $C = C_0, \ldots, C_n$ is the sequence of curves generated by Q. Let d_i denote the degree of C_i, $S_i = \mathrm{Sing}(C_i)$, and $N_i = \mathrm{Ngr}(C_i)$. Also, for simplicity, when we work with a point P in either S_i or N_i we will denote by m_P its multiplicity on the corresponding curve.

Theorem 3.11. *Let C, C_i, S_i, N_i be as above.*

(1) $\mathrm{genus}(C) = \frac{1}{2} \cdot [(d_n - 1)(d_n - 2) - \sum_{P \in S_n} m_P(m_P - 1)]$.

(2) *For every i, $0 \le i < n$,*

$$(d_i - 1)(d_i - 2) - \sum_{P \in N_i} r_P(r_P - 1) = (d_{i+1} - 1)(d_{i+1} - 2) - \sum_{P \in N_{i+1}} r_P(r_P - 1).$$

(3) $\mathrm{genus}(C) = \frac{1}{2} \cdot [((d - 1)(d - 2) - \sum_{P \in N_0} m_P(m_P - 1)]$.

Proof. (1) The r.h.s. is equal to $\mathrm{genus}(C_n)$, since C_n has only ordinary singularities. C_n is birationally equivalent to C, so $\mathrm{genus}(C) = \mathrm{genus}(C_n)$.

(2) Let $S_i = \{P_1, P_2, P_3, \ldots, P_s\}$, where $P_1 = (1 : 0 : 0), P_2 = (0 : 1 : 0), P_3 = (0 : 0 : 1)$. By abuse of notation we include all the fundamental points in S_i,

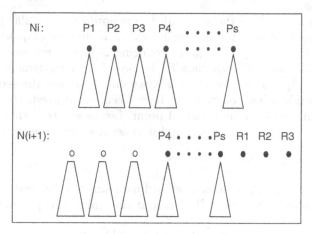

Fig. 3.1. Neighborhood graphs

even if they are not singular points of the curve C_i. That, however, does not affect the count in the equation. The points in S_{i+1} are the singular neighboring points of P_1, P_2, P_3 at their first neighborhood, the transformed points $Q_{i+1}(P_k)$ of P_k, $4 \leq k \leq s$, and possibly three new ordinary singularities R_1, R_2, R_3 resulting from the fundamental points, compare Theorem 3.9(4). Again, w.l.o.g. we include R_j in S_{i+1}, even if it is a simple point. The quadratic transformation does not affect the character and multiplicity of P_k, $4 \leq k \leq s$, so we identify P_k and $Q_{i+1}(P_k)$, $4 \leq k \leq s$. The points R_i, $1 \leq i \leq 3$, do not have any neighboring singularities. See Fig. 3.1 for a sketch of the neighborhoods.

So the equation is equivalent to

$$(d_i - 1)(d_i - 2) - \sum_{j=1}^{3} m_{P_j}(m_{P_j} - 1) = (d_{i+1} - 1)(d_{i+1} - 2) - \sum_{j=1}^{3} m_{R_j}(m_{R_j} - 1).$$

But this follows immediately from the relations

$$d_{i+1} = 2d_i - \sum_{j=1}^{3} r_{P_j}, \qquad m_{R_j} = d_i - \sum_{\substack{k=1 \\ k \neq j}}^{3} m_{P_k}, \quad 1 \leq j \leq 3.$$

(3) The statement follows immediately from (1) and (2).   □

The method described above for determining a birationally equivalent curve with only ordinary singularities is a global one. However, the problem can be solved by local analysis of the curve at the nonordinary singularities. This results in a computationally better way of determining the genus of a curve C. In the local method we do not compute the whole sequence of transform curves of the given curve C. Instead, we act in an equivalent

way: Let $\{P_1, \ldots, P_s\}$ be the set of all the nonordinary singular points of $\mathcal{C}$.
It is clear that for every P_k there always exists a sequence of quadratic trans-
formations $\mathcal{Q}_{P_k} = (\mathcal{Q}_{1,k}, \ldots, \mathcal{Q}_{n_k,k})$ reducing $\mathcal{C}$ to a curve having only ordinary singularities and such that P_k is moved to a fundamental point by
the action of $\mathcal{Q}_{1,k}$. Then, for every P_k, we only compute the sequence $\mathcal{Q}_{P_k}$
till all the neighboring points of P_k have been determined, that is till an-
other $P_{k'}$ is moved to a fundamental point. Let us say that this sequence is
$\mathcal{Q}_{P_k}^* = (\mathcal{Q}_{1,k}, \ldots, \mathcal{Q}_{r_k,k})$, $r_k \le n_k$, and it generates the sequence of curves

$$\mathcal{C} \xrightarrow{\;\mathcal{Q}_{1,k}\;} \mathcal{C}_{1,P_k} \xrightarrow{\;\mathcal{Q}_{2,k}\;} \cdots \xrightarrow{\;\mathcal{Q}_{r_k,k}\;} \mathcal{C}_{r_k,P_k},$$

where in general $\mathcal{C}_{r_k}(P_k)$ can have nonordinary singularities, but these are not
singular neighboring points of P_k. Then at the end of this process we have

$$\mathcal{C} \longrightarrow \mathcal{C}_{1,P_1} \longrightarrow \cdots \longrightarrow \mathcal{C}_{r_1,P_1},$$
$$\vdots$$
$$\mathcal{C} \longrightarrow \mathcal{C}_{s,P_s} \longrightarrow \cdots \longrightarrow \mathcal{C}_{r_s,P_s}.$$

Theorem 3.12. *Let $P_1, \ldots, P_s$ be the singularities of the irreducible projective
curve $\mathcal{C}$ of degree d. Let $S = \{P_1, \ldots, P_s\} \cup N(P_1) \cup \ldots \cup N(P_s)$, where $N(P_k)$
is the set of all the neighboring singularities of P_k w.r.t. $\mathcal{Q}_{P_k}^*$ as above. For
every $P \in S$ let m_P denote the multiplicity of P in the corresponding curve.
Then*
$$\text{genus}(\mathcal{C}) = \frac{1}{2}[(d-1)(d-2) - \sum_{P \in S} m_P(m_P - 1)].$$

Proof. Taking into account the result stated in Theorem 3.11(3), it is enough
to note that the multiplicity of a neighboring point does not depend on the
reduction process of other singularities.   □

Thus, the genus of an irreducible algebraic plane curve $\mathcal{C}$ can be determined computationally by analyzing the multiplicities of the singularities and
neighboring singularities of $\mathcal{C}$. This analysis can be achieved either by a global
transformation to a curve with only ordinary singularities, or by the local process of resolving one nonordinary singularity at a time. We summarize this
result in an algorithm for determining the genus of a curve.

Algorithm GENUS
Given the defining polynomial $F \in K[x, y, z]$ of an irreducible projective
curve $\mathcal{C}$ of degree d. The algorithm computes $g = \text{genus}(\mathcal{C})$.

1. Determine, using the quadratic transformation techniques explained
   above, the neighborhood graph $\mathcal{N} = \text{Ngr}(\mathcal{C})$ of the curve $\mathcal{C}$, computing
   also the multiplicity m_P of every point P in $\mathcal{N}$.
2. Set $g = \frac{1}{2}[(d-1)(d-2) - \sum_{P \in \mathcal{N}} m_P(m_P - 1)]$.
3. Return g.

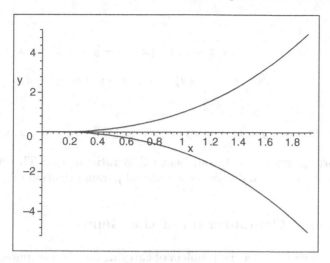

Fig. 3.2. $C_{*,z}$

Example 3.13. Let C be the quintic over $\mathbb{C}$ (see Fig. 3.2) of equation

$$F(x, y, z) = y^2 z^3 - x^5.$$

We have that

$$\text{Sing}(C) = \{(0 : 1 : 0), (0 : 0 : 1)\},$$

where $(0 : 1 : 0)$ is a triple nonordinary point, and $(0 : 0 : 1)$ is a double nonordinary point. Furthermore, the neighboring graph of C is:

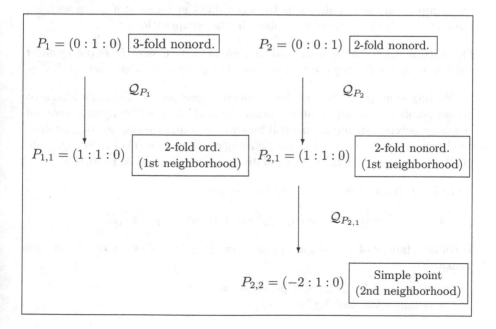

where

$$\mathcal{Q}_{P_1} = \{x = yz, y = xz, z = xy\} \circ \{x = x + y, y = z, z = x - y\}$$

$$\mathcal{Q}_{P_2} = \{x = yz, y = xz, z = xy\} \circ \{x = x + y, y = -x + y, z = z\}$$

and

$$\mathcal{Q}_{P_{2,1}} = \{x = yz, y = xz, z = xy\} \circ \{x = x + y + z, y = y - x, z = y\}.$$

Therefore, $\mathrm{genus}(\mathcal{C}) = 0$, and hence $\mathcal{C}$ is rational (see Theorem 4.63). In Example 3.13, we will determine a rational parametrization of $\mathcal{C}$.

3.3 Symbolic Computation of the Genus

In this section we consider the problem of carrying out the computation of the genus of a curve $\mathcal{C}$ in the smallest possible subfield of K. Thus, throughout this subsection we assume that $\mathbb{K}$ is the smallest computable subfield of K containing the coefficients of the defining polynomial of $\mathcal{C}$, i.e. $\mathbb{K}$ is the ground field of $\mathcal{C}$; see Definition 3.14. In addition, we also assume that $\mathbb{L}$ is a subfield of K such that $\mathbb{K} \subset \mathbb{L} \subset K$. For technical reasons, we assume that $\deg(\mathcal{C}) > 1$. Obviously the genus of a line is 0.

In a direct method one would introduce algebraic extensions of the ground field $\mathbb{K}$ during the computation. However, in [Noe83] M. Noether proved that the computation can be carried out without ever extending the ground field. In the following we present a method based on the notion of a family of conjugate points which allows us to determine the genus of a curve without directly introducing algebraic numbers in the computations.

Definition 3.14. *The* ground field *of a plane curve $\mathcal{C}$ over K is the smallest subfield of K containing the coefficients of any defining polynomial of $\mathcal{C}$.*

We introduce the notion of a family of conjugate points. The basic idea is to collect points whose coordinates depend algebraically on all conjugate roots of the same polynomial $m(t)$. This will imply that computations on such families can be carried out by using only the defining polynomial $m(t)$ of these algebraic numbers, without ever having to isolate any of the individual roots of $m(t)$.

Definition 3.15. *The set of projective points*

$$\mathcal{F} = \{(p_1(\alpha) : p_2(\alpha) : p_3(\alpha)) \mid m(\alpha) = 0\} \subset \mathbb{P}^2(K)$$

is called a family of s conjugate points over $\mathbb{L}$ *if the following conditions are satisfied:*

(1) $p_1, p_2, p_3, m \in \mathbb{L}[t]$, and $\gcd(p_1, p_2, p_3) = 1$
(2) m is squarefree and $\deg(m) = s$,

(3) $\deg(p_i) < \deg(m)$ *for* $i = 1, 2, 3$,

(4) $\mathcal{F}$ *contains exactly s different points of* $\mathbb{P}^2(K)$.

We denote such a family by $\{(p_1(t) : p_2(t) : p_3(t))\}_{m(t)}$. *The polynomial* $m(t)$ *is called the* generating polynomial *of* $\mathcal{F}$.

Note that condition (4) in Definition 3.15 is necessary. For instance, the family $\mathcal{F} = \{(t^2 : t^2 - 1 : t^2 + 1)\}_{t^3 - t}$ satisfies conditions (1), (2), (3) but $\mathcal{F} = \{(0 : -1 : 1), (1 : 0 : 2)\}$ contains only two points.

Definition 3.16. *We say that a family $\mathcal{F}$ of conjugate points over $\mathbb{L}$ is a family of conjugate r-fold points on $\mathcal{C}$ over $\mathbb{L}$ iff* $\operatorname{mult}_P(\mathcal{C}) = r$ *for all* $P \in \mathcal{F}$.

A simple way for finding families of conjugate points on $\mathcal{C}$ is to intersect $\mathcal{C}$ with a line over $\mathbb{L}$. We consider an example.

Example 3.17. Let $\mathcal{C}$ be the quintic over $\mathbb{C}$ of equation

$$F(x, y, z) = -4y^2z^3 + y^4z - \frac{1}{3}x^2y^3 + \frac{8}{3}x^2z^3 - 4xy^2z^2 + 8xz^4 + 4z^5 + x^2y^2z.$$

Thus, the ground field is $\mathbb{Q}$. Intersecting $\mathcal{C}$ with the line $x = 0$ we get the family

$$\mathcal{F}_1 = \{(0 : 1 : t)\}_{2t^2 - 1}.$$

It is easy to check that $\mathcal{F}_1$ is a family of 2 double points of $\mathcal{C}$ over $\mathbb{Q}$, namely

$$\{(0 : \pm\sqrt{2} : 1)\}.$$

Similarly, if we intersect with the line $x = y$ we get a family of five simple points of $\mathcal{C}$ over $\mathbb{Q}$:

$$\mathcal{F}_2 = \{(1 : 1 : t)\}_{4t^5 + 8t^4 - \frac{4}{3}t^3 - 4t^2 + 2t - \frac{1}{3}}.$$

In order to compute with families of conjugate points we need not explicitly consider the individual points, but we can instead compute modulo generating polynomial. This is shown in the next lemma.

Lemma 3.18. *Let $\mathcal{F} = \{(p_1(t) : p_2(t) : p_3(t))\}_{m(t)}$ be a family of s conjugate points over $\mathbb{L}$, and let $F \in \mathbb{K}[x, y, z]$ be the defining polynomial of $\mathcal{C}$. Then the following statements are equivalent*

(1) $\mathcal{F}$ is a family of conjugate r-fold points on $\mathcal{C}$ over $\mathbb{L}$,

(2) r is the greatest non-negative integer such that all partial derivatives of F of order less than r vanish at $(p_1(t) : p_2(t) : p_3(t))$ modulo $m(t)$.

Proof. (2) implies (1) is trivial. In order to prove that (1) implies (2), let G be any partial derivative of F of order less than r. For every root α of $m(t)$, let $P_\alpha = (p_1(\alpha) : p_2(\alpha) : p_3(\alpha))$. Then, since $\operatorname{mult}_{P_\alpha}(\mathcal{C}) = r$, one has that $G(P_\alpha) = 0$. Thus $m(t)$ divides $G(p_1(t), p_2(t), p_3(t))$. On the other hand, since P_α is an r-fold point $\mathcal{C}$, there exists a partial derivative of F of order r not vanishing at P_α. Therefore, there exists at least one partial derivative of F of order r not vanishing at $(p_1(t), p_2(t), p_3(t))$ modulo $m(t)$. $\qquad\square$

The important property of families of conjugate points is that they can be used to generate linear systems of curves without extending the ground field.

Lemma 3.19. *Let $\mathcal{F}$ be a family of s conjugate points (singular or not) over a field $\mathbb{L}$, and let $k, \ell \in \mathbb{N}$. Then, $\mathbb{L}$ is the ground field of all curves in the linear system*

$$\mathcal{H}\left(k, \sum_{P \in \mathcal{F}} \ell P\right).$$

Proof. Let $\mathcal{H}$ denote the linear system considered in the statement of the theorem. If $\mathcal{H} = \emptyset$ the result is trivial. If $\dim(\mathcal{H}) \geq 0$ the result follows from Lemma 3.18. $\qquad\square$

Our final goal is to compute the genus of $\mathcal{C}$. Thus, first we have to analyze whether singularities can be structured in families of conjugate points and afterwards we will have to deal with the neighboring points. For this purpose, we introduce the following notion (see [FaS90]).

Definition 3.20. *We say that the irreducible affine curve $\mathcal{C}$ defined by the polynomial $f(x, y)$ is in* regular position *w.r.t. x iff*

(1) the coefficient of $y^{\deg(\mathcal{C})}$ in f is not zero,

(2) if $f(x_0, y_i) = \dfrac{\partial f}{\partial x}(x_0, y_i) = 0$ for $i = 0, 1$ then $y_0 = y_1$.

Remarks. (1) Note that condition (1) in Definition 3.20 is equivalent to $(0 : 1 : 0) \notin \mathcal{C}^*$ and condition (2) is equivalent to requiring that two different ramification points (see [Ful89], pp. 265) are not on the same vertical line; in particular no vertical line contains two singularities of the curve.

(2) For any irreducible projective curve $\mathcal{C}$, one can always find a suitable change of coordinates such that $\mathcal{C}_{*,z}$ is in regular position. In [FaS90] a deterministic algorithm for finding such a change of coordinates is given.

(3) The conditions in Definition 3.20 can be checked algorithmically. $\qquad\square$

Lemma 3.21. *Let $\mathcal{F} = \{(p_1(t) : p_2(t) : p_3(t))\}_{m(t)}$ be a family of conjugate r-fold points of $\mathcal{C}$ over $\mathbb{L}$. If $m(t)$ is irreducible over $\mathbb{L}$, then all points in $\mathcal{F}$ have the same character (compare Definition 2.6).*

Proof. First we observe that all points of $\mathcal{F}$ are either affine or points at infinity. This follows immediately from the irreducibility of $m(t)$ and condition (3) in Definition 3.15. Let us assume w.l.o.g. that the points in $\mathcal{F}$ are affine. For every root α of $m(t)$, let $T(\alpha, x, y, z)$ be the polynomial in x, y, z defining the tangents of $\mathcal{C}$ at $P_\alpha = (p_1(\alpha) : p_2(\alpha) : p_3(\alpha))$. Then the character of P_α depends on the discriminant $D(\alpha, y)$ of $T(\alpha, x, y, 1)$ w.r.t. x being identically zero or not. Thus, since $m(t)$ is irreducible, the character depends on the divisibility of $D(t, y)$ by $m(t)$, and therefore all points in $\mathcal{F}$ have the same character. $\qquad\square$

Theorem 3.22. *Let C be an irreducible projective curve such that $C_{*,z}$ is in regular position w.r.t. either x or y. Then $\mathrm{Sing}(C)$ can be decomposed as a finite union of families of conjugate points over $\mathbb{K}$ such that all points in the same family have the same multiplicity and character.*

Proof. Let us assume w.l.o.g. that $C_{*,z}$ is in regular position w.r.t. x, and let the defining polynomial of C be expressed as

$$F(x,y,z) = f_d(x,y) + \cdots + f_0(x,y)z^d,$$

where f_i is a form of degree i. By Lemma 3.21 it is enough to prove that singularities can be distributed in conjugate families over $\mathbb{K}$ of the same multiplicity. We analyze separately the cases of affine singularities and singularities at infinity.

(1) First we deal with the singularities at infinity. For $1 \leq i \leq d$, we introduce polynomials

$$\overline{M}_i(x,y) = \gcd\left(f_d, \frac{\partial^i f_d}{\partial x^i}, \frac{\partial^i f_d}{\partial x^{i-1}y}, \ldots, \frac{\partial^i f_d}{\partial y^i} \right),$$

Note that linear factors of $\overline{M}_i(x,y)$ correspond to singularities at infinity of multiplicity at least $i+1$. Furthermore, since $C_{*,z}$ is in regular position w.r.t. x, $(0:1:0) \notin C^*$. Therefore, the singularities at infinity of multiplicity at least $i+1$ are of the form $(\alpha:1:0)$ where α is a root of $\overline{M}_i(t,1)$. Thus, if

$$M_i(t) = \frac{\overline{M}_i(t,1)}{\gcd\left(\overline{M}_i(t,1), \dfrac{\partial \overline{M}_i(t,1)}{\partial t} \right)}, \qquad N_i(t) = \frac{M_i(t)}{M_{i+1}(t)},$$

then the singularities at infinity of multiplicity $i+1$ are in the family $\{(t:1:0)\}_{N_i(t)}$. Now, for each irreducible factor $m(t)$ of $N_i(t) \in \mathbb{K}[t]$ over $\mathbb{K}$, we consider the family

$$\{(t:1:0)\}_{m(t)}.$$

Note that this is a family of $(i+1)$-fold points on C over $\mathbb{K}$, and its cardinality equals the degree of $m(t)$.

(2) For the affine singularities let $f(x,y) = F(x,y,1)$, and for $1 \leq i \leq d$ let us consider the polynomials

$$\overline{B}_i(x) = \gcd\left(\mathrm{res}_y\left(f, \frac{\partial^i f}{\partial x^i} \right), \mathrm{res}_y\left(f, \frac{\partial^i f}{\partial x^{i-1}y} \right), \ldots, \mathrm{res}_y\left(f, \frac{\partial^i f}{\partial y^i} \right) \right),$$

$$B_i(x) = \frac{\overline{B}_i}{\gcd\left(\overline{B}_i, \dfrac{\partial \overline{B}_i}{\partial x} \right)}, \qquad A_i = \frac{B_i}{B_{i+1}}.$$

The x-coordinates of the $(i+1)$-fold affine points of C are exactly the roots of A_i. Furthermore, since $C_{*,z}$ is in regular position w.r.t. x, for every root α of A_i the y–coordinate of the corresponding singularity is the unique solution of

$$\left\{ f(\alpha, y) = 0, \frac{\partial f}{\partial x}(\alpha, y) = 0 \right\}.$$

Thus, if we take any irreducible factor $m(t)$ of A_i over $\mathbb{K}$, and $\mathbb{L}$ denotes the extension of $\mathbb{K}$ by a root α of $m(t)$, i.e., $\mathbb{L} = \mathbb{K}(\alpha) = \mathbb{K}[t]/\langle m(t) \rangle$, we have that

$$\gcd_{\mathbb{L}[y]} \left(f(t,y), \frac{\partial f}{\partial x}(t,y) \right)$$

is linear; say it is $b(t)y - a(t)$.

Therefore, the y-coordinate can be expressed rationally in terms of t. Summarizing, for every irreducible factor m of A_i over $\mathbb{K}$ we get a family of singularities of multiplicity $i + 1$ of the form

$$\{(c(t) : a(t) : b(t))\}_{m(t)},$$

where $c(t)$ is the remainder of $tb(t)$ modulo $m(t)$. Let us finally see that the above family is in fact a family of conjugate points over $\mathbb{K}$. Note that, by construction, conditions (1),(2) and (3) in Definition 3.15 are satisfied. In order to check condition (4), let P_α be the point generated by a root α of $m(t)$, and let us assume that there exist two different roots α, β of $m(t)$ such that $P_\alpha = P_\beta$. The polynomial $b(t)$ is not identically zero because it is the leading coefficient of the gcd. Thus, since $\deg(b) < \deg(m)$ and m is irreducible, one has that $b(\alpha)$ and $b(\beta)$ are not zero. Therefore,

$$P_\alpha = (c(\alpha)/b(\alpha) : a(\alpha)/b(\alpha) : 1) = (\alpha : a(\alpha)/b(\alpha) : 1),$$

and

$$P_\beta = (c(\beta)/b(\beta) : a(\beta)/b(\beta) : 1) = (\beta : a(\beta)/b(\beta) : 1).$$

Hence, $\alpha = \beta$.   □

Corollary 3.23. *The singularities of an irreducible projective curve with ground field $\mathbb{K}$ can be decomposed as a finite union of families of conjugate points over $\mathbb{K}$ such that all points in the same family have the same multiplicity and character.*

Proof. Let C be an irreducible projective curve with ground field $\mathbb{K}$, and let $\mathcal{L}$ be a change of projective coordinates over $\mathbb{K}$ such that $\mathcal{L}(C)_{*,z}$ is in regular position. By Theorem 3.22 the singularities of $\mathcal{L}(C)_{*,z}$ can be decomposed as a finite union of families of conjugate points over $\mathbb{K}$ such that all points in the same family have the same multiplicity and character. Therefore, applying $\mathcal{L}^{-1}$ to the coordinates of each family we get the result.   □

Definition 3.24. *The decomposition of* Sing(C) *into families of conjugate points given in the Corollary 3.23 is called a* standard decomposition *of the singular locus of* C, *and we denote it by* $\mathcal{D}(\mathrm{Sing}(C))$.

The ideas described in the proof of Theorem 3.22 immediately yield an algorithm for computing a standard decomposition of the singularities.

Algorithm STANDARD-DECOMPOSITION-SINGULARITIES

Given the defining polynomial $F \in \mathbb{K}[x, y, z]$ of a nonlinear irreducible projective curve C of degree d, the algorithm computes a standard decomposition $SD = \mathcal{D}(\mathrm{Sing}(C))$.

1. Apply a change of projective coordinates $\mathcal{L}$ over $\mathbb{K}$ such that the affine version of $G(x, y, z) := F(\mathcal{L}(x, y, z))$ is in regular position w.r.t. x. Set $SD = \emptyset$.

2. Let $g(x, y) := G(x, y, 0)$. For $1 \leq i \leq d$, and while $\overline{M}_i(x, y)$ is not constant compute

   2.1. $\overline{M}_i(x, y) = \gcd\left(g, \dfrac{\partial^i g}{\partial x^i}, \dfrac{\partial^i g}{\partial x^{i-1} y}, \ldots, \dfrac{\partial^i g}{\partial y^i}\right)$.

   2.2. $M_i(t) = \dfrac{\overline{M}_i(t, 1)}{\gcd\left(\overline{M}_i(t, 1), \dfrac{\partial \overline{M}_i(t, 1)}{\partial t}\right)}$, and $N_i(t) = \dfrac{M_i(t)}{M_{i+1}(t)}$.

   2.3. For every $N_i(t)$ and for every irreducible factor $m(t)$ of $N_i(t)$ over $\mathbb{K}$ do
   $$SD = SD \cup \{\mathcal{L}^{-1}((t : 1 : 0))\}_{m(t)}.$$

3. Let $f(x, y) = G(x, y, 1)$. For $1 \leq i \leq d$, and while $\overline{B}_i(x, y)$ is not constant compute

   3.1. $\overline{B}_i(x) = \gcd\left(\mathrm{res}_y\left(f, \dfrac{\partial^i f}{\partial x^i}\right), \mathrm{res}_y\left(f, \dfrac{\partial^i f}{\partial x^{i-1} y}\right), \ldots, \mathrm{res}_y\left(f, \dfrac{\partial^i f}{\partial y^i}\right)\right)$

   3.2. $B_i(x) = \dfrac{\overline{B}_i}{\gcd\left(\overline{B}_i, \dfrac{\partial \overline{B}_i}{\partial x}\right)}$, and $A_i(x) = \dfrac{B_i}{B_{i+1}}$.

   3.3. For every A_i and for every irreducible factor $m(t)$ of $A_i(t)$ over $\mathbb{K}$ do

       3.3.1. Compute $\gcd_{\mathbb{L}[y]}\left(f(t, y), \dfrac{\partial f}{\partial x}(t, y)\right) := b(t)y - a(t)$ where $\mathbb{L}$ is the extension of $\mathbb{K}$ by a root α of $m(t)$.

       3.3.2. Let $c(t)$ be the remainder of $tb(t)$ by $m(t)$ and
   $$SD = SD \cup \{\mathcal{L}^{-1}((c(t) : a(t) : b(t)))\}_{m(t)}.$$

4. Return SD.

Example 3.25. Let $\mathcal{C}$ be the irreducible curve defined by

$$F = \tfrac{359}{12}xy^2z^2 + 2yz^4 + \tfrac{187}{12}y^3z^2 + xz^4 + \tfrac{67}{3}x^2yz^2 + \tfrac{117}{4}y^5 + 9x^5 + 6x^3z^2$$

$$+ \tfrac{393}{4}xy^4 + 145x^2y^3 + 115x^3y^2 + 49x^4y \ .$$

First we observe that F is in regular position w.r.t. x. In Step 2.2., we get the polynomials

$$N_1(t) = 3 + 4t + 2t^2, \quad N_2(t) = 1.$$

Thus, the singularities of $\mathcal{C}$ at infinity are double points and are collected in the family (see Step 2.3.)

$$SD_1 = \{(t : 1 : 0)\}_{3+4t+2t^2}.$$

In Step 3.2. we get the polynomials

$$A_1(t) = (1 + 3t^2)(1 + t^2), \quad A_2(t) = 1.$$

For $m(t) = 1 + 3t^2$, we obtain the family of double affine points

$$SD_2 = \{(t : 0 : 1)\}_{1+3t^2},$$

and for $m(t) = 1 + t^2$ we obtain the family of double affine points

$$SD_3 = \{(t : -t : 1)\}_{1+t^2}.$$

Thus, all singularities of $\mathcal{C}$ are double points and the singular locus of $\mathcal{C}$ has the decomposition

$$\mathcal{D}(\mathrm{Sing}(\mathcal{C})) = SD_1 \cup SD_2 \cup SD_3.$$

Therefore, genus$(\mathcal{C}) = 0$.

In the next theorem we prove that neighboring points can also be structured in families of conjugate points.

Theorem 3.26. *Let $\mathcal{C}$ be an irreducible projective curve, and let $\mathcal{F} \in \mathcal{D}$ $(\mathrm{Sing}(\mathcal{C}))$ be a conjugate family of nonordinary singularities on $\mathcal{C}$. The singularities at each neighborhood of $\mathcal{F}$ can be decomposed as a finite union of families of conjugate points over $\mathbb{K}$ such that all points in the same family have the same multiplicity and character as neighboring points.*

Proof. Let $F(x, y, z) \in \mathbb{K}[x, y, z]$ be the defining polynomial of $\mathcal{C}$, let d be the degree of $\mathcal{C}$, and let

$$\mathcal{F} = \{(p_1(t) : p_2(t) : p_3(t))\}_{m(t)}.$$

Let $P_t = (p_1(t) : p_2(t) : p_3(t))$ be a generic element of $\mathcal{F}$. Let $\mathbb{L}$ be the extension of $\mathbb{K}$ by a root α of $m(t)$. We apply a change of projective coordinates

$\mathcal{L}_t$, defined over $\mathbb{L}$, such that P_t is moved to $(0 : 0 : 1)$, no tangent of $\mathcal{L}_t(\mathcal{C})$ at P_t is an irregular line, and no other point on an irregular line is singular on $\mathcal{L}_t(\mathcal{C})$. Let $\mathcal{Q}_t$ be the composition of $\mathcal{L}_t$ with the standard quadratic transformation. Let $F'(t, x, y, z)$ be the defining equation of the transformed curve $\mathcal{C}'_t$ of $\mathcal{C}$ under $\mathcal{Q}_t$, and let $T'(t, x, y)$ be the squarefree part of $F'(t, x, y, 0)$ after crossing out the factors of the form x_1^ℓ and y_2^ℓ. Then, the first neighborhood of P_t w.r.t. $\mathcal{Q}_t$ is

$$\{(h : 1 : 0)\}_{T'(t,h,1)}.$$

Furthermore, applying the reasoning in step (1) of the proof of Theorem 3.22, the above family can be decomposed into families of conjugate points over $\mathbb{L}$ such that all elements in the same family have the same multiplicity and character as points in $\mathcal{C}'_t$. Thus, the first neighborhood of P_t w.r.t. $\mathcal{Q}_t$ can be expressed as

$$\bigcup_{i \in I} \{(h : 1 : 0)\}_{m_i(t,h)},$$

where within a family multiplicity and character are the same.

Multiple separable field extensions can always be rewritten as simple field extensions. Hence, the family is now expressed as a family over $\mathbb{L}$.

Repeating this process through all levels of neighborhoods and for all families of nonordinary singularities in $\mathcal{D}(\mathrm{Sing}(\mathcal{C}))$, we reach a representation, in families of conjugate points, of the neighborhood graph of $\mathcal{C}$.   □

Definition 3.27. *The decomposition of* $\mathrm{Ngr}(\mathcal{C})$ *into families of conjugate points described in the proof of Theorem 3.26 is called a* standard decomposition *of the neighborhood graph of* $\mathcal{C}$, *and we denote it by* $\mathcal{D}(\mathrm{Ngr}(\mathcal{C}))$.

The ideas described above provide an algorithm for computing a standard decomposition of the neighborhood graph (see Exercise 3.8).

Exercises

3.1. Prove that $\mathcal{L}(D)$ is a linear space over K.

3.2. Let φ be a nonzero rational function on a nonsingular algebraic curve, $\mathcal{C}$, in $\mathbb{P}^2(K)$. Prove the following statements:

(i) φ has as many zeros as poles on $\mathcal{C}$.
(ii) A function φ is said to have valence r if it takes every value $c \in K$ exactly r-times. If φ is not a constant in $K(\mathcal{C})$ then φ has valence equal to the number of poles of φ on $\mathcal{C}$.
(iii) Let ϕ be a nonzero rational function on $\mathcal{C}$. Then, $\div(\varphi \cdot \phi) = \div(\varphi) + \div(\phi), \div(\varphi^{-1}) = -\div(\varphi)$. Hence $\div(\varphi/\phi) = \div(\varphi) - \div(\phi)$.

3.3. Compute the neighboring graph of the curve defined by $y + x^2 - 2xy^3 + y^6$ and determine its genus.

3.4. Let $f(x, y)$ define an affine algebraic curve $\mathcal{C}$ over the field K and assume that at least one of the coefficients of f is 1. Let L be the algebraic extension of the prime field of K by the coefficients of f. Show that L is the ground field of $\mathcal{C}$.

3.5. Compute a standard decomposition of the singularities of the cardiod

$$F(x, y, z) = (x^2 + 4yz + y^2)^2 - 16(x^2 + y^2)z^2$$

and determine the genus of the cardioid.

3.6. Find a curve of degree 8 with all its singularities being triple points, and collected in only one family.

3.7. Check that the curve defined implicitly by the polynomial

$$
\begin{aligned}
F = {} & 385262029422463\, z^4 x - 13271226736003144\, z^4 y - 51195290397407\, z^3 x^2 \\
& - 141398711398438\, z^3 y^2 - 236012324707\, z^2 x^3 - 314714123403525377\, z^5 \\
& + 47129502337494\, x\, z^3 y - 526034082419\, x^2 z^2 y + 852907374437\, x\, z^2 y^2 \\
& - 567607920\, x^2 z\, y^2 + 4968341530\, x\, z\, y^3 + 256764427424\, z^2 y^3 \\
& - 1220539178\, x^3 z\, y - 187208023\, z\, x^4 + 4861447511\, z\, y^4 - 52488 x^5 \\
& - 12754584\, y^5 - 4723920\, x^3 y^2 - 14171760\, x^2 y^3 - 787320\, x^4 y \\
& - 21257640\, x\, y^4
\end{aligned}
$$

has six double points in a unique family.

3.8. Describe an algorithm for computing a standard decomposition of the neighboring graph of an irreducible projective curve.

4

Rational Parametrization

Summary. Chapter 4 is the central chapter of the book. In this chapter, we focus on rational or parametric curves, we study different problems related to this type of curves and we show how to algorithmically parametrize a rational curve. The chapter consists of three conceptual blocks. The first one (Sect. 4.1) is devoted to the notion of rational parametrization of a curve, and to the study of the class of rational curves, i.e., curves having a rational parametrization. In the second block of the chapter (Sects. 4.2–4.5), we assume that a rational parametrization of a curve is provided and we consider various problems related to such a rational parametrization. The material in this part of the chapter follows the ideas in [SeW01a]. In Sect. 4.2 the injectivity of the parametrization is studied, in Sect. 4.3 we analyze the number of times the points on the curve are traced via the parametrization, in Sect. 4.4 the inversion problem for proper parametrizations is studied, and in Sect. 4.5 the implicitization question is addressed. The third block of the chapter (Sects. 4.6–4.8) deals with the problem of algorithmically deciding whether a given curve is rational, and in the affirmative case, of actually computing a rational parametrization of the curve. The material in this part of the chapter follows the ideas in [SeW91], which are based on [AbB87a, AbB87b, AbB88, AbB89]. In Sect. 4.6 we study the simple case of curves parametrizable by lines, in Sect. 4.7 these ideas are extended to the general case, and in Sect. 4.8, once the theoretical and algorithmic ideas have been developed, we show how to carry out all these algorithms symbolically.

Alternatively, a parametrization algorithm can be constructed from methods in [VaH94]. Also, the reader interested in the parametrization problem for surfaces may see [Sch98a].

Throughout this chapter, unless explicitly stated otherwise, we use the following notation. K is an algebraically closed field of characteristic 0. We consider either affine or projective plane algebraic curves. In addition, if $\mathcal{C}$ is an affine rational curve, and $\mathcal{P}(t)$ is a rational affine parametrization of C over K (see Definition 4.1), we write its components either as

$$\mathcal{P}(t) = \left(\frac{\chi_{11}(t)}{\chi_{12}(t)}, \frac{\chi_{21}(t)}{\chi_{22}(t)} \right),$$

where $\chi_{ij}(t) \in K[t]$ and $\gcd(\chi_{1i}, \chi_{2i}) = 1$, or as

$$\mathcal{P}(t) = (\chi_1(t), \chi_2(t)),$$

where $\chi_i(t) \in K(t)$. Similarly, rational projective parametrizations (see Definition 4.2) are expressed as

$$\mathcal{P}(t) = (\chi_1(t), \chi_2(t), \chi_3(t)),$$

where $\chi_i(t) \in K[t]$ and $\gcd(\chi_1, \chi_2, \chi_3) = 1$.

Furthermore, associated with a given parametrization $\mathcal{P}(t)$ we consider the polynomials

$$G_1^{\mathcal{P}}(s,t) = \chi_{11}(s)\chi_{12}(t) - \chi_{12}(s)\chi_{11}(t), \quad G_2^{\mathcal{P}}(s,t) = \chi_{21}(s)\chi_{22}(t) - \chi_{22}(s)\chi_{21}(t)$$

as well as the polynomials

$$H_1^{\mathcal{P}}(t,x) = x \cdot \chi_{12}(t) - \chi_{11}(t), \quad H_2^{\mathcal{P}}(t,y) = y \cdot \chi_{22}(t) - \chi_{21}(t).$$

The polynomials $G_i^{\mathcal{P}}$ will play an important role in Sect. 4.3 in deciding whether a parametrization $\mathcal{P}(t)$ is proper by means of the tracing index; i.e., in studying whether the parametrization is injective for almost all parameter values. The polynomials $H_i^{\mathcal{P}}$ will be used in Sect. 4.5 for the implicitization problem.

4.1 Rational Curves and Parametrizations

Some plane algebraic curves can be expressed by means of rational parametrizations, i.e., pairs of univariate rational functions that, except for finitely many exceptions, represent all the points on the curve. For instance, the parabola $y = x^2$ can also be described as the set $\{(t, t^2) \mid t \in \mathbb{C}\}$; in this case, all affine points on the parabola are given by the parametrization (t, t^2). Also, the tacnode curve (see Exercise 2.9 and Fig. 4.1) defined in $\mathbb{A}^2(\mathbb{C})$ by the polynomial

$$f(x,y) = 2x^4 - 3x^2 y + y^2 - 2y^3 + y^4$$

can be represented, for instance, as

$$\left\{ \left(\frac{t^3 - 6t^2 + 9t - 2}{2t^4 - 16t^3 + 40t^2 - 32t + 9}, \frac{t^2 - 4t + 4}{2t^4 - 16t^3 + 40t^2 - 32t + 9} \right) \middle| t \in \mathbb{C} \right\}.$$

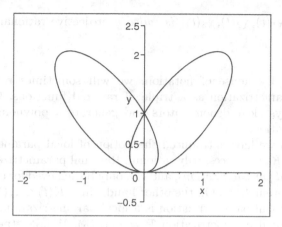

Fig. 4.1. Tacnode curve

In this case, all points on the tacnode are reachable by this pair of rational functions with the exception of the origin.

However, not all plane algebraic curves can be rationally parametrized, as we will see in Example 4.3. In this section, we introduce the notion of rational or parametrizable curves and in this whole chapter we study the main properties and characterizations of this type of curves. In fact, we will see that the rationality of curves can be characterized by means of the genus, and therefore the algorithmic methods described in Chap. 3 will be used.

Definition 4.1. *The affine curve C in $\mathbb{A}^2(K)$ defined by the square-free polynomial $f(x, y)$ is* rational *(or* parametrizable*) if there are rational functions $\chi_1(t), \chi_2(t) \in K(t)$ such that*

(1) for almost all $t_0 \in K$ (i.e. for all but a finite number of exceptions) the point $(\chi_1(t_0), \chi_2(t_0))$ is on C, and
(2) for almost every point $(x_0, y_0) \in C$ there is a $t_0 \in K$ such that $(x_0, y_0) = (\chi_1(t_0), \chi_2(t_0))$.

In this case $(\chi_1(t), \chi_2(t))$ is called an affine rational parametrization *of C. We say that $(\chi_1(t), \chi_2(t))$ is in* reduced form *if the rational functions $\chi_1(t)$, and $\chi_2(t)$ are in reduced form; i.e., if for $i = 1, 2$ the gcd of the numerator and the denominator of χ_i is trivial.*

Definition 4.2. *The projective curve C in $\mathbb{P}^2(K)$ defined by the square-free homogeneous polynomial $F(x, y, z)$ is* rational *(or* parametrizable*) if there are polynomials $\chi_1(t), \chi_2(t), \chi_3(t) \in K[t]$, $\gcd(\chi_1, \chi_2, \chi_3) = 1$, such that*

(1) for almost all $t_0 \in K$ the point $(\chi_1(t_0) : \chi_2(t_0) : \chi_3(t_0))$ is on C, and
(2) for almost every point $(x_0 : y_0 : z_0) \in C$ there is a $t_0 \in K$ such that $(x_0 : y_0 : z_0) = (\chi_1(t_0) : \chi_2(t_0) : \chi_3(t_0))$.

In this case, $(\chi_1(t), \chi_2(t), \chi_3(t))$ *is called a* projective rational parametrization *of* $\mathcal{C}$.

Remarks. (1) By abuse of notation, we will sometimes refer to a projective parametrization as a triple of rational functions. Of course, we could always clear denominators and generate a polynomial projective parametrization.

(2) In Sect. 2.5 we have introduced the notion of local parametrization of a curve over K, not necessarily rational. Rational parametrizations are also called *global parametrizations*, and can only be achieved for curves of genus 0 (see Theorem 4.11). On the other hand, since $K(t) \subset K((t))$, it is clear that any global parametrization is a local parametrization. Consider the global rational parametrization $\mathcal{P} = (\chi_1, \chi_2)$. W.l.o.g. (perhaps after a linear change of parameter) we may assume that $t = 0$ is not a root of the denominators. By interpreting the numerators and denominators of a global parametrization as formal power series, and inverting the denominators, we get exactly a local parametrization $\chi_1 = A(t), \chi_2 = B(t)$ with center $(\chi_1(0), \chi_2(0))$, as introduced in Sect. 2.5,

(3) The notion of rational parametrization can be stated by means of rational maps. More precisely, let $\mathcal{C}$ be a rational affine curve and $\mathcal{P}(t) \in K(t)^2$ a rational parametrization of $\mathcal{C}$. By Definition 4.1, the parametrization $\mathcal{P}(t)$ induces the rational map

$$\mathcal{P} : \mathbb{A}^1(K) \longrightarrow \mathcal{C}$$
$$t \longmapsto \mathcal{P}(t),$$

and $\mathcal{P}(\mathbb{A}^1(K))$ is a dense (in the Zariski topology) subset of $\mathcal{C}$. Sometimes, by abuse of notation, we also call this rational map a rational parametrization of $\mathcal{C}$.

(4) Every rational parametrization $\mathcal{P}(t)$ defines a monomorphism from the field of rational functions $K(\mathcal{C})$ to $K(t)$ as follows (see proof of Theorem 4.9):

$$\varphi : K(\mathcal{C}) \longrightarrow K(t)$$
$$R(x, y) \longmapsto R(\mathcal{P}(t)). \qquad \square$$

Example 4.3. An example of an irreducible curve which is not rational is the projective cubic $\mathcal{C}$, defined over $\mathbb{C}$, by $x^3 + y^3 = z^3$. Suppose that $\mathcal{C}$ is rational, and let $(\chi_1(t), \chi_2(t), \chi_3(t))$ be a projective parametrization of $\mathcal{C}$. Observe that not all components of the parametrization can be constant. Then

$$\chi_1^3 + \chi_2^3 - \chi_3^3 = 0.$$

Differentiating this equation w.r.t. t we get

$$3 \cdot (\chi_1' \chi_1^2 + \chi_2' \chi_2^2 - \chi_3' \chi_3^2) = 0.$$

W.l.o.g. assume that χ_2 is not constant, so $\chi_2 \neq 0$ and $\chi_2' \neq 0$. $\chi_1^2, \chi_2^2, \chi_3^2$ are a solution of the system of homogeneous linear equations with coefficient matrix

$$\begin{pmatrix} \chi_1 & \chi_2 & -\chi_3 \\ \chi_1' & \chi_2' & -\chi_3' \end{pmatrix}.$$

By fundamental line operations we reduce this coefficient matrix to

$$\begin{pmatrix} \chi_2\chi_1' - \chi_2'\chi_1 & 0 & \chi_2'\chi_3 - \chi_2\chi_3' \\ 0 & \chi_2\chi_1' - \chi_2'\chi_1 & \chi_3'\chi_1 - \chi_3\chi_1' \end{pmatrix}.$$

So

$$(\chi_1^2 : \chi_2^2 : \chi_3^2) = (\chi_2\chi_3' - \chi_3\chi_2' : \chi_3\chi_1' - \chi_1\chi_3' : \chi_1\chi_2' - \chi_2\chi_1').$$

Since χ_1, χ_2, χ_3 are relatively prime, this proportionality implies

$$\chi_1^2 \,|\, (\chi_2\chi_3' - \chi_3\chi_2'), \quad \chi_2^2 \,|\, (\chi_3\chi_1' - \chi_1\chi_3'), \quad \chi_3^2 \,|\, (\chi_1\chi_2' - \chi_2\chi_1').$$

Suppose $\deg(\chi_1) \geq \deg(\chi_2), \deg(\chi_3)$. Then, we get that the first divisibility implies $2\deg(\chi_1) \leq \deg(\chi_2) + \deg(\chi_3) - 1$, a contradiction. Similarly, we see that $\deg(\chi_2) \geq \deg(\chi_1), \deg(\chi_3)$ and $\deg(\chi_3) \geq \deg(\chi_1), \deg(\chi_2)$ are impossible. Thus, there can be no parametrization of $\mathcal{C}$.

Definitions 4.1 and 4.2 are stated for general affine and projective curves, respectively. However, in the next theorem we show that only irreducible curves can be parametrizable.

Theorem 4.4. *Any rational curve is irreducible.*

Proof. We prove this for affine curves, the proof for projective curves is similar and is left to the reader. Let $\mathcal{C}$ be a rational affine curve parametrized by a rational parametrization $\mathcal{P}(t)$. First observe that the ideal of $\mathcal{C}$ consists of the polynomials vanishing at $\mathcal{P}(t)$, i.e.,

$$I(\mathcal{C}) = \{h \in K[x, y] \,|\, h(\mathcal{P}(t)) = 0\}.$$

Indeed, if $h \in I(\mathcal{C})$ then $h(P) = 0$ for all $P \in \mathcal{C}$. In particular h vanishes on all points of $\mathcal{C}$ generated by the parametrization, and hence $h(\mathcal{P}(t)) = 0$. Conversely, let $h \in K[x, y]$ be such that $h(\mathcal{P}(t)) = 0$. Therefore, h vanishes on all points of the curve generated by $\mathcal{P}(t)$, i.e., on all points of $\mathcal{C}$ with finitely many exceptions. So, since $\mathcal{C}$ is the Zariski closure of the image of $\mathcal{P}$, it vanishes on $\mathcal{C}$, i.e., $h \in I(C)$ (see Appendix B).

Finally, in order to prove that $\mathcal{C}$ is irreducible, we prove that $I(\mathcal{C})$ is prime (see Appendix B). Let $h_1 \cdot h_2 \in I(\mathcal{C})$. Then $h_1(\mathcal{P}(t)) \cdot h_2(\mathcal{P}(t)) = 0$. Thus, either $h_1(\mathcal{P}(t)) = 0$ or $h_2(\mathcal{P}(t)) = 0$. Therefore, either $h_1 \in I(\mathcal{C})$ or $h_2 \in I(\mathcal{C})$.
□

The rationality of a curve does not depend on its embedding into an affine or projective plane. So, in the sequel, we may choose freely between projective and affine situations, whatever we find more convenient.

Lemma 4.5. *Let C be an irreducible affine curve and C^* its corresponding projective curve. Then C is rational if and only if C^* is rational. Furthermore, a parametrization of C can be computed from a parametrization of C^* and vice versa.*

Proof. Let

$$(\chi_1(t), \chi_2(t), \chi_3(t))$$

be a parametrization of C^*. Observe that $\chi_3(t) \neq 0$, since the curve C^* can have only finitely many points at infinity. Hence,

$$\left(\frac{\chi_1(t)}{\chi_3(t)}, \frac{\chi_2(t)}{\chi_3(t)} \right)$$

is a parametrization of the affine curve C.

Conversely, a rational parametrization of C can always be extended to a parametrization of C^* by normalizing the z-coordinate to 1 and clearing denominators.   □

Definition 4.1 clearly implies that associated with any rational plane curve there is a pair of univariate rational functions over K, not both simultaneously constant, which is a parametrization of the curve. The converse is also true. That is, associated with any pair of univariate rational functions over K, not both simultaneously constant, there is a rational plane curve C such that the image of the parametrization is dense in C. The implicit equation of this curve C is directly related to a resultant. In the following lemma we state this property. Later, in Sect. 4.5, we give a geometric interpretation to the integer r that appears in Lemma 4.6, proving that it counts the number of times the curve is traced when one gives values to the parameter of the parametrization.

Lemma 4.6. *Let C be an affine rational curve over K, $f(x, y)$ its the defining polynomial, and*

$$\mathcal{P}(t) = (\chi_1(t), \chi_2(t))$$

a rational parametrization of C. Then, there exists $r \in \mathbb{N}$ such that

$$\mathrm{res}_t(H_1^{\mathcal{P}}(t, x), H_2^{\mathcal{P}}(t, y)) = (f(x, y))^r.$$

Proof. Let $\chi_i(t) = \frac{\chi_{i1}(t)}{\chi_{i2}(t)}$, and let

$$h(x, y) = \mathrm{res}_t(H_1^{\mathcal{P}}(t, x), H_2^{\mathcal{P}}(t, y)).$$

First we observe that $H_1^{\mathcal{P}}$ and $H_2^{\mathcal{P}}$ are irreducible, because $\chi_1(t)$ and $\chi_2(t)$ are in reduced form. Hence $H_1^{\mathcal{P}}$ and $H_2^{\mathcal{P}}$ do not have common factors. Therefore, $h(x, y)$ is not the zero polynomial. Furthermore, h cannot be a constant polynomial either. Indeed: let $t_0 \in K$ be such that $\chi_{12}(t_0)\chi_{22}(t_0) \neq 0$. Then $H_1^{\mathcal{P}}(t_0, \mathcal{P}(t_0)) = H_2^{\mathcal{P}}(t_0, \mathcal{P}(t_0)) = 0$. So $h(\mathcal{P}(t_0)) = 0$, and since h is not the zero polynomial it cannot be constant.

Now, we consider the square-free part $h'(x,y)$ of $h(x,y)$ and the plane curve $\mathcal{C}$ defined by $h'(x,y)$ over K. Let us see that $\mathcal{P}(t)$ parametrizes $\mathcal{C}$. For this purpose, we check the conditions introduced in Definition 4.1.

1. Let $t_0 \in K$ be such that $\chi_{12}(t_0)\chi_{22}(t_0) \neq 0$. Reasoning as above, we see that $h(\mathcal{P}(t_0)) = 0$. So $h'(\mathcal{P}(t_0)) = 0$, and hence $\mathcal{P}(t_0)$ is on $\mathcal{C}$.

2. Let c_1, c_2 be the leading coefficients of $H_1^{\mathcal{P}}, H_2^{\mathcal{P}}$ w.r.t. t, respectively. Note that $c_1 \in K[x], c_2 \in K[y]$ are of degree at most 1. For every (x_0, y_0) on $\mathcal{C}$ such that $c_1(x_0) \neq 0$ or $c_2(y_0) \neq 0$ (note that there is at most one point in K^2 where c_1 and c_2 vanish simultaneously), we have $h(x_0, y_0) = 0$. Thus, since h is a resultant, there exists $t_0 \in K$ such that $H_1^{\mathcal{P}}(t_0, x_0) = H_2^{\mathcal{P}}(t_0, y_0) = 0$. Also, observe that $\chi_{12}(t_0) \neq 0$ since otherwise the first component of the parametrization would not be in reduced form. Similarly, $\chi_{22}(t_0) \neq 0$. Thus, $(x_0, y_0) = \mathcal{P}(t_0)$. Therefore, almost all points on $\mathcal{C}$ are generated by $\mathcal{P}(t)$.

Now by Theorem 4.4 it follows that $h'(x,y)$ is irreducible. Therefore, there exists $r \in \mathbb{N}$ such that $h(x,y) = (h'(x,y))^r$. $\quad\square$

Sometimes it is useful to apply equivalent characterizations of the concept of rationality. In Theorems 4.7, 4.9, 4.10, and 4.11 some such equivalent characterizations are established.

Theorem 4.7. *An irreducible curve $\mathcal{C}$, defined by $f(x,y)$, is rational if and only if there exist rational functions $\chi_1(t), \chi_2(t) \in K(t)$, not both constant, such that $f(\chi_1(t), \chi_2(t)) = 0$. In this case, $(\chi_1(t), \chi_2(t))$ is a rational parametrization of $\mathcal{C}$.*

Proof. Let $\mathcal{C}$ be rational. So there exist rational functions $\chi_1, \chi_2 \in K(t)$ satisfying conditions (1) and (2) in Definition 4.1. Obviously not both rational functions χ_i are constant, and clearly $f(\chi_1(t), \chi_2(t)) = 0$.

Conversely, let $\chi_1, \chi_2 \in K(t)$, not both constant, be such that $f(\chi_1(t), \chi_2(t))$ is identically zero. Let $\mathcal{D}$ be the irreducible plane curve defined by $(\chi_1(t), \chi_2(t))$ (see Lemma 4.6). Then $\mathcal{C}$ and $\mathcal{D}$ are both irreducible, because of Theorem 4.4, and have infinitely many points in common. Thus, by Bézout's Theorem (Theorem 2.48) one concludes that $\mathcal{C} = \mathcal{D}$. Hence, $(\chi_1(t), \chi_2(t))$ is a parametrization of $\mathcal{C}$. $\quad\square$

An alternative characterization of rationality in terms of field theory is given in Theorem 4.9. This theorem can be seen as the geometric version of Lüroth's Theorem. Lüroth's Theorem appears in basic text books on algebra such as [Jac74], [Jac80], or [VaW70]. Here we do not give a proof of this result.

Theorem 4.8 (Lüroth's Theorem). *Let* $\mathbb{L}$ *be a field (not necessarily alge-braically closed),* t *a transcendental element over* $\mathbb{L}$. *If* $\mathbb{K}$ *is a subfield of* $\mathbb{L}(t)$ *strictly containing* $\mathbb{L}$, *then* $\mathbb{K}$ *is* $\mathbb{L}$*-isomorphic to* $\mathbb{L}(t)$.

Theorem 4.9. *An irreducible affine curve* $\mathcal{C}$ *is rational if and only if the field of rational functions on* $\mathcal{C}$, *i.e.* $K(\mathcal{C})$, *is isomorphic to* $K(t)$ *(t a transcendental element).*

Proof. Let $f(x, y)$ be the defining polynomial of $\mathcal{C}$, and let $\mathcal{P}(t)$ be a parametr-ization of $\mathcal{C}$. We consider the map

$$\varphi_{\mathcal{P}} : \ K(\mathcal{C}) \ \longrightarrow \ K(t)$$
$$R(x, y) \ \longmapsto \ R(\mathcal{P}(t)).$$

First we observe that $\varphi_{\mathcal{P}}$ is well-defined. Let $\frac{p_1}{q_1}, \frac{p_2}{q_2}$, where $p_i, q_i \in K[x, y]$, be two different expressions of the same element in $K(\mathcal{C})$. Then f divides $p_1 q_2 - q_1 p_2$. In addition, by Theorem 4.7, $f(\mathcal{P}(t))$ is identically zero, and therefore $p_1(\mathcal{P}(t))q_2(\mathcal{P}(t)) - q_1(\mathcal{P}(t))p_2(\mathcal{P}(t))$ is also identically zero. Further-more, since $q_1 \neq 0$ in $K(\mathcal{C})$, we have $q_1(\mathcal{P}(t)) \neq 0$. Similarly $q_2(\mathcal{P}(t)) \neq 0$. Therefore, $\varphi_{\mathcal{P}}(\frac{p_1}{q_1}) = \varphi_{\mathcal{P}}(\frac{p_2}{q_2})$.

Now, since $\varphi_{\mathcal{P}}$ is not the zero homomorphism, the map $\varphi_{\mathcal{P}}$ defines an iso-morphism of $K(\mathcal{C})$ onto a subfield of $K(t)$ that properly contains K. Thus, by Lüroth's Theorem, this subfield, and $K(\mathcal{C})$ itself, must be isomorphic to $K(t)$.

Conversely, let $\psi : K(\mathcal{C}) \to K(t)$ be an isomorphism and $\chi_1(t) = \psi(x), \chi_2(t) = \psi(y)$. Clearly, since the image of ψ is $K(t)$, χ_1 and χ_2 can-not both be constant. Furthermore

$$f(\chi_1(t), \chi_2(t)) = f(\psi(x), \psi(y)) = \psi(f(x, y)) = 0.$$

Hence, by Theorem 4.7, the pair $(\chi_1(t), \chi_2(t))$ is a rational parametrization of $\mathcal{C}$. □

Remarks. From the proof of Theorem 4.9 we see that every parametrization $\mathcal{P}(t)$ induces a monomorphism $\varphi_{\mathcal{P}}$ from $K(\mathcal{C})$ to $K(t)$. We will refer to $\varphi_{\mathcal{P}}$ as the *monomorphism induced* by $\mathcal{P}(t)$.

In the following theorem we see how rationality can also be established by means of rational maps.

Theorem 4.10. *An affine algebraic curve* $\mathcal{C}$ *is rational if and only if it is birationally equivalent to* K *(i.e., the affine line* $\mathbb{A}^1(K)$*).*

Proof. By Theorem 2.38 one has that $\mathcal{C}$ is birationally equivalent to K if and only if $K(\mathcal{C})$ is isomorphic to $K(t)$. Thus, by Theorem 4.9 we get the desired result. □

The following theorem states that only curves of genus 0 can be rational. In fact, all irreducible conics are rational, and an irreducible cubic is rational if and only if it has a double point.

Theorem 4.11. *If an algebraic curve C is rational then* genus$(C) = 0$.

Proof. By the remark after Definition 3.4 the genus is invariant under birational maps. Hence the result follows from Theorem 4.10.   □

In Sect. 4.7 (see Theorem 4.63) we will demonstrate that also the converse is true, namely that every curve of genus 0 is rational.

4.2 Proper Parametrizations

Although the implicit representation for a plane curve is unique, up to multiplication by nonzero constants, there exist infinitely many different parametrizations of the same rational curve. For instance, for every $i \in \mathbb{N}$, (t^i, t^{2i}) parametrizes the parabola $y = x^2$. Obviously (t, t^2) is the parametrization of lowest degree in this family and it generates every point on the parabola only once. Such parametrizations are called proper parametrizations (see Definition 4.12).

The parametrization algorithms presented in this book always output proper parametrizations. Furthermore, there are algorithms for determining whether a given parametrization of a plane curve is proper, and if that is not the case, for transforming it to a proper one. In Sect. 6.1 we will describe these methods.

In this section, we introduce the notion of proper parametrization and we study some of their main properties. For this purpose, in the following we assume that C is an affine rational plane curve, and $\mathcal{P}(t)$ is an affine rational parametrization of C.

Definition 4.12. *An affine parametrization $\mathcal{P}(t)$ of a rational curve C is proper if the map*

$$\mathcal{P} : \mathbb{A}^1(K) \longrightarrow C$$
$$t \longmapsto \mathcal{P}(t)$$

is birational, or equivalently, if almost every point on C is generated by exactly one value of the parameter t.

We define the inversion *of a proper parametrization $\mathcal{P}(t)$ as the inverse rational mapping of $\mathcal{P}$, and we denote it by $\mathcal{P}^{-1}$.*

Lemma 4.13. *Every rational curve can be properly parametrized.*

Proof. From Theorem 4.10 one deduces that every rational curve C is birationally equivalent to $\mathbb{A}^1(K)$. Therefore, every rational curve can be properly parametrized.   □

The notion of properness can also be stated algebraically in terms of fields of rational functions. From Theorem 2.38 we deduce that a rational parametrization $\mathcal{P}(t)$ is proper if and only if the induced monomorphism $\varphi_{\mathcal{P}}$ (see Remark to Theorem 4.9)

$$\varphi_{\mathcal{P}} : \begin{array}{l} K(\mathcal{C}) \longrightarrow K(t) \\ R(x,y) \longmapsto R(\mathcal{P}(t)). \end{array}$$

is an isomorphism. Therefore, $\mathcal{P}(t)$ is proper if and only if the mapping $\varphi_{\mathcal{P}}$ is surjective, that is, if and only if $\varphi_{\mathcal{P}}(K(\mathcal{C})) = K(\mathcal{P}(t)) = K(t)$. More precisely, we have the following theorem.

Theorem 4.14. *Let $\mathcal{P}(t)$ be a rational parametrization of a plane curve $\mathcal{C}$. Then, the following statements are equivalent:*

(1) $\mathcal{P}(t)$ is proper.
(2) The monomorphism $\varphi_{\mathcal{P}}$ induced by $\mathcal{P}$ is an isomorphism.
(3) $K(\mathcal{P}(t)) = K(t)$.

Remarks. We have introduced the notion of properness for affine parametrizations. For projective parametrizations the notion of properness can be introduced in a similar way by requiring the rational map, associated with the projective parametrization, to be birational. Moreover, if $\mathcal{C}$ is an irreducible affine curve and $\mathcal{C}^\star$ is its projective closure, then $K(\mathcal{C}) = K(\mathcal{C}^\star)$. Thus, taking into account Theorem 4.14 one has that the properness of affine and projective parametrizations are equivalent.

Now, we characterize proper parametrizations by means of the degree of the corresponding rational curve. To state this result, we first introduce the notion of degree of a parametrization.

Definition 4.15. *Let $\chi(t) \in K(t)$ be a rational function in reduced form. If $\chi(t)$ is not zero, the degree of $\chi(t)$ is the maximum of the degrees of the numerator and denominator of $\chi(t)$. If $\chi(t)$ is zero, we define its degree to be -1. We denote the degree of $\chi(t)$ as $\deg(\chi(t))$. Rational functions of degree 1 are called* linear.

Obviously the degree is multiplicative with respect to the composition of rational functions. Furthermore, invertible rational functions are exactly the linear rational functions (see Exercise 4.1).

Definition 4.16. *We define the* degree *of an affine rational parametrization $\mathcal{P}(t) = (\chi_1(t), \chi_2(t))$ as the maximum of the degrees of its rational components; i.e.*

$$\deg(\mathcal{P}(t)) = \max \{\deg(\chi_1(t)), \deg(\chi_2(t))\}.$$

We start this study with a lemma that shows how proper and improper parametrizations of an affine plane curve are related.

Lemma 4.17. *Let $\mathcal{P}(t)$ be a proper parametrization of an affine rational curve $\mathcal{C}$, and let $\mathcal{P}'(t)$ be any other rational parametrization of $\mathcal{C}$.*

(1) There exists a nonconstant rational function $R(t) \in K(t)$ such that $\mathcal{P}'(t) = \mathcal{P}(R(t))$.

(2) $\mathcal{P}'(t)$ is proper if and only if there exists a linear rational function $L(t) \in K(t)$ such that $\mathcal{P}'(t) = \mathcal{P}(L(t))$.

Proof. (1) We consider the following diagram:

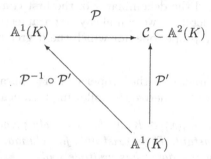

Then, since $\mathcal{P}$ is a birational mapping, it is clear that $R(t) = \mathcal{P}^{-1}(\mathcal{P}'(t)) \in K(t)$.

(2) If $\mathcal{P}'(t)$ is proper, then from the diagram above we see that $\varphi := \mathcal{P}^{-1} \circ \mathcal{P}'$ is a birational mapping from $\mathbb{A}^1(K)$ onto $\mathbb{A}^1(K)$. Hence, by Theorem 2.38 one has that φ induces an automorphism $\tilde{\varphi}$ of $K(t)$ defined as:

$$\tilde{\varphi} : K(t) \longrightarrow K(t)$$
$$t \longmapsto \varphi(t).$$

Therefore, since K-automorphisms of $K(t)$ are the invertible rational functions (see e.g., [VaW70]), we see that $\tilde{\varphi}$ is our linear rational function.

Conversely, let ψ be the birational mapping from $\mathbb{A}^1(K)$ onto $\mathbb{A}^1(K)$ defined by the linear rational function $L(t) \in K(t)$. Then, it is clear that $\mathcal{P}' = \mathcal{P} \circ \psi : \mathbb{A}^1(K) \to \mathcal{C}$ is a birational mapping, and therefore $\mathcal{P}'(t)$ is proper. $\square$

Lemma 4.17 seems to suggest that a parametrization of prime degree is proper. But in fact, this is not true, as can easily be seen from the parametrization (t^2, t^2) of a line. Exercise 4.2 asks whether the line is the only curve for which primality of a parametrization does not imply properness.

Proper parametrizations can always be normalized such that in every component of the parametrization the degrees of the numerator and denominator agree. This will be useful later.

Lemma 4.18. *Every rational curve $\mathcal{C}$ has a proper parametrization $\mathcal{P}(t) = (\chi_1(t), \chi_2(t))$ such that if $\chi_i(t)$ is nonzero, then $\deg(\chi_{i1}) = \deg(\chi_{i2})$.*

Proof. By Lemma 4.13 we know that $\mathcal{C}$ has a proper rational parametrization, say $\mathcal{P}'(t)$. Note that if the i−th component of a parametrization is zero, then it is zero for every parametrization. Let us assume w.l.o.g that χ_1 is nonzero.

By Lemma 4.17, any linear reparametrization of a proper parametrization is again proper. If 0 is a root of none of the numerator and denominator of $\chi_1(t)$, then $\mathcal{P}'(\frac{1}{t})$ is still proper and the requirement on the degree is fulfilled. If 0 is a root of any of the numerator or denominator of $\chi_1(t)$, we consider the proper parametrization $\mathcal{P}'(t+a)$, where a is not a root of any of the numerator and denominator. This a always exists since $\chi_1(t)$ is nonzero. Now, observe that the numerator and the denominator of the first component of $\mathcal{P}'(t + a)$ do not vanish at 0. Therefore, we can always reparametrize the initial proper parametrization into a proper one, for which the degree requirement holds.

□

Before we can characterize the properness of a parametrization via the degree of the curve, we first derive the following technical property.

Lemma 4.19. *Let* $p(x), q(x) \in K[x]^\star$ *be relatively prime such that at least one of them is nonconstant. There exist only finitely many values* $a \in K$ *such that the polynomial* $p(x) - aq(x)$ *has multiple roots.*

Proof. Let us consider the polynomial $f(x, y) = p(x) - yq(x) \in K[x, y]$. Since $\gcd(p, q) = 1$ and $p(x), q(x)$ are nonzero, the polynomial f is irreducible. Now we study the existence of roots of the discriminant of f w.r.t. y. Let $g(x, y) = \frac{\partial f}{\partial x}$. Note that g is nonzero, since at least one of the two polynomials $p(x)$ and $q(x)$ is not constant. Since $\deg(g) < \deg(f)$ and f is irreducible, we get $\gcd(f, g) = 1$. So $\mathrm{discr}_x(f) \neq 0$. Hence the result follows immediately.   □

Corollary 4.20. *Let* $p(x), q(x) \in K[x]^\star$ *be relatively prime such that at least one of them is nonconstant, and let* $R(y)$ *be the resultant*

$$R(y) = \mathrm{res}_x(p(x) - yq(x), p'(x) - yq'(x)).$$

Then, for all $b \in K$ *such that* $R(b) \neq 0$*, the polynomial* $p(x) - bq(x)$ *is squarefree.*

The next theorem characterizes the properness of a parametrization by means of the degree of the implicit equation of the curve.

Theorem 4.21. *Let* $\mathcal{C}$ *be an affine rational curve defined over* K *with defining polynomial* $f(x, y) \in K[x, y]$*, and let* $\mathcal{P}(t) = (\chi_1(t), \chi_2(t))$ *be a parametrization of* $\mathcal{C}$*. Then* $\mathcal{P}(t)$ *is proper if and only if*

$$\deg(\mathcal{P}(t)) = \max\{\deg_x(f), \deg_y(f)\}.$$

Furthermore, if $\mathcal{P}(t)$ *is proper and* $\chi_1(t)$ *is nonzero, then* $\deg(\chi_1(t)) = \deg_y(f)$*; similarly, if* $\chi_2(t)$ *is nonzero then* $\deg(\chi_2(t)) = \deg_x(f)$*.*

Proof. First we prove the result for the special case of parametrizations having a constant component; i.e., for horizontal or vertical lines. Afterwards, we consider the general case. Let $\mathcal{P}(t)$ be a parametrization such that one of its two components is constant, say $\mathcal{P}(t) = (\chi_1(t), \lambda)$ for some $\lambda \in K$. Then the curve $\mathcal{C}$ is the line of equation $y = \lambda$. Hence, by Lemma 4.17 (2) and because (t, λ) parametrizes $\mathcal{C}$ properly, we get that all proper parametrizations of $\mathcal{C}$ are of the form $(\frac{at+b}{ct+d}, \lambda)$, where $a, b, c, d, \in K$ and $ad - bc \neq 0$. Therefore, $\deg(\chi_1) = 1$, and the theorem clearly holds.

Now we consider the general case, i.e., $\mathcal{C}$ is not a horizontal or vertical line. Let $\mathcal{P}(t)$ be proper and in reduced form, such that none of its components is constant. Then we prove that $\deg(\chi_2(t)) = \deg_x(f)$, and analogously one can prove that $\deg(\chi_1(t)) = \deg_y(f)$. From these relations we immediately get that $\deg(\mathcal{P}(t)) = \max\{\deg_x(f), \deg_y(f)\}$. Let $\chi_2(t) = \chi_{21}(t)/\chi_{22}(t)$. We define $\mathcal{S}$ as the subset of K containing

(a) all the second coordinates of those points on $\mathcal{C}$ that are either not generated by $\mathcal{P}(t)$, or more than once by different values of t,
(b) those $b \in K$ such that the polynomial $\chi_{21}(t) - b\chi_{22}(t)$ has multiple roots,
(c) $\mathrm{lc}(\chi_{21})/\mathrm{lc}(\chi_{22})$, where " lc" denotes the leading coefficient,
(d) those $b \in K$ such that the polynomial $f(x, b)$ has multiple roots,
(e) the roots of the leading coefficient of $f(x, y)$ w.r.t. x.

We claim that $\mathcal{S}$ is finite. Indeed: Since $\mathcal{P}(t)$ is a proper parametrization, there are only finitely many values satisfying (a). According to Lemma 4.19 there are only finitely many field elements satisfying (b). The argument for (c) is trivial. An element $b \in K$ satisfies (d) if and only if b is the second coordinate of a singular point of $\mathcal{C}$ or the line $y = b$ is tangent to the curve at some simple point (see Theorem 2.50(6)). By Theorem 2.10, $\mathcal{C}$ has only finitely many singular points, and $y = b$ is tangent to $\mathcal{C}$ at some point (a, b) if (a, b) is a solution of the system $\{f = 0, \frac{\partial f}{\partial x} = 0\}$. However, by Bézout's Theorem (Theorem 2.48), this system has only finitely many solutions. So only finitely many field elements satisfy (d). Since the leading coefficient of $f(x, y)$ w.r.t. x is a nonzero univariate polynomial, only finitely many field elements satisfy (e). Therefore, $\mathcal{S}$ is finite.

Now we take an element $b \in K \setminus \mathcal{S}$ and we consider the intersection of $\mathcal{C}$ and the line of equation $y = b$. Because of condition (e) the degree of $f(x, b)$ is exactly $\deg_x(f(x, y))$, say $m := \deg_x(f(x, y))$. Furthermore, by (d), $f(x, b)$ has m different roots, say $\{r_1, \ldots, r_m\}$. So, there are m different points on $\mathcal{C}$ having b as a second coordinate, namely $\{(r_i, b)\}_{i=1,\ldots,m}$, and they can be generated by $\mathcal{P}(t)$ because of (a).

On the other hand, we consider the polynomial $M(t) = \chi_{21}(t) - b\chi_{22}(t)$. We note that $\deg_t(M) \geq m$, since every point (r_i, b) is generated by some value of the parameter t. But, since every point $(a, b) \in \mathcal{C}$ is generated exactly once by $\mathcal{P}$ (see condition (a)) and M cannot have multiple roots, we get that $\deg_t(M) = m = \deg_x(f(x, y))$. Now, since b is not the quotient of the leading coefficients of χ_{21} and χ_{22} (because of (c)), we finally see that $\deg_x(f(x, y)) = \deg(M) = \max\{\deg(\chi_{21}), \deg(\chi_{22})\}$.

Conversely, let $\mathcal{P}(t)$ be a parametrization of $\mathcal{C}$ such that $\deg(\mathcal{P}(t)) = \max\{\deg_x(f), \deg_y(f)\}$, and let $\mathcal{P}'(t)$ be any proper parametrization of $\mathcal{C}$.

Then, by Lemma 4.17(1), there exists $R(t) \in K(t)$ such that $\mathcal{P}'(R(t)) = \mathcal{P}(t)$. $\mathcal{P}'(t)$ is proper, so $\deg(\mathcal{P}'(t)) = \max\{\deg_x(f), \deg_y(f)\} = \deg(\mathcal{P}(t))$. Therefore, since the degree is multiplicative with respect to composition, $R(t)$ must be of degree 1, and hence invertible. Thus, by Lemma 4.17(2), $\mathcal{P}(t)$ is proper.   □

The next corollary follow from Theorem 4.21 and Lemma 4.17(1).

Corollary 4.22. *Let C be a rational affine plane curve defined by $f(x, y) \in K[x, y]$. Then the degree of any rational parametrization of C is a multiple of $\max\{\deg_x(f), \deg_y(f)\}$.*

Example 4.23. We consider the rational quintic C defined by the polynomial $f(x, y) = y^5 + x^2 y^3 - 3\,x^2 y^2 + 3\,x^2 y - x^2$. By Theorem 4.21, any proper rational parametrization of C must have a first component of degree 5, and a second component of degree 2. It is easy to check that

$$\mathcal{P}(t) = \left(\frac{t^5}{t^2 + 1}, \frac{t^2}{t^2 + 1} \right)$$

properly parametrizes C. Note that $f(\mathcal{P}(t)) = 0$.

For a generalization of the Theorem 4.21 to the surface case see [PDS05].

4.3 Tracing Index

In Sect. 2.2 we have introduced the notion of degree of a dominant rational map between varieties (i.e., irreducible algebraic sets). In this section, we investigate the degree of a special type of rational maps, namely those induced by rational parametrizations of curves. That is, if $\mathcal{P}(t)$ is an affine rational parametrization of C, we study the degree of the dominant rational map $\mathcal{P}$: $\mathbb{A}(K) \longrightarrow C: t \mapsto \mathcal{P}(t)$. Later, in Sect. 4.5, we will see that the degree of the rational map induced by the parametrization plays a role in the implicitization problem.

In addition, we will work with the fibres of the map $\mathcal{P}$. We will denote by $\mathcal{F}_{\mathcal{P}}(P)$ the fibre of a point $P \in C$; that is

$$\mathcal{F}_{\mathcal{P}}(P) = \mathcal{P}^{-1}(P) = \{t \in K \mid \mathcal{P}(t) = P\}.$$

In Theorem 2.43 we have seen that the degree of a dominant rational map between two varieties of the same dimension is the cardinality of the fiber of a generic element. Therefore, in the case of the mapping $\mathcal{P}$, this implies that almost all points of C are generated via $\mathcal{P}(t)$ by the same number of parameter values, and this number is the degree. Thus, intuitively speaking, the degree measures the number of times the parametrization traces the curve when the parameter takes values in K. Taking into account this intuitive meaning of the notion of degree, we will also call the degree of the mapping $\mathcal{P}$ the tracing index of $\mathcal{P}(t)$.

Definition 4.24. *Let C be an affine rational curve, and let $\mathcal{P}(t)$ be a rational parametrization of C. Then the* tracing index *of $\mathcal{P}(t)$, denoted by* index$(\mathcal{P}(t))$, *is the degree of $\mathcal{P} : \mathbb{A}(K) \longrightarrow C$, $t \mapsto \mathcal{P}(t)$; i.e.,* index$(\mathcal{P}(t))$ *is a natural number such that almost all points on C are generated, via $\mathcal{P}(t)$, by exactly* index$(\mathcal{P}(t))$ *parameter values.*

4.3.1 Computation of the Index of a Parametrization

Theorem 4.25. *Let $\mathcal{P}(t)$ be a parametrization in reduced form. Then for almost all $\alpha \in K$ we have*

$$\mathrm{card}(\mathcal{F}_{\mathcal{P}}(\mathcal{P}(\alpha))) = \deg_t(\gcd(G_1^{\mathcal{P}}(\alpha, t), G_2^{\mathcal{P}}(\alpha, t))).$$

Proof. Let $\chi_i = \chi_{i1}/\chi_{i2}$, in reduced form, be the i-th component of $\mathcal{P}(t)$. Let S be the set of all $\alpha \in K$ such that either $\mathcal{P}(\alpha)$ is not defined or both polynomials $G_1^{\mathcal{P}}(\alpha, t)$ and $G_2^{\mathcal{P}}(\alpha, t)$ have multiple roots. First, we see that S is a finite set. Indeed: clearly there exist only finitely many values such that $\mathcal{P}(t)$ is not defined. Now, we assume w.l.o.g. that $\chi_1(t)$ is nonconstant. Let α be such that $\chi_{12}(\alpha)\chi_{22}(\alpha) \neq 0$. If $G_1^{\mathcal{P}}(\alpha, t)$ has multiple roots, then $H_1^{\mathcal{P}}(t, \chi_1(\alpha)) = 1/\chi_{12}(\alpha)G_1^{\mathcal{P}}(\alpha, t)$ also has multiple roots. But by Lemma 4.19 this can only happen for finitely many values of α. Therefore, S is finite.

Now, let $\alpha \in K \setminus S$. We observe that every element of the fibre $\mathcal{F}_{\mathcal{P}}(\mathcal{P}(\alpha))$ is a common root of $G_1^{\mathcal{P}}(\alpha, t)$ and $G_2^{\mathcal{P}}(\alpha, t)$. On the other hand, let β be a root of $\gcd(G_1^{\mathcal{P}}(\alpha, t), G_2^{\mathcal{P}}(\alpha, t))$. Note that $\gcd(G_1^{\mathcal{P}}(\alpha, t), G_2^{\mathcal{P}}(\alpha, t))$ is defined since not both components of $\mathcal{P}(t)$ are constant, and therefore at least one of the polynomials $G_i^{\mathcal{P}}(\alpha, t)$ is not zero. Let us assume that χ_1 is not constant. Then $\chi_{12}(\beta) \neq 0$, since otherwise $\chi_{12}(\alpha)\chi_{11}(\beta) = 0$. But $\chi_{12}(\alpha) \neq 0$ and hence $\chi_{11}(\beta) = 0$, which is impossible because $\gcd(\chi_{11}, \chi_{12}) = 1$. Similarly, if χ_2 is not constant, we get that $\chi_{22}(\beta) \neq 0$. Note that if some χ_i is constant the result is obtained trivially. Thus, $\beta \in \mathcal{F}_{\mathcal{P}}(P(\alpha))$. Therefore, since $G_1^{\mathcal{P}}(\alpha, t)$ and $G_2^{\mathcal{P}}(\alpha, t)$ do not have multiple roots, the cardinality of the fibre is the degree of the gcd. $\square$

Theorem 4.25 implies that almost all points $(x_\alpha, y_\alpha) = \mathcal{P}(\alpha) \in C$ are generated more than once if and only if $\deg_t(\gcd(G_1^{\mathcal{P}}(\alpha, t), G_2^{\mathcal{P}}(\alpha, t))) > 1$. In Lemma 4.27 we will see that the degree of this gcd is preserved under almost all specializations of the variable s. First we state the following result on gcds. Let φ_a denote the natural evaluation homomorphism of $K[x, y]$ into $K[y]$, i.e., for $a \in K$,

$$\varphi_a : K[x, y] \longrightarrow K[y]$$
$$f(x, y) \longmapsto f(a, y).$$

Lemma 4.26. *Let $f, g \in K[x, y]^*$, $f = \bar{f} \cdot \gcd(f, g)$, $g = \bar{g} \cdot \gcd(f, g)$. Let $a \in K$ be such that not both leading coefficients of f and g w.r.t. y vanish at a.*

(1) $\deg_y(\gcd(\varphi_a(f), \varphi_a(g))) \geq \deg_y(\varphi_a(\gcd(f, g)) = \deg_y(\gcd(f, g)).$

(2) If the resultant w.r.t. y of $\bar{f}$ and $\bar{g}$ does not vanish at a, then

$$\gcd(\varphi_a(f), \varphi_a(g)) = \varphi_a(\gcd(f, g)).$$

Proof. Let $h = \gcd(f, g)$. Since not both leading coefficients (w.r.t. y) of f and g vanish under φ_a, also the leading coefficient of h cannot vanish under φ_a. So $\deg_y(\varphi_a(h)) = \deg_y(h)$. Furthermore, $\varphi_a(f) = \varphi_a(\bar{f})\varphi_a(h)$ and $\varphi_a(g) = \varphi_a(\bar{g})\varphi_a(h)$.

(1) $\varphi_a(h)$ divides $\gcd(\varphi_a(f), \varphi_a(g))$, so

$$\deg_y(\gcd(\varphi_a(f), \varphi_a(g))) \geq \deg_y(\varphi_a(h)) = \deg_y(h).$$

(2) We have

$$\gcd(\varphi_a(f), \varphi_a(g)) = \gcd(\varphi_a(\bar{f}), \varphi_a(\bar{g})) \cdot \varphi_a(h).$$

If $\gcd(\varphi_a(f), \varphi_a(g)) \neq \varphi_a(h)$, then $\gcd(\varphi_a(\bar{f}), \varphi_a(\bar{g})) \neq 1$. Hence, the resultant w.r.t. y of $\varphi_a(\bar{f}), \varphi_a(\bar{g})$ is zero. Therefore, since φ_a is a ring homomorphism, one obtains that

$$0 = \text{res}_y(\varphi_a(\bar{f}), \varphi_a(\bar{g})) = \varphi_a(\text{res}_y(\bar{f}, \bar{g})).$$

This, however, is excluded by the assumptions.   □

Lemma 4.27. *Let $\mathcal{P}(t)$ be a rational parametrization in reduced form. Then for almost all values $\alpha \in K$ of s we have*

$$\deg_t(\gcd(G_1^{\mathcal{P}}(s, t), G_2^{\mathcal{P}}(s, t))) = \deg_t(\gcd(G_1^{\mathcal{P}}(\alpha, t), G_2^{\mathcal{P}}(\alpha, t))).$$

Proof. We distinguish two cases. First, we assume that no component of $\mathcal{P}(t)$ is constant, so $G_1^{\mathcal{P}}(s, t)$ and $G_2^{\mathcal{P}}(s, t)$ cannot be zero. Thus, if $G = \gcd(G_1^{\mathcal{P}}, G_2^{\mathcal{P}})$ and $G_1^{\mathcal{P}} = \overline{G_1^{\mathcal{P}}} \cdot G, G_2^{\mathcal{P}} = \overline{G_2^{\mathcal{P}}} \cdot G$, then $T(s) = \text{res}_t(\overline{G_1^{\mathcal{P}}}, \overline{G_2^{\mathcal{P}}}) \in K[s]$ is not identically zero. Therefore, $T(s)$ and the leading coefficients of $G_1^{\mathcal{P}}$ and $G_2^{\mathcal{P}}$, w.r.t. t, can only vanish at finitely many values. From Lemma 4.26 (2) we get $\varphi_\alpha(\gcd(G_1^{\mathcal{P}}, G_2^{\mathcal{P}})) = \gcd(\varphi_\alpha(G_1^{\mathcal{P}}), \varphi_\alpha(G_2^{\mathcal{P}}))$ for almost all $\alpha \in K$.
Second, if any component of the parametrization $\mathcal{P}(t)$ is constant, we obviously have $\varphi_\alpha(\gcd(G_1^{\mathcal{P}}, G_2^{\mathcal{P}})) = \gcd(\varphi_\alpha(G_1^{\mathcal{P}}), \varphi_\alpha(G_2^{\mathcal{P}}))$.
So, for almost all $\alpha \in K$,

$$\deg_t(\gcd(\varphi_\alpha(G_1^{\mathcal{P}}), \varphi_\alpha(G_2^{\mathcal{P}}))) = \deg_t(\varphi_\alpha(\gcd(G_1^{\mathcal{P}}, G_2^{\mathcal{P}}))) \leq \deg_t(\gcd(G_1^{\mathcal{P}}, G_2^{\mathcal{P}})).$$

On the other hand, by Lemma 4.26 (1), for almost all $\alpha \in K$,

$$\deg_t(\gcd(\varphi_\alpha(G_1^{\mathcal{P}}), \varphi_\alpha(G_2^{\mathcal{P}}))) \geq \deg_t(\gcd(G_1^{\mathcal{P}}, G_2^{\mathcal{P}})).$$

Thus, for almost all $\alpha \in K$, $\deg_t(\gcd(\varphi_\alpha(G_1^{\mathcal{P}}), \varphi_\alpha(G_2^{\mathcal{P}}))) = \deg_t(\gcd(G_1^{\mathcal{P}}, G_2^{\mathcal{P}})).$
    □

Theorem 4.28. *Let $\mathcal{P}(t)$ be a parametrization in reduced form of the curve $\mathcal{C}$. Then*

$$\text{index}(\mathcal{P}(t)) = \deg_t(\gcd(G_1^{\mathcal{P}}(s,t), G_2^{\mathcal{P}}(s,t)).$$

Proof. The result follows from Theorem 4.25, Lemma 4.27, and Theorem 2.43.

$\square$

Now from Lemma 4.26, Theorem 4.28, and the proof of Lemma 4.27 we get the following corollary.

Corollary 4.29. *Let $\mathcal{P}(t)$ be a parametrization in reduced form, and let $G^{\mathcal{P}}(s,t) = \gcd(G_1^{\mathcal{P}}, G_2^{\mathcal{P}})$. We define $T(s) = \text{res}_t(\frac{G_1^{\mathcal{P}}}{G^{\mathcal{P}}}, \frac{G_2^{\mathcal{P}}}{G^{\mathcal{P}}})$ if $\mathcal{P}$ does not have constant components, and $T(s) = 1$ otherwise. Then, for $\alpha \in K$ such that $\chi_{12}(\alpha)\chi_{22}(\alpha)T(\alpha) \neq 0$, and such that α is not a common root of the leading coefficients of $G_1^{\mathcal{P}}$ and $G_2^{\mathcal{P}}$ w.r.t. t, we have*

(1) $\text{card}(\mathcal{F}_{\mathcal{P}}(\mathcal{P}(\alpha))) = \deg_t(G^{\mathcal{P}}(\alpha,t)) = \deg_t(G^{\mathcal{P}}(s,t))$,
(2) $\mathcal{F}_{\mathcal{P}}(\mathcal{P}(\alpha)) = \{\beta \in K \mid G^{\mathcal{P}}(\alpha,\beta) = 0\}$.

$\square$

Since a parametrization is proper if and only if it defines a birational mapping between the affine line and the curve, it is clear that a parametrization is proper if and only if its tracing index is 1.

Theorem 4.30. *A rational parametrization is proper if and only if its tracing index is 1, i.e. if and only if $\deg_t(\gcd(G_1^{\mathcal{P}}, G_2^{\mathcal{P}})) = 1$.*

The previous results can be used to derive the following algorithm for computing the tracing index of a given parametrization. This algorithm can also be used for checking the properness of a parametrization.

Algorithm TRACING INDEX

Given a rational parametrization $\mathcal{P}(t)$ in reduced form, the algorithm computes $\text{index}(\mathcal{P}(t))$, and decides whether the parametrization is proper.

1. Compute the polynomials $G_1^{\mathcal{P}}(s,t), G_2^{\mathcal{P}}(s,t)$.
2. Determine $G^{\mathcal{P}}(s,t) := \gcd(G_1^{\mathcal{P}}, G_2^{\mathcal{P}})$.
3. $\ell := \deg_t(G^{\mathcal{P}}(s,t))$.
4. If $\ell = 1$ then return "$\mathcal{P}(t)$ is proper and $\text{index}(\mathcal{P}(t)) = 1$" else return "$\mathcal{P}(t)$ is not proper and $\text{index}(\mathcal{P}(t)) = \ell$"

We illustrate the algorithm by an example.

Example 4.31. Let $\mathcal{P}(t)$ be the rational parametrization

$$\mathcal{P}(t) = \left(\frac{(t^2 - 1)\,t}{t^4 - t^2 + 1}, \frac{(t^2 - 1)\,t^2}{t^6 - 3\,t^4 + 3\,t^2 - 1 - 2\,t^3} \right).$$

In Step 1 the polynomials

$$G_1^{\mathcal{P}}(s,t) = s^3t^4 - st^4 + s^2t^3 - s^4t^3 - t^3 - s^3t^2 + st^2 + s^4t - s^2t + t + s^3 - s$$

$$G_2^{\mathcal{P}}(s,t) = s^4t^6 - s^2t^6 - s^6t^4 + 2\,s^3t^4 + t^4 - 2\,s^4t^3 + 2\,s^2t^3 + s^6t^2 - 2\,s^3t^2$$
$$-t^2 - s^4 + s^2,$$

are generated. Their gcd, computed in Step 2, is $G^{\mathcal{P}}(s,t) = st^2 - s^2t + t - s$. Thus, $\text{index}(\mathcal{P}(t)) = 2$, and therefore the parametrization is not proper.

For a generalization of these results to the surface case see [PDS04].

4.3.2 Tracing Index Under Reparametrizations

In order to study the behavior of the index under reparametrizations we first prove a technical lemma where we show that, in the case of a single nonconstant rational function $R(t)$, the degree w.r.t. t of $R(t)$ is the degree of the rational map from K to K induced by $R(t)$.

Lemma 4.32. *Let* $R(t) = p(t)/q(t) \in K(t)$ *be nonconstant and in reduced form. Let* $R : K \to K$ *be the rational map induced by* $R(t)$. *Then* $\text{card}(R^{-1}(a)) = \deg(R(t))$ *for almost all* $a \in K$.

Proof. Let W_0 be the nonempty open subset of K where R is defined, and let V_0 be the subset of points $a \in K$ such that $p(t) - aq(t)$ is square-free, and such that $\deg(p(t) - aq(t)) = \deg(R(t))$. From Lemma 4.19 we get that V_0 is open and nonempty. Furthermore, since R is nonconstant, $R(W_0)$ is also a nonempty open set (see Exercise 4.5). We consider the set $U = V_0 \cap R(W_0)$. So also U is a nonempty open set. We show that $\text{card}(R^{-1}(a)) = \deg(R(t))$ for all $a \in U$. Indeed: take $a \in U$. Then $R^{-1}(a)$ is nonempty. Moreover, since $\gcd(p,q) = 1$, $p(t) - aq(t)$ is square-free, and $\deg(p(t) - aq(t)) = \deg(R(t))$. Then, $\text{card}(R^{-1}(a)) = \deg(R(t))$. □

Theorem 4.33. *Let* $\mathcal{P}(t)$ *be a rational parametrization, and* $R(t) \in K(t) \setminus K$. *Then*

$$\text{index}(\mathcal{P}(R(t))) = \deg(R(t)) \cdot \text{index}(\mathcal{P}(t)).$$

Proof. The statement follows from Lemmas 2.42 and 4.32. □

Corollary 4.34. *Let* $\mathcal{C}$ *be an affine rational curve defined over* K *by* $f(x,y)$, *and let* $\mathcal{P}(t) = (\chi_1(t), \chi_2(t))$ *be a parametrization of* $\mathcal{C}$. *If* $\chi_1(t)$ *is nonzero then* $\deg_y(f) = \frac{\deg(\chi_1(t))}{\text{index}(\mathcal{P})}$; *similarly if* $\chi_2(t)$ *is nonzero then* $\deg_x(f) = \frac{\deg(\chi_2(t))}{\text{index}(\mathcal{P})}$.

Proof. By Lemmas 4.13 and 4.17, there exists a proper parametrization $\mathcal{Q}(t) = (\xi_1(t), \xi_2(t))$ of $\mathcal{C}$, and $R(t) \in K(t) \setminus K$ such that $\mathcal{P}(t) = \mathcal{Q}(R(t))$. By Theorem 4.33

$$\text{index}(\mathcal{P}(t)) = \deg(R(t)) \cdot \text{index}(\mathcal{Q}(t)) = \deg(R(t)).$$

Moreover, $\deg(\chi_i(t)) = \deg(R(t)) \cdot \deg(\xi_i(t))$. Now, the result follows from Theorem 4.21. □

In Theorem 4.35 we show the relation between the index of a parametrization, the degree of a parametrization and the degree of the curve.

Theorem 4.35. *Let C be an affine rational curve defined by $f(x, y) \in K[x, y]$, let $n = \max\{\deg_x(f), \deg_y(f)\}$, and let $\mathcal{P}(t)$ be a rational parametrization of C. Then,*

$$\mathrm{index}(\mathcal{P}(t)) = \frac{\deg(\mathcal{P}(t))}{n}.$$

Proof. Because of Lemma 4.13 there exists a proper parametrization $\mathcal{P}'(t)$ of C, and because of Lemma 4.17 there exists $R(t) \in K(t) \setminus K$ such that $\mathcal{P}(t) = \mathcal{P}'(R(t))$. From Theorem 4.33 and the fact that $\mathcal{P}'(t)$ is proper we get that

$$\mathrm{index}(\mathcal{P}(t)) = \deg(R(t)) \cdot \mathrm{index}(\mathcal{P}'(t)) = \deg(R(t)).$$

Furthermore, since the degree of rational functions under composition is multiplicative, we arrive at $\deg(\mathcal{P}(t)) = \deg(R(t)) \cdot \deg(\mathcal{P}'(t))$. Thus

$$\mathrm{index}(\mathcal{P}(t)) = \frac{\deg(\mathcal{P}(t))}{\deg(\mathcal{P}'(t))}.$$

Applying Theorem 4.21 we see that $\deg(\mathcal{P}'(t)) = n$, which completes the proof.
□

4.4 Inversion of Proper Parametrizations

In Theorems 4.14, 4.21, and 4.30 we have deduced various different criteria for deciding the properness of a parametrization. Now, we show how to compute the inverse map of a proper rational parametrization. Let $\mathcal{P}(t)$ be a proper parametrization of an affine rational curve C. Then the *inversion problem* consists of computing the inverse rational mapping of the birational map (compare Definition 4.12)

$$\mathcal{P} : \mathbb{A}^1(K) \longrightarrow C.$$

More precisely, we want to compute the rational map

$$\varphi : \quad C \quad \longrightarrow \mathbb{A}^1(K)$$
$$(x, y) \longmapsto \varphi(x, y) \, ,$$

satisfying

(1) $\varphi \circ \mathcal{P} = \mathrm{id}_{\mathbb{A}^1(K)}$, i.e. $\varphi(\mathcal{P}(t)) = t$, and
(2) $\mathcal{P} \circ \varphi = \mathrm{id}_C$, i.e. $\chi_{i2}(\varphi)x - \chi_{i1}(\varphi) = 0 \mod I(C)$ for $i = 1, 2$,

where χ_{i1}/χ_{i2} is the i-th component of $\mathcal{P}(t)$. In this case φ is the inverse $\mathcal{P}^{-1}$ we are looking for.

So the inversion problem is essentially an elimination problem, and therefore elimination techniques such as Gröbner bases can be applied. Here we give a different approach to the problem based on the computation of gcds over the function field of the curve. A generalization to surfaces of these ideas can be found in [PDSS02]. For a more general statement of the problem, namely inversion of birational maps, see [Sch98b]. Alternative methods for inverting proper parametrizations can be found in [BuD06], [ChG92b], and [GSA84].

In addition, in order to check whether a rational function is the inverse of a given parametrization, it is enough to test one the two conditions given above. A proof of this fact, for the general case of hypersurfaces, can be found in [PDSS02]. Thus, in the sequel, we will choose freely one of the conditions to check the rational invertibility of a parametrization.

Lemma 4.36. *Let*

$$\mathcal{P} : \mathbb{A}^1(K) \longrightarrow \mathcal{C} \subset \mathbb{A}^2(K)$$
$$t \longmapsto (\chi_1(t), \chi_2(t))$$

be a rational parametrization of a plane curve $\mathcal{C}$*, and let*

$$\mathcal{U} : \quad \mathcal{C} \longrightarrow \mathbb{A}^1(K)$$
$$(x, y) \longmapsto \mathcal{U}(x, y)$$

be a rational map, where the denominators of $\mathcal{U}$ *do not belong to the ideal of* $\mathcal{C}$*. The following statements are equivalent:*

(1) $\mathcal{U}$ *is the inverse of* $\mathcal{P}$*.*
(2) $\mathcal{P}(\mathcal{U}(P)) = P$ *for almost all points* $P \in \mathcal{C}$*.*
(3) $\mathcal{U}(\mathcal{P}(t)) = t$ *for almost all values* $t \in K$*.*   □

First we observe that $K(\mathcal{C})[t]$ is a Euclidean domain. Furthermore, since we know how to computationally perform the arithmetic in the coordinate ring $\Gamma(\mathcal{C})$ (see Sect. 2.2), we know how to compute gcds in $K(\mathcal{C})[t]$. Moreover, since $I(\mathcal{C})$ is principal, all computations can be carried out by means of remainders w.r.t. the defining polynomial. Alternatively we may use the parametrization $\mathcal{P}(t)$ to check whether a class in the quotient ring $\Gamma(\mathcal{C})$ is zero. Of course, this second approach avoids the use of the implicit equation but representatives of the classes are not reduced.

Theorem 4.37. *Let* $\mathcal{P}(t)$ *be a proper parametrization in reduced form with nonconstant components of a rational curve* $\mathcal{C}$*. Let* $H_1^{\mathcal{P}}(t, x), H_2^{\mathcal{P}}(t, y)$ *be considered as polynomials in* $K(\mathcal{C})[t]$*. Then,*

$$\deg_t (\gcd_{K(\mathcal{C})[t]} (H_1^{\mathcal{P}}, H_2^{\mathcal{P}})) = 1.$$

Moreover, the single root of this gcd is the inverse of $\mathcal{P}$*.*

Proof. Let $R(t) = \gcd_{K(\mathcal{C})[t]}(H_1^{\mathcal{P}}, H_2^{\mathcal{P}})$, and let φ be the inverse of $\mathcal{P}$. Then φ is a root of $R(t)$, and therefore $\deg_t(R) \geq 1$. Now, since R is the gcd, there exist polynomials $M_i(x, y, t) \in K(\mathcal{C})[t]$ such that $H_i^{\mathcal{P}}(x, y, t) = M_i(x, y, t)R(x, y, t) \bmod I(\mathcal{C})$. Thus, if f defines $\mathcal{C}$, the above equality can be written in $K[x, y, t]$ as:

$$N_i(x, y)H_i^{\mathcal{P}}(x, y, t) = M_i^*(x, y, t)S(x, y, t) + A(x, y)f(x, y) \ ,$$

where $\deg_t(S) = \deg_t(R)$, and neither N_i nor all coefficients of M_i^* w.r.t. t, nor the leading coefficient of S w.r.t. t belong to $I(\mathcal{C})$. Thus, substituting $\mathcal{P}(s)$ into this formula and clearing denominators, we see that $\deg_t(S) \leq \deg_t(\gcd(G_1^{\mathcal{P}}(s, t), G_1^{\mathcal{P}}(s, t)))$. Now, by Theorem 4.30, we get that $\deg_t(R) = \deg_t(S) \leq 1$. $\qquad\square$

In the following we outline an algorithm for inverting a proper parametrization, based on Theorem 4.37.

Algorithm INVERSE

Given an affine rational parametrization $\mathcal{P}(t)$, in reduced form, the algorithm decides whether the parametrization is proper, and in the affirmative case it determines the inverse of the mapping $\mathcal{P}$.

1. Apply algorithm TRACING INDEX to check whether $\mathcal{P}(t)$ is proper. If $\mathcal{P}(t)$ is not proper then return "**not proper**" and **exit**.
2. Compute $H_1^{\mathcal{P}}(t, x)$ and $H_2^{\mathcal{P}}(t, y)$.
3. Determine $M(x, y, t) = \gcd_{K(\mathcal{C})[t]}(H_1^{\mathcal{P}}, H_2^{\mathcal{P}})$. By Theorem 4.37 $M(x, y, t)$ is linear in t; let us say

$$M(x, y, t) = D_1(x, y)t - D_0(x, y) \ .$$

4. Return "**the inverse is** $\frac{D_0(x,y)}{D_1(x,y)}$."

Example 4.38. Let $\mathcal{C}$ be the plane curve over $\mathbb{C}$ defined by the rational parametrization

$$\mathcal{P}(t) = \left(\frac{t^3 + 1}{t^2 + 3}, \frac{t^3 + t + 1}{t^2 + 1} \right).$$

It is easy to check, applying algorithm TRACING INDEX, that $\text{index}(\mathcal{P}(t)) = 1$ and therefore $\mathcal{P}(t)$ is proper. Furthermore, the implicit equation of $\mathcal{C}$ is

$$\begin{aligned} f(x, y) = {}&-4\,x^2y^3 + 4\,xy^3 - 2\,y^3 + 4\,x^3y^2 - 8\,x^2y^2 + 4\,xy^2 + 3\,y^2 + 4\,x^3y \\ &-3\,x^2y - 11\,xy + 13\,x^3 + 8\,x^2 + 3\,x - 1 \ . \end{aligned}$$

For a method for computing the implicit equation see Theorem 4.39. In Step 2 we consider the polynomials in $K(\mathcal{C})[t]$

$$H_1^{\mathcal{P}}(t, x) = -t^3 + xt^2 + 3\,x - 1, \quad H_2^{\mathcal{P}}(t, y) = -t^3 + yt^2 - t + y - 1.$$

In Step 3, we determine $\gcd_{\mathbb{C}(\mathcal{C})[t]}(H_1^{\mathcal{P}}, H_2^{\mathcal{P}})$. The polynomial remainder sequence of $H_1^{\mathcal{P}}$ and $H_2^{\mathcal{P}}$ is:

$$R_0(t) = -t^3 + xt^2 + 3x - 1$$

$$R_1(t) = -t^3 + yt^2 - t + y - 1$$

$$R_2(t) = (x - y)t^2 + t + 3x - y$$

$$R_3(t) = \frac{2x^2 - 3yx - 1 + y^2}{(-x+y)^2}t + \frac{(-2y-1)x^2 + (2y^2 + 2y - 3)x - y^2 + y}{(-x+y)^2}$$

$$R_4(t) = 0.$$

Thus, $\gcd_{\mathbb{C}(\mathcal{C})[t]}(H_1^{\mathcal{P}}, H_2^{\mathcal{P}})$

$$= \frac{\left(2x^2 - 3yx - 1 + y^2\right)}{(-x+y)^2}t + \frac{-y^2 + 2y^2x + 2yx - 3x - 2yx^2 - x^2 + y}{(-x+y)^2}.$$

Therefore, the inverse mapping is:

$$\mathcal{P}^{-1}(x, y) = -\frac{-y^2 + 2y^2x + 2yx - 3x - 2yx^2 - x^2 + y}{2x^2 - 3yx - 1 + y^2}.$$

4.5 Implicitization

Given an affine rational parametrization $\mathcal{P}(t)$, the *implicitization problem* consists of computing the defining polynomial for the Zariski closure of the set

$$S = \{\mathcal{P}(t) \mid t \in K \text{ such that } \mathcal{P}(t) \text{ is defined}\}.$$

Therefore, the problem consists of finding the smallest algebraic set in $\mathbb{A}^2(K)$ containing S. Note also, that if we are given a projective rational parametrization the implicitization problem is the same since the defining polynomial of the projective curve is the homogenization of the defining polynomial of the affine curve.

The problem can be solved by general elimination techniques such as Gröbner bases ([AdL94] and [CLO97]). This approach is valid not only for curves but for the more general case of parametric varieties in $\mathbb{A}(K)^n$. Also, for surfaces, different approaches can be found in [BCD03], [ChG92a], [Gon97], [Kot04], [SGD97]. However, for the case of plane curves, the implicit equation can be found by means of gcd's and resultants alone. For instance, applying Lemma 4.6, the defining polynomial of the curve parametrized by $\mathcal{P}(t)$ can be obtained by computing the square-free part of a resultant. Moreover, if properness is guaranteed, Theorem 4.39 shows that the implicit equation can be computed by a single resultant. This result can be found in [SGD97], [SeW89], or in [SeW01a]. In addition to these results, in Theorem 4.41 we see that in this resultant the implicit equation appears to the power of the tracing index. Similar results on implicitization can be found in [ChG92a] and [CLO97].

Theorem 4.39. *Let $\mathcal{P}(t)$ be a proper parametrization in reduced form of a rational affine plane curve $\mathcal{C}$. Then, the defining polynomial of $\mathcal{C}$ is the resultant*

$$\operatorname{res}_t(H_1^{\mathcal{P}}(t,x), H_2^{\mathcal{P}}(t,y)).$$

Proof. Let $\mathcal{P}(t) = (\chi_1(t), \chi_2(t))$. We know from Theorem 4.21 that, if $f(x,y)$ is the implicit equation of $\mathcal{C}$, then $\deg_y(f) = \deg(\chi_1(t))$, and $\deg_x(f) = \deg(\chi_2(t))$. The polynomials $H_i^{\mathcal{P}}$ can be written as

$$H_1^{\mathcal{P}}(t,x) = a_m(x)t^m + \cdots + a_0(x), \qquad \text{where } m = \deg_y(f),$$

$$H_2^{\mathcal{P}}(t,y) = b_n(y)t^n + \cdots + b_0(y), \qquad \text{where } n = \deg_x(f),$$

where $\deg_x(a_i) \leq 1$ and $\deg_y(b_i) \leq 1$.

Let $R(x,y)$ be the resultant of $H_1^{\mathcal{P}}$ and $H_2^{\mathcal{P}}$ with respect to t, and let A be the Sylvester matrix of $H_1^{\mathcal{P}}, H_2^{\mathcal{P}} \in K(x,y)[t]$ seen as univariate polynomials in t:

$$A = \begin{pmatrix} a_m(x) \cdots & \cdots & \cdots a_0(x) & & \\ & \ddots & & & \ddots & \\ & & a_m(x) \cdots & \cdots & \cdots a_0(x) \\ b_n(y) \cdots & \cdots & \cdots b_0(y) & & \\ & \ddots & & & \ddots & \\ & & b_n(y) \cdots & \cdots & \cdots b_0(y) \end{pmatrix}.$$

Therefore, since only the entries in the first n rows depend on x, and this dependence on x is linear, $\deg_x(R) \leq n$. Analogously, $\deg_y(R) \leq m$. On the other hand, it is known that $f(x,y)$ is a factor of $R(x,y)$ (compare Lemma 4.6). Thus, $\deg_x(f) = \deg_x(R)$ and $\deg_y(f) = \deg_y(R)$. Therefore, up to a constant, $f(x,y) = R(x,y)$. □

We finish this section showing how Lemma 4.6, Theorem 4.39, and the notion of tracing index of a parametrization (compare Definition 4.24) are related. Basically, the result follows from the next lemma on resultants, which is valid for an arbitrary field.

Lemma 4.40. *Let $A, B \in \mathbb{L}[t]$ be nonconstant polynomials over a field $\mathbb{L}$:*

$$A(t) = a_m t^m + \cdots + a_0, \quad B(t) = b_n t^n + \cdots + b_0, \quad a_m b_n \neq 0$$

and let $R(t) = \frac{M(t)}{N(t)} \in \mathbb{L}(t)$ be a nonconstant rational function in reduced form, such that $\deg(M - \beta N) = \deg(R)$ for every root β of $A(t)B(t)$. Let $A'(t)$ and $B'(t)$ be the polynomials

$$A'(t) = a_m M(t)^m + a_{m-1} M(t)^{m-1} N(t) + \cdots + a_0 N(t)^m,$$

$$B'(t) = b_n M(t)^n + b_{n-1} M(t)^{n-1} N(t) + \cdots + b_0 N(t)^n.$$

Then, if b' is the leading coefficient of B',

$$\mathrm{res}_t(A', B') = \frac{(b')^{m(\deg(R)-\deg(N))}}{b_n^{m\,\deg(R)}} \mathrm{res}_t(A, B)^{\deg(R)} \cdot \mathrm{res}_t(B', N)^m.$$

Proof. Let B decompose over the algebraic closure of $\mathbb{L}$ as

$$B(t) = b_n \prod_{i=1}^{n} (t - \beta_i).$$

Since $B'(t) = N^n \cdot B(R)$ one has that

$$B'(t) = b_n \prod_{i=1}^{n} (M(t) - b_i N(t)).$$

Therefore, since $\deg(M - \beta_i N) = \deg(R)$ for every $i \in \{1, \ldots, n\}$, we have $\deg(B') = n \cdot \deg(R)$. In particular, since R is nonconstant, B' is not a constant polynomial. Similarly we see that $\deg(A') = m \cdot \deg(R)$, and that A' is also a nonconstant polynomial.

Now, observe that if $r = \deg(R)$, every root β_i of B generates r roots $\{\beta_{i,1}, \ldots, \beta_{i,r}\}$ of $B'(t)$, namely the roots of $M(t) - \beta_i N(t)$. Moreover, if α is a root of B' then $N(\alpha) \neq 0$, since otherwise one gets that $M(\alpha) = 0$, which is impossible because of $\gcd(M, N) = 1$. Therefore,

$$\beta_i = \frac{M(\beta_{i,j})}{N(\beta_{i,j})} = R(\beta_{i,j}), \quad j = 1, \ldots, r.$$

Let $S = \mathrm{res}_t(A, B)$, $S' = \mathrm{res}_t(A', B')$ and $S'' = \mathrm{res}_t(B', N)$. From the relation $A' = N^m \cdot A(R)$ we get

$$S' = (b')^{mr} \prod_{B'(\alpha)=0} A'(\alpha) = (b')^{mr} \prod_{i=1}^{n} \prod_{j=1}^{r} A'(\beta_{i,j}) = (b')^{mr} \prod_{i=1}^{n} A(\beta_i)^r \prod_{j=1}^{r} N(\beta_{i,j})^m.$$

Furthermore, if $k = \deg(N)$, we have

$$S = b_n^m \prod_{i=1}^{n} A(\beta_i), \qquad S'' = (b')^k \prod_{i=1}^{n} \prod_{j=1}^{r} N(\beta_{i,j}).$$

Thus,

$$S' = \frac{(b')^{mr}}{b_n^{rm}} S^r \prod_{i=1}^{n} \prod_{j=1}^{r} N(\beta_{i,j})^m = \frac{(b')^{mr-km}}{b_n^{rm}} S^r \cdot (S'')^m. \qquad \square$$

Theorem 4.41. *Let $\mathcal{P}(t)$ be a parametrization in reduced form of an affine rational plane curve $\mathcal{C}$, and let $f(x,y)$ be the defining polynomial of $\mathcal{C}$. Then for some nonzero constant c we have*

$$\operatorname{res}_t(H_1^{\mathcal{P}}(t,x), H_2^{\mathcal{P}}(t,y)) = c \cdot (f(x,y))^{\operatorname{index}(\mathcal{P})}.$$

Proof. If $\mathcal{C}$ is a line parallel to one of the axes, let us say $y = a$, then $\mathcal{P}(t) = (\frac{\chi_{11}(t)}{\chi_{12}(t)}, a)$. By Lemma 4.32 $\operatorname{index}(\mathcal{P}) = \deg(\mathcal{P})$. Therefore,

$$\operatorname{res}_t(H_1^{\mathcal{P}}(t), H_2^{\mathcal{P}}(t))$$

$$= \operatorname{res}_t(x \cdot \chi_{12}(t) - \chi_{11}(t), y - a) = (y - a)^{\deg(\mathcal{P}(t))} = (y - a)^{\operatorname{index}(\mathcal{P})}.$$

Let us now assume that the irreducible curve $\mathcal{C}$ is not a line parallel to one of the axes, i.e. its defining polynomial depends on both variables x, y. By Lemma 4.18 there is a proper parametrization of $\mathcal{C}$ in which the degrees of numerator and denominator at each component agree. So let

$$\mathcal{P}'(t) = \left(\frac{\xi_{11}(t)}{\xi_{12}(t)}, \frac{\xi_{21}(t)}{\xi_{22}(t)}\right)$$

be a proper parametrization, in reduced form, of $\mathcal{C}$ where $\deg(\xi_{i1}) = \deg(\xi_{i2})$. By Lemma 4.17 there exists a nonconstant rational function $R(t)$ such that $\mathcal{P}(t) = \mathcal{P}'(R(t)) = \left(\frac{\chi_{11}(t)}{\chi_{12}(t)}, \frac{\chi_{21}(t)}{\chi_{22}(t)}\right)$. Let $R(t) = \frac{M(t)}{N(t)}$ be in reduced form. We consider the polynomials

$$H_1^{\mathcal{P}}(t) = x \cdot \chi_{12}(t) - \chi_{11}(t), \quad H_2^{\mathcal{P}}(t) = y \cdot \chi_{22}(t) - \chi_{21}(t),$$

$$H_1^{\mathcal{P}'}(t) = x \cdot \xi_{12}(t) - \xi_{11}(t), \quad H_2^{\mathcal{P}'}(t) = y \cdot \xi_{22}(t) - \xi_{21}(t).$$

Note that $H_i^{\mathcal{P}}, H_i^{\mathcal{P}'} \in (\mathbb{K}[x,y])[t]$.

We structure the remaining part of the proof in the following way:

(1) we relate the polynomials $H_i^{\mathcal{P}}$ and $\overline{H}_i^{\mathcal{P}'}$ (the result of substituting the rational function R into $H_i^{\mathcal{P}'}$),
(2) we extract common factors in these relations,
(3) we derive a nontrivial relation between $\operatorname{res}_t(H_1^{\mathcal{P}}, H_2^{\mathcal{P}})$ and $\operatorname{res}_t(\overline{H}_1^{\mathcal{P}'}, \overline{H}_2^{\mathcal{P}'})$,
(4) these resultants contain powers of the defining polynomial of $\mathcal{C}$. We express the exponent as $\operatorname{index}(\mathcal{P})$.

So let us deal with step (1). Let

$$\xi_{i1}(t) = \sum_{j=0}^{n_i} a_{i,j}t^j, \quad \xi_{i2}(t) = \sum_{j=0}^{n_i} b_{i,j}t^j, \quad H_i^{\mathcal{P}'}(t) = \sum_{j=0}^{m_i} h_{i,j}t^j, \quad \text{for } i = 1, 2.$$

Observe that $m_i = n_i$. For $i = 1, 2$, we introduce the new polynomials

$$\overline{\xi}_{i1}(t) = \sum_{j=0}^{n_i} a_{i,j} M(t)^j N(t)^{n_i-j}, \quad \overline{\xi}_{i2}(t) = \sum_{j=0}^{n_i} b_{i,j} M(t)^j N(t)^{n_i-j},$$

$$\overline{H}_i^{\mathcal{P}'}(t) = \sum_{j=0}^{m_i} h_{i,j} M(t)^j N(t)^{m_i-j},$$

which result from $\xi_{i1}, \xi_{i2}, H_i^{\mathcal{P}'}$ by substituting $R(t)$ for t and clearing denominators. In order to apply Lemma 4.40 to the nonconstant polynomials $H_1^{\mathcal{P}'}(t), H_2^{\mathcal{P}'}(t) \in K(x, y)[t]$ and the rational function $R(t)$, let us see that $\deg(M(t) - \beta N(t)) = \deg(R)$ for every root β of $H_1^{\mathcal{P}'}(t) \cdot H_2^{\mathcal{P}'}(t)$. Indeed, if β is such that $\deg(M(t) - \beta N(t)) < \deg(R)$ then $\beta \in K$. Therefore, either $H_1^{\mathcal{P}'}(\beta) = 0$ or $H_2^{\mathcal{P}'}(\beta) = 0$ and $\beta \in K$. This implies that either $\gcd(\xi_{11}, \xi_{12}) \neq 1$ or $\gcd(\xi_{21}, \xi_{22}) \neq 1$, which is impossible. The application of Lemma 4.40 leads to

$$\mathrm{res}_t(\overline{H}_1^{\mathcal{P}'}, \overline{H}_2^{\mathcal{P}'}) =$$

$$\frac{(b')^{m_1(\deg(R)-\deg(N))}}{h_{2,m_2}^{\deg(R)m_1}} \cdot \mathrm{res}_t(H_1^{\mathcal{P}'}, H_2^{\mathcal{P}'})^{\deg(R)} \cdot \mathrm{res}_t(\overline{H}_2^{\mathcal{P}'}, N)^{m_1}, \tag{4.1}$$

where b' is the leading coefficient of $\overline{H}_2^{\mathcal{P}'}$ w.r.t. t. In addition, since $\mathcal{P}(t) = \mathcal{P}'(R(t))$, we have

$$\chi_{j1}(t) \cdot \xi_{j2}(R(t)) = \xi_{j1}(R(t)) \cdot \chi_{j2}(t), \quad j = 1, 2.$$

Thus,

$$\chi_{j1}(t) \cdot H_j^{\mathcal{P}'}(R(t)) = \xi_{j1}(R(t)) \cdot H_j^{\mathcal{P}}(t), \quad j = 1, 2,$$

and (note that $m_j = n_j$)

$$\chi_{j1}(t)\overline{H}_j^{\mathcal{P}'}(t) = \overline{\xi}_{j1}(t) H_j^{\mathcal{P}}(t), \quad \chi_{j1}(t)\overline{\xi}_{j2}(t) = \overline{\xi}_{j1}(t)\chi_{j2}(t), \quad \text{for } j = 1, 2.$$

Next we deal with step (2). We prove that $\gcd(\chi_{11}, \chi_{21}) = \gcd(\overline{\xi}_{11}, \overline{\xi}_{21})$. Indeed: from the line above and the fact that the numerators and denominators in the parametrization are relatively prime we deduce $\chi_{j1}|\overline{\xi}_{j1}$ and thus $\gcd(\chi_{11}, \chi_{21})| \gcd(\overline{\xi}_{11}, \overline{\xi}_{21})$. In order to prove that $\gcd(\overline{\xi}_{11}, \overline{\xi}_{21})$ divides $\gcd(\chi_{11}, \chi_{21})$, we first see that $\gcd(\overline{\xi}_{j1}, \overline{\xi}_{j2}) = 1$. Let a be a common root of $\overline{\xi}_{j1}$ and $\overline{\xi}_{j2}$. Note that by definition of $\overline{\xi}_{j1}$ it follows that $N(a) \neq 0$, since otherwise it would imply that $M(a) = 0$, which is impossible since $\gcd(M, N) = 1$. Therefore, taking into account that $\overline{\xi}_{j1} = N^{n_j}\xi_{j1}(R)$, $\overline{\xi}_{j2} = N^{n_j}\xi_{j2}(R)$, one deduces that $\xi_{j1}(R(a)) = \xi_{j2}(R(a)) = 0$ which is impossible since $\gcd(\xi_{j1}, \xi_{j2}) = 1$. So we have $\gcd(\overline{\xi}_{j1}, \overline{\xi}_{j2}) = 1$, from which we get by a similar reasoning as above that $\gcd(\overline{\xi}_{11}, \overline{\xi}_{21})$ divides $\gcd(\chi_{11}, \chi_{21})$.

As a consequence of this remark we can extract this gcd from the equalities above and express them as:

$$\chi_{j1}^*(t)\overline{H}_j^{\mathcal{P}'}(t) = \overline{\xi}_{j1}^*(t)H_j^{\mathcal{P}}(t), \quad \chi_{j1}^*(t)\overline{\xi}_{j2}(t) = \overline{\xi}_{j1}^*(t)\chi_{2j}(t), \quad \text{for } j = 1, 2,$$

where $\gcd(\chi_{11}^*, \chi_{21}^*) = \gcd(\overline{\xi}_{11}^*, \overline{\xi}_{21}^*) = 1$.

Now we come to step (3). Observe that

$$\mathrm{res}_t(\chi_{11}^*\overline{H}_1^{\mathcal{P}'}, \chi_{21}^*\overline{H}_2^{\mathcal{P}'}) = \mathrm{res}_t(\overline{\xi}_{11}^*H_1^{\mathcal{P}}, \overline{\xi}_{21}^*H_2^{\mathcal{P}}).$$

So,

$$\mathrm{res}_t(\chi_{11}^*, \chi_{21}^*) \cdot \mathrm{res}_t(\chi_{11}^*, \overline{H}_2^{\mathcal{P}'}) \cdot \mathrm{res}_t(\overline{H}_1^{\mathcal{P}'}, \chi_{21}^*) \cdot \mathrm{res}_t(\overline{H}_1^{\mathcal{P}'}, \overline{H}_2^{\mathcal{P}'})$$

$$= \mathrm{res}_t(\overline{\xi}_{11}^*, \overline{\xi}_{21}^*) \cdot \mathrm{res}_t(\overline{\xi}_{11}^*, H_2^{\mathcal{P}}) \cdot \mathrm{res}_t(H_1^{\mathcal{P}}, \overline{\xi}_{21}^*(t)) \cdot \mathrm{res}_t(H_1^{\mathcal{P}}, H_2^{\mathcal{P}}).$$

Let us see that none of the factors involving χ_{j1}^* or $\overline{\xi}_{j1}^*$ vanishes. Since χ_{11}^*, χ_{21}^* are relatively prime, their resultant does not vanish. Analogously for $\overline{\xi}_{j1}^*$. In order to see that the remaining factors do not vanish, we prove that if $L(t) \in K[t]^*$ then $\gcd(L, H_i^{\mathcal{P}}) = \gcd(L, \overline{H}_i^{\mathcal{P}'}) = 1$; note that since we have assumed that $\mathcal{C}$ is not a line parallel to the axes, none of the polynomial $\overline{\xi}_{ij}^*, \chi_{ij}^*$ can be zero. Indeed: if the gcd is not trivial there exists $a \in K$ such that, for instance, $H_i^{\mathcal{P}}(a) = 0$. But this implies that $\gcd(\chi_{i1}, \chi_{i2}) \neq 1$, which is impossible. Also, if $\overline{H}_i^{\mathcal{P}'}(a) = 0$, from its definition it follows that $N(a) \neq 0$. Therefore, since $\overline{H}_i^{\mathcal{P}'}(t) = N^{m_i}H_i^{\mathcal{P}'}(R(t))$, one would deduce that $H_i^{\mathcal{P}'}(R(a)) = 0$, and hence $\gcd(\xi_{i1}, \xi_{i2}) \neq 1$, which is impossible.

Taking into account this fact, the previous equality on resultants can be written as

$$T_1(y)T_2(x)\mathrm{res}_t(\overline{H}_1^{\mathcal{P}'}, \overline{H}_2^{\mathcal{P}'}) = T_1'(y)T_2'(x)\mathrm{res}_t(H_1^{\mathcal{P}}, H_2^{\mathcal{P}}), \tag{4.2}$$

where T_i, T_i' are univariate nonzero polynomials over K. Now, combining (4.1) and (4.2) we get

$$T_1(y)T_2(x)\left(\frac{(b')^{m_1(\deg(R)-\deg(N))}}{h_{2,m_2}^{\deg(R)m_1}}\mathrm{res}_t(H_1^{\mathcal{P}'}, H_2^{\mathcal{P}'})^{\deg(R)} \cdot \mathrm{res}_t(\overline{H}_2^{\mathcal{P}'}, N)^{m_1}\right)$$

$$= T_1'(y)T_2'(x)\mathrm{res}_t(H_1^{\mathcal{P}}, H_2^{\mathcal{P}}).$$

Finally we come to step (4). If $f(x, y)$ is the implicit equation of $\mathcal{C}$, from Lemma 4.6 and Theorem 4.39 we see that there exists $\ell \in \mathbb{N}$ such that

$$T_1(y)T_2(x)\left(\frac{(b')^{m_1(\deg(R)-\deg(N))}}{h_{2,m_2}^{\deg(R)m_1}}f(x, y)^{\deg(R)} \cdot \mathrm{res}_t(\overline{H}_2^{\mathcal{P}'}, N)^{m_1}\right)$$

$$= T_1'(y)T_2'(x)f(x, y)^\ell.$$

Moreover, since $b', h_{2,m_2} \in K[y]^*$ and $\mathrm{res}_t(\overline{H}_2^{\mathcal{P}'}, N)^{m_1} \in K[y]^*$ (note that we have already proved that the gcd of $\overline{H}_2^{\mathcal{P}'}$ and a nonzero polynomial depending only on t is trivial) the above equality can be rewritten as

$$U_1(y)U_2(x)f(x,y)^{\deg(R)} = U_1'(y)U_2'(x)f(x,y)^\ell$$

for some nonzero polynomials U_i, U_i'. Therefore, since $f(x,y)$ is irreducible and it depends on both variables x, y (note that we are assuming that $\mathcal{C}$ is not a line parallel to the axes), we conclude that $\deg(R) = \ell$. Furthermore, from Theorem 4.33 we get that

$$\mathrm{index}(\mathcal{P}(t)) = \mathrm{index}(\mathcal{P}'(R(t)) = \deg(R) \cdot \mathrm{index}(\mathcal{P}'(t)) = \deg(R),$$

which finishes the proof. □

4.6 Parametrization by Lines

In this section we treat some straight-forward cases in which we can easily parametrize implicitly given algebraic curves. This approach will be generalized in Sect. 4.7. The basic idea consists in using a pencil of lines through a suitable point on the curve such that by computing an intersection point of a generic element of the pencil with the curve one determines a parametrization of the curve. Of course every line $\mathcal{L}$ can be rationally parametrized, in fact by a pencil of lines with a base point not on $\mathcal{L}$. In the following we will not consider lines.

4.6.1 Parametrization of Conics

Only irreducible curves can be rational (see Theorem 4.4). So let $\mathcal{C}$ be an irreducible conic defined by the quadratic polynomial

$$f(x,y) = f_2(x,y) + f_1(x,y) + f_0(x,y),$$

where $f_i(x,y)$ are homogeneous of degree i. Let us first assume w.l.o.g. that $\mathcal{C}$ passes through the origin, so $f_0(x,y) = 0$. Let $\mathcal{H}(t)$ be the linear system $\mathcal{H}(1, O)$ of lines through the origin (compare Sect. 2.4), the elements of $\mathcal{H}(t)$ being parametrized by their slope t. So the defining polynomial of $\mathcal{H}(t)$ is

$$h(x,y,t) = y - tx.$$

Now, we compute the intersection points of a generic element of $\mathcal{H}(t)$ and $\mathcal{C}$. That is, we solve the system

$$\begin{cases} y = tx \\ f(x,y) = 0 \end{cases}$$

w.r.t. the variables x, y. The solutions are

$$O = (0,0) \quad \text{and} \quad Q(t) = \left(-\frac{f_1(1,t)}{f_2(1,t)}, -\frac{t \cdot f_1(1,t)}{f_2(1,t)} \right).$$

Note that $f_1(x, y)$ is not identically zero, since $\mathcal{C}$ is an irreducible curve. Therefore, Q depends on the parameter t. Furthermore, $f(Q(t)) = 0$, so by Theorem 4.7 $Q(t)$ is a parametrization of $\mathcal{C}$.

Theorem 4.42. *The irreducible projective conic $\mathcal{C}$ defined by the polynomial $F(x, y, z) = f_2(x, y) + f_1(x, y)z$ (f_i a form of degree i, respectively), has the rational projective parametrization*

$$\mathcal{P}(t) = (-f_1(1,t), -tf_1(1,t), f_2(1,t)).$$

Corollary 4.43. *Every irreducible conic is rational.*

So after a suitable change of coordinates, Theorem 4.42 yields a parametrization of the irreducible conic $\mathcal{C}$. We summarize this process in the following algorithm.

Algorithm CONIC-PARAMETRIZATION.
Given the defining polynomial $F(x, y, z)$ of an irreducible projective conic $\mathcal{C}$, the algorithm computes a rational parametrization.

1. Determine a point $(a : b : 1) \in \mathcal{C}$.
2. $g(x, y) = F(x + a, y + b, 1)$. Let $g_2(x, y)$ and $g_1(x, y)$ be the homogeneous components of $g(x, y)$ of degree 2 and 1, respectively.
3. Return $\mathcal{P}(t) = (-g_1(1,t) + ag_2(1,t), -tg_1(1,t) + bg_2(1,t), g_2(1,t))$.

Remarks. Note that, because of the geometric construction, the output parametrization of algorithm CONIC-PARAMETRIZATION is proper. Moreover, if $\mathcal{P}_{*,z}(t)$ is the affine parametrization of $\mathcal{C}_{*,z}$ derived from $\mathcal{P}(t)$, and $(a : b : 1)$ is the point on $\mathcal{C}$ used in the algorithm, then its inverse can be expressed as

$$\mathcal{P}_{*,z}^{-1}(x, y) = \frac{y - b}{x - a}. \qquad \square$$

Example 4.44. Let $\mathcal{C}$ be the ellipse defined by

$$F(x, y, z) = x^2 + 2y^2 - z^2.$$

We apply algorithm CONIC-PARAMETRIZATION. In Step (1) we take the point $(1 : 0 : 1)$ on $\mathcal{C}$. Then, performing Step (2), we get $g(x, y) = x^2 + 2x + 2y^2$. So, a parametrization of $\mathcal{C}$ is

$$\mathcal{P}(t) = (-1 + 2t^2, -2t, 1 + 2t^2).$$

4.6.2 Parametrization of Curves with a Point of High Multiplicity

Obviously, this approach can be immediately generalized to the situation where we have an irreducible projective curve C of degree d with a $(d-1)$–fold point P. W.l.o.g. we assume that $P = (0 : 0 : 1)$. So the defining polynomial of C is of the form

$$F(x, y, z) = f_d(x, y) + f_{d-1}(x, y)z ,$$

where f_i is a form of degree i, respectively. Of course, there can be no other singularity of C, since otherwise the line passing through the two singularities would intersect C more than d times.

As above, let $\mathcal{H}(t)$ be the linear system of lines $\mathcal{H}(1, O)$ through $O = (0 : 0 : 1)$. Intersecting C with an element of $\mathcal{H}(t)$ we get the origin as an intersection point of multiplicity at least $d - 1$. Reasoning as in the case of conics, we see that

$$Q(t) = (-f_{d-1}(1, t), -t \cdot f_{d-1}(1, t), f_d(1, t)).$$

is a rational parametrization of the curve C. We summarize this in the following theorem.

Theorem 4.45. *Let C be an irreducible projective curve of degree d defined by the polynomial $F(x, y, z) = f_d(x, y) + f_{d-1}(x, y)z$ (f_i a form of degree i, resp.), i.e. having a $(d-1)$–fold point at $(0 : 0 : 1)$. Then C is rational and a rational parametrization is*

$$\mathcal{P}(t) = (-f_{d-1}(1, t), -tf_{d-1}(1, t), f_d(1, t)).$$

Corollary 4.46. *Every irreducible curve of degree d with a $(d-1)$-fold point is rational.*

So after a suitable change of coordinates Theorem 4.45 yields a parametrization of the irreducible curve C. We summarize this process in the following algorithm.

Algorithm PARAMETRIZATION-BY-LINES.
Given the defining polynomial $F(x, y, z)$ of an irreducible projective curve C of degree d, having a $(d-1)$–fold point, the algorithm computes a rational parametrization of C.

1. If $d = 1$, then proceed as in Remark to Definition 4.48. If $d > 1$, compute the $(d-1)$–fold point P of C. W.l.o.g., perhaps after renaming the variables, let $P = (a : b : 1)$.
2. $g(x, y) := F(x + a, y + b, 1)$. Let $g_d(x, y)$ and $g_{d-1}(x, y)$ be the homogeneous components of $g(x, y)$ of degree d and $d-1$, respectively.
3. Return $\mathcal{P}(t) = (-g_{d-1}(1, t) + ag_d(1, t), -tg_{d-1}(1, t) + bg_d(1, t), g_d(1, t))$.

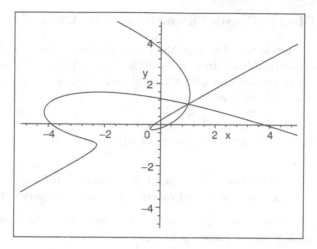

Fig. 4.2. Quartic C

Remarks. Note that, because of the underlying geometric construction, the parametrization computed by algorithm PARAMETRIZATION-BY-LINES is proper. Furthermore, if $\mathcal{P}_{\star,z}(t)$ is the affine parametrization of $C_{\star,z}$ derived from $\mathcal{P}(t)$, then its inverse can be computed as follows. W.l.o.g., perhaps after renaming the variables, let $P = (a : b : 1)$ be the singularity of the curve. Then

$$\mathcal{P}_{\star,z}^{-1}(x,y) = \frac{y-b}{x-a}.$$

Example 4.47. Let C be the affine quartic curve defined by (see Fig. 4.2)

$$f(x,y) = 1 + x - 15\,x^2 - 29\,y^2 + 30\,y^3 - 25\,xy^2 + x^3y + 35\,xy + x^4 - 6\,y^4 + 6\,x^2y.$$

C has an affine triple point at $(1,1)$. We apply algorithm PARAMETRIZATION-BY-LINES to parametrize C. In Step 2, we compute the polynomial

$$g(x,y) = 5\,x^3 + 6\,y^3 - 25\,xy^2 + x^3y + x^4 - 6\,y^4 + 9\,x^2y,$$

and determining the homogeneous forms of $g(x,y)$, we get the rational parametrization of C

$$\mathcal{P}(t) = \left(\frac{4 + 6\,t^3 - 25\,t^2 + 8\,t + 6\,t^4}{-1 + 6\,t^4 - t}, \frac{4\,t + 12\,t^4 - 25\,t^3 + 9\,t^2 - 1}{-1 + 6\,t^4 - t} \right).$$

Furthermore, taking into account the remark to the algorithm we have that

$$\mathcal{P}^{-1}(x,y) = \frac{y-1}{x-1}.$$

4.6.3 The Class of Curves Parametrizable by Lines

A natural question is whether only the rational curves considered previously are those parametrizable by lines. In order to answer this question, first of all, we must be more precise and give a formal definition of what we mean by a curve parametrizable by lines.

Definition 4.48. *The irreducible projective curve C is* parametrizable by lines *if there exists a linear system of curves $\mathcal{H}$ of degree 1 such that*

(1) $\dim(\mathcal{H}) = 1$,
(2) the intersection of a generic element in $\mathcal{H}$ and C contains a nonconstant point whose coordinates depend rationally on the free parameter of $\mathcal{H}$.

We say that an irreducible affine curve is parametrizable by lines *if its projective closure is parametrizable by lines.*

Remarks. 1. Note that in Definition 4.48 we have not required that the base point of $\mathcal{H}$ is on the curve. Later, we will see that in fact the base point must lie on C, unless C is a line.
  2. Any line is parametrizable by lines (see Exercise 4.12).
  3. Note that an affine curve parametrizable by lines is in fact rational. Moreover, the implicit equation of C vanishes on the generic intersection point depending rationally on the parameter. So, by Theorem 4.7, this generic point is a rational parametrization of C. Furthermore, if the irreducibility condition in Definition 4.48 is not imposed, then the curve has a rational component (see Exercises 4.13 and 4.14).
  4. Let C be an affine curve such that its associated projective curve $C^\star$ is parametrizable by the linear system of lines $\mathcal{H}(t)$ of equation $L_1(x, y, z) - tL_2(x, y, z)$. Then, the affine parametrization of C, generated by $\mathcal{H}(t)$, is proper and $\frac{L_1(x,y,1)}{L_2(x,y,1)}$ is its inverse (see Exercise 4.15). In fact, $\mathcal{H}(t)$ is a pencil of lines (Definition 2.53) and its base point is $L_1 \cap L_2$.

Theorem 4.49. *Let C be an irreducible projective plane curve of degree $d > 1$. The following statements are equivalent:*

(1) C is parametrizable by a pencil of lines $\mathcal{H}(t)$.
(2) C has a point of multiplicity $d - 1$ which is the base point of $\mathcal{H}(t)$.

Proof. That (2) implies (1) follows from Definition 4.48 and Theorem 4.45. Conversely, let $L_1(x, y, z) - tL_2(x, y, z)$ be the defining polynomial of $\mathcal{H}(t)$, let $\mathcal{P}(t)$ be the proper parametrization derived from $\mathcal{H}(t)$, and let Q be the base point of $\mathcal{H}(t)$. Since $d > 1$, for almost all $t_0 \in K$, $\mathcal{H}(t_0)$ intersects C in at least two points, and one of them is $\mathcal{P}(t_0)$. First we prove that $\mathcal{H}(t_0) \cap C = \{\mathcal{P}(t_0), Q\}$ for almost all $t_0 \in K$. Let $P \in [\mathcal{H}(t_0) \cap C] \setminus \mathcal{P}(t_0)$. If P is reachable by $\mathcal{P}(t)$, then there exists $t_1 \in K$, $t_1 \neq t_0$, such that $\mathcal{P}(t_1) = P$. This implies that $P \in \mathcal{H}(t_1) \cap \mathcal{H}(t_0)$. Therefore, $P = Q$. If P is not reachable, the inverse of $\mathcal{P}(t)$ is not defined at P, and hence $L_2(P) = 0$. But, since $P \in \mathcal{H}(t_0)$, then $L_1(P) = 0$. Thus P is in all the lines of the system of lines $\mathcal{H}(t)$, so $P = Q$.

Now, since $\mathcal{C}$ is irreducible, it has only finitely many singularities. Thus $\text{mult}_{\mathcal{P}(t_0)}(\mathcal{C}, \mathcal{H}(t_0)) = 1$ for almost all $t_0 \in K$. This implies, by Bézout's Theorem, that $\text{mult}_Q(\mathcal{C}, \mathcal{H}(t_0)) = d - 1$ for almost all $t_0 \in K$. Therefore, $d - 1 = \text{mult}_Q(\mathcal{C})$, i.e. the base point of $\mathcal{H}(t)$ is a point on $\mathcal{C}$ of multiplicity $d - 1$. Thus, (1) implies (2). $\qquad\qquad\square$

We have seen that the inverse of an affine parametrization generated by the algorithm PARAMETRIZATION-BY-LINES is linear. In the next theorem we see that this phenomenon also characterizes the curves parametrizable by lines.

Theorem 4.50. *Let $\mathcal{C}$ be an irreducible affine plane curve. The following statements are equivalent:*

(1) $\mathcal{C}$ is parametrizable by lines.
(2) There exists a proper affine parametrization of $\mathcal{C}$ with a linear inverse.
(3) The inverse of any proper affine parametrization of $\mathcal{C}$ is linear.

Proof. Let d be the degree of $\mathcal{C}$. If $d = 1$ the result is trivial. Let us assume that $d > 1$. If (1) holds, by Theorem 4.49 we know that $\mathcal{C}^*$ has a $(d-1)$–fold point. Therefore, applying algorithm PARAMETRIZATION-BY-LINES one gets a proper affine parametrization of $\mathcal{C}$ with linear inverse. Thus, (2) holds.

We prove now that (2) implies (3). Let $\mathcal{P}(t)$ be a proper affine parametrization with linear inverse, and let $\mathcal{P}'(t)$ be any other proper affine parametrization of $\mathcal{C}$. Because of Lemma 4.17 (2) there exists a linear rational function $L(t)$ such that $\mathcal{P}'(t) = \mathcal{P}(L(t))$. Therefore, $\mathcal{P}'^{-1} = L^{-1} \circ \mathcal{P}^{-1}$ is also linear.

Finally, we prove that (3) implies (1). Let $\mathcal{P}(t)$ be a proper affine parametrization of $\mathcal{C}$ with a rational inverse of the form $(ax + by + c)/(a'x + b'y + c')$. Let $\mathcal{P}^*(t)$ be the projective parametrization generated by $\mathcal{P}(t)$. Then, we consider the pencil of lines $\mathcal{H}(t)$ defined by $H(x, y, z, t) = (ax + by + cz) - (a'x + b'y + c'z)t$. Clearly, $H(\mathcal{P}^*(t), t) = 0$. Thus, $\mathcal{P}^*(t) \in \mathcal{H}(t) \cap \mathcal{C}^*$. Therefore, $\mathcal{C}^*$ is parametrizable by lines. $\qquad\qquad\square$

4.7 Parametrization by Adjoint Curves

In Theorem 4.11 we saw that only curves of genus 0 have any chance of being rationally parametrizable. In this section we conclude that the curves of genus 0 are exactly the rational curves.

In Theorem 4.49 we have seen that, in general, rational curves can not be parametrized by lines. In fact, we have proved that a rational curve $\mathcal{C}$ of degree $d \geq 2$ is parametrizable by lines if and only if it has a $(d-1)$–fold point. In order to treat the general case, we develop here a method based on the notion of adjoint curves that, intuitively speaking, is a generalization of the idea underlining the parametrization by lines method. The method described in this section follows basically the approach in [Wal50] and [SeW91]. There are alternative parametrization methods such as [VaH97] based on the computation of the anticanonical divisor, or [Sch92] where adjoints of high degree are used.

Throughout this section, $\mathcal{C}$ will be an irreducible projective curve of degree $d > 2$ and genus 0. Note that this is not a loss of generality, because we have seen in the previous section that lines and irreducible conics can be parametrized by lines. Before showing how adjoints are defined and how they can be used to solve the parametrization problem, we first generalize the notion of parametrization by lines. We need to guarantee that every curve in the parametrizing system $\mathcal{H}$ intersects $\mathcal{C}$ in finitely many points. This is trivial when we parametrize by lines, but in the generalization it leads to an additional condition.

Definition 4.51. *A linear system of curves $\mathcal{H}$ parametrizes $\mathcal{C}$ iff*

(1) $\dim(\mathcal{H}) = 1$,
(2) the intersection of a generic element in $\mathcal{H}$ and $\mathcal{C}$ contains a nonconstant point whose coordinates depend rationally on the free parameter in $\mathcal{H}$,
(3) $\mathcal{C}$ is not a component of any curve in $\mathcal{H}$.

In this case we say that $\mathcal{C}$ is parametrizable by $\mathcal{H}$.

Lemma 4.52. *Let $\mathcal{H}(t)$ be a linear system of curves parametrizing $\mathcal{C}$, then there exists only one nonconstant intersection point of a generic element of $\mathcal{H}(t)$ and $\mathcal{C}$ depending on t, and it is a proper parametrization of $\mathcal{C}$.*

Proof. By condition (2) in Definition 4.51 we know that there exists a nonconstant point $\mathcal{P}(t)$ in $\mathcal{H}(t) \cap \mathcal{C}$ depending rationally on t. Let us see that $\mathcal{P}(t)$ is a proper parametrization of $\mathcal{C}$. It is clear that the defining polynomial of $\mathcal{C}$ vanishes at it. Thus, $\mathcal{P}(t)$ is a parametrization of $\mathcal{C}$. In order to see that it is proper, we find the inverse of the affine parametrization $\mathcal{P}_{*,z}(t)$ of $\mathcal{C}_{*,z}$ generated by $\mathcal{P}(t)$. Let $H(t, x, y, z) = H_0(x, y, z) - tH_1(x, y, z)$ be the defining polynomial of $\mathcal{H}(t)$. Then, $H(t, \mathcal{P}(t)) = 0$. Moreover, $H_1(\mathcal{P}(t)) \neq 0$, because otherwise we would have that $H_0(\mathcal{P}(t)) = 0$, which is impossible because of condition (3) in Definition 4.51. Therefore, $M = H_0/H_1$ is defined at $\mathcal{P}(t)$ and $M(\mathcal{P}(t)) = t$. Thus, by Lemma 4.36, $M(x, y, 1)$ is the inverse of $\mathcal{P}_{*,z}(t)$.

Finally, let us see that $\mathcal{P}(t)$ is unique. Let $\mathcal{Q}(t)$ be another intersection point depending rationally on t. By the argument above, we know that both are proper rational parametrizations, and that $\mathcal{P}_{*,z}^{-1}(t) = \mathcal{Q}_{*,z}^{-1}(t)$. Thus, $\mathcal{P}(t) = \mathcal{Q}(t)$. $\quad\square$

Now let us see how to actually compute a parametrization from a parametrizing linear system of curves. For this purpose, for a polynomial G in $K[x, y, z][t]$ we use the notation $\mathrm{pp}_t(G)$ to denote the primitive part of G w.r.t. t, i.e. G divided by the gcd of its coefficients.

Theorem 4.53. *Let $F(x, y, z)$ be the defining polynomial of C, and let $H(t, x, y, z)$ be the defining polynomial of a linear system $\mathcal{H}(t)$ parametrizing C. Then, the proper parametrization $\mathcal{P}(t)$ generated by $\mathcal{H}(t)$ is the solution in $\mathbb{P}^2(K(t))$ of the system of algebraic equations*

$$\left.\begin{array}{l} \mathrm{pp}_t(\mathrm{res}_y(F, H)) = 0 \\ \mathrm{pp}_t(\mathrm{res}_x(F, H)) = 0 \end{array}\right\}.$$

Proof. Let $\{P_1, \ldots, P_s, \mathcal{P}(t)\}$ be the intersection points of $\mathcal{H}(t)$ and C. By Lemma 4.52 we know that $P_i \in \mathbb{P}^2(K)$ and $\mathcal{P}(t) \in \mathbb{P}^2(K(t))$. Let $P_i = (a_i : b_i : c_i)$ and $\mathcal{P}(t) = (\chi_1(t), \chi_2(t), \chi_3(t))$. Condition (3) in Definition 4.51 implies that $\mathrm{res}_y(F, H)$ and $\mathrm{res}_x(F, H)$ are not identically zero. Furthermore, from Bézout's Theorem we get that

$$\mathrm{res}_y(F, H) = (\chi_3(t)x - \chi_1(t)z)^\beta \prod_{i=1}^s (c_i x - a_i z)^{\alpha_i}$$

$$\mathrm{res}_x(F, H) = (\chi_3(t)y - \chi_2(t)z)^{\beta'} \prod_{i=1}^s (c_i y - b_i z)^{\alpha_i'}$$

for some $\alpha_i, \alpha_i', \beta, \beta' \in \mathbb{N}$. So, obviously, the parametrization is determined by the primitive parts of these resultants. □

The following theorem gives sufficient conditions for a linear system of curves to be a parametrizing system.

Theorem 4.54. *Let $\mathcal{H}$ be a linear system of curves of degree k and let $\mathcal{B}$ be the set of base points of $\mathcal{H}$ (cf. Definition 2.54). If*

(1) $\dim(\mathcal{H}) = 1$,
(2) $\sum_{P \in \mathcal{B}} \mathrm{mult}_P(C, C') = dk - 1$ for almost all curves $C' \in \mathcal{H}$, and
(3) C is not a component of any curve in $\mathcal{H}$,

then $\mathcal{H}$ parametrizes C.

Proof. We just have to prove that condition (2) in the statement of the theorem implies condition (2) in Definition 4.51. By condition (3) we know that C is not a component of any curve in $\mathcal{H}$. Thus, by Bézout's Theorem and condition (2) we see that $(C' \cap C) \setminus \mathcal{B}$ consists of a single point for almost all $C' \in \mathcal{H}$. Therefore, this point depends rationally on the parameter of $\mathcal{H}$. □

Now, the natural question is how to determine parametrizing linear systems of curves. We will show that adjoints provide an answer to this question. Adjoint curves can be defined for reducible curves. However, since our final goal is to work with rational curves, we will only consider irreducible curves. For the reducible case we refer to [BrK86],[Ful89],[Wal50].

Before we introduce the notion of adjoint curves and establish some of their important properties, we remind the reader of some of the notation introduced in Sect. 3.2 concerning the blowing up of curves:

1. Sing($\mathcal{C}$) denotes the singular locus of $\mathcal{C}$.
2. Ngr($\mathcal{C}$) denotes the neighboring graph of $\mathcal{C}$, i.e. Ngr($\mathcal{C}$) comprises the singularities and neighboring singularities of $\mathcal{C}$.
3. For $P \in$ Sing($\mathcal{C}$), Ngr$_P$($\mathcal{C}$) denotes the subgraph of Ngr($\mathcal{C}$) with root at P.
4. For $P \in$ Ngr($\mathcal{C}$) we denote by $\mathcal{Q}_P$ the sequence of quadratic transformations and linear transformations generating the neighborhood where P belongs to. Moreover, for any projective curve $\mathcal{C}'$ we denote by $\mathcal{Q}_P(\mathcal{C}')$ the quadratic transform of $\mathcal{C}'$ by $\mathcal{Q}_P$.

Definition 4.55. *A projective curve $\mathcal{C}'$ is an* adjoint curve *of the irreducible projective curve $\mathcal{C}$ iff* $\text{mult}_P(\mathcal{Q}_P(\mathcal{C}')) \geq \text{mult}_P(\mathcal{Q}_P(\mathcal{C})) - 1$ *for every $P \in$* Ngr($\mathcal{C}$). *We say that $\mathcal{C}'$ is an* adjoint curve of degree k *of $\mathcal{C}$, if $\mathcal{C}'$ is an adjoint of $\mathcal{C}$ and* $\deg(\mathcal{C}') = k$.

All algebraic conditions required in the definition of adjoint curves are linear. Therefore if one fixes the degree, the set of all adjoint curves of $\mathcal{C}$ is a linear system of curves (see Sect. 2.4). In fact, if $\mathcal{C}$ has only ordinary singularities, then the set of adjoint curves of degree k of $\mathcal{C}$ is the linear system generated by the effective divisor

$$\sum_{P \in \text{Sing}(\mathcal{C})} (\text{mult}_P(\mathcal{C}) - 1)P.$$

This remark motivates the following definition.

Definition 4.56. *The set of all adjoints of $\mathcal{C}$ of degree k, $k \in \mathbb{N}$, is called the* system of adjoints *of $\mathcal{C}$ of degree k. We denote this system by $\mathcal{A}_k(\mathcal{C})$.*

Theorem 4.57. *Let $\mathcal{C}$ be a projective curve of degree d and genus 0, and let $k \geq d - 2$, then $\mathcal{A}_k(\mathcal{C}) \neq \emptyset$.*

Proof. The full linear system of curves of degree k has dimension $k(k + 3)/2$ (cf. Sect. 2.4). Since genus($\mathcal{C}$) $= 0$, the number of linear conditions required by $\mathcal{A}_k(\mathcal{C})$ is

$$\sum_{P \in \text{Ngr}(\mathcal{Q}_P(\mathcal{C}))} \frac{\text{mult}_P(\mathcal{Q}_P(\mathcal{C}))(\text{mult}_P(\mathcal{Q}_P(\mathcal{C})) - 1)}{2} = \frac{(d - 1)(d - 2)}{2}.$$

Therefore,

$$\dim(\mathcal{A}_k(\mathcal{C})) \geq \frac{k(k + 3)}{2} - \frac{(d - 1)(d - 2)}{2}$$

(compare to Theorem 2.59 for the case of curves with only ordinary singularities). Now, if $k \geq d-2$, then $\dim(\mathcal{A}_k(\mathcal{C})) \geq d-2 > 0$ and hence $\mathcal{A}_k(\mathcal{C}) \neq \emptyset$. $\square$

In [Noe83], Sect. 50, the dimension of the linear system of adjoints of an irreducible curve is determined. Applying this result one has the following result.

Theorem 4.58. *Let C be a projective curve of degree d and genus 0, and let $k \geq d - 2$, then*

$$\dim(\mathcal{A}_k(C)) = \frac{k(k+3)}{2} - \frac{(d-1)(d-2)}{2}.$$

Now, we proceed to show how from linear systems of adjoint curves we may generate parametrizing linear systems. For this purpose, we first prove two preliminary lemmas. In the first one, if C_1, C_2 are projective curves defined respectively by the forms F_1, F_2, we denote by $C_1 C_2$ the curve defined by $F_1 F_2$, and by $\lambda C_1 + \mu C_2$ the curve defined by $\lambda F_1 + \mu F_2$ where $\lambda, \mu \in K$, assuming that the corresponding polynomial is not identically zero.

Lemma 4.59. *Let C be an irreducible projective curve of degree d, let $k \in \{d, d-1, d-2\}$, let $\mathcal{F} \subset C \setminus \mathrm{Sing}(C)$ be a finite set and let*

$$\mathcal{H}_k := \mathcal{A}_k(C) \cap \mathcal{H}(k, \sum_{P \in \mathcal{F}} P).$$

Then the following hold:

(1) If $k = d$, for every $C' \in \mathcal{H}_d$, and for almost all $(\lambda, \mu) \in K^2$ we have $\mu C' + \lambda C \in \mathcal{H}_d$, and $\mu C' + \lambda C$ does not have multiple components.

(2) If we take a fixed $k \in \{d-1, d-2\}$, then for every $C' \in \mathcal{H}_k$, for every projective curve $\mathcal{M}$ of degree $d - k$, and for almost all $(\lambda, \mu) \in K^2$ we have

$$\mu \mathcal{M} C' + \lambda C \in \mathcal{H}_d \cap \mathcal{H}(d, \sum_{P \in \mathcal{M} \cap C} P),$$

and $\mu \mathcal{M} C' + \lambda C$ does not have multiple components.

Proof. If $\mathcal{H}_k = \emptyset$, then there is nothing to prove. Let us assume that $\mathcal{H}_k \neq \emptyset$. Let F, G, M be the defining polynomials of $C, C', \mathcal{M}$, respectively.

In order to prove Statement (1), we first observe that if $C' = C$ the result trivially holds for $\lambda, \mu \in K$ such that $\lambda + \mu \neq 0$. Let us assume that $C' \neq C$. We observe that $C', C \in \mathcal{H}_d$. Therefore, since $\mathcal{H}_d$ is a projective linear variety, if λ, μ are such that $\mu G + \lambda F$ is not identically zero, then $\mu C' + \lambda C \in \mathcal{H}_d$. Moreover, since $C' \neq C$, for all $(\lambda, \mu) \in \Omega_1 := K^2 \setminus \{(0,0)\}$ we have that $\mu G + \lambda F$ is not identically zero. Let us prove the second part of Statement (1). For this purpose, we take the polynomial $A(\lambda, \mu, x, y, z) := \mu G + \lambda F$, where λ, μ are considered as formal parameters. Let us see that A is irreducible as a polynomial in $\mathbb{K}[\lambda, \mu, x, y, z]$. Indeed, if it factors, since A is linear in $\{\lambda, \mu\}$, one factor belongs to $\mathbb{K}[x, y, z]$. But this implies that F is either reducible or $F = G$ up to constant, which is impossible since F is irreducible and we have assumed that $C' \neq C$. Moreover, taking into account that F is irreducible and nonlinear (lines have been excluded), A does depend on $\{x, y, z\}$, and hence A can be seen as a nonconstant polynomial in $K[\lambda, \mu, x, y][z]$. Now, because of the irreducibility of A, one has that A is primitive w.r.t. z, and it is square-free. Therefore, by Theorem 8.1, p. 338, in [GCL92], the discriminant of A w.r.t. z is not identically zero. Thus, computing this discriminant, we find a nonempty open Zariski subset Ω_2 of K^2, such that $A(\lambda_0, \mu_0, x, y, z)$ is squarefree for every $(\lambda_0, \mu_0) \in \Omega_2$. So, for every $(\lambda, \mu) \in \Omega_1 \cap \Omega_2$, which is nonempty because K^2 is irreducible, we have that $\mu \mathcal{C}' + \lambda \mathcal{C} \in \mathcal{H}_d$, and $\mu \mathcal{C}' + \lambda \mathcal{C}$ does not have multiple components.

Let us prove Statement (2). F is irreducible and $k < d$, so $\mu MG + \lambda F$ is identically zero if and only if $\lambda = \mu = 0$. We prove that if $(\lambda, \mu) \neq (0, 0)$ then $\mu \mathcal{M}\mathcal{C}' + \lambda \mathcal{C} \in \mathcal{H}_d \cap \mathcal{H}(d, \sum_{P \in \mathcal{M} \cap \mathcal{C}} P)$. Indeed:

(i) Let us see that $\mu \mathcal{M}\mathcal{C}' + \lambda \mathcal{C} \in \mathcal{H}(d, \sum_{P \in \mathcal{F}} P)$. Clearly, $\mathcal{C} \in \mathcal{H}(d, \sum_{P \in \mathcal{F}} P)$. Moreover, by hypothesis, $\mathcal{C}' \in \mathcal{H}(k, \sum_{P \in \mathcal{F}} P)$, hence $\text{mult}_P(\mathcal{C}') \geq 1$ for $P \in \mathcal{F}$. Furthermore, $\text{mult}_P(\mathcal{M}\mathcal{C}') = \text{mult}_P(\mathcal{M}) + \text{mult}_P(\mathcal{C}') \geq 1$ for $P \in \mathcal{F}$ (see Exercise 2.10). So, since $\deg(\mathcal{M}\mathcal{C}') = d$, one gets that $\mathcal{M}\mathcal{C}' \in \mathcal{H}(d, \sum_{P \in \mathcal{F}} P)$. Now, the statement follows from the linearity of $\mathcal{H}(d, \sum_{P \in \mathcal{F}} P)$; note that $\mu MG + \lambda F$ is not identically zero.

(ii) Reasoning similarly as in (i) we deduce that $\mu \mathcal{M}\mathcal{C}' + \lambda \mathcal{C} \in \mathcal{H}(d, \sum_{P \in \mathcal{M} \cap \mathcal{C}} P)$.

(iii) Let us see that $\mu \mathcal{M}\mathcal{C}' + \lambda \mathcal{C} \in \mathcal{A}_d(\mathcal{C})$. First, we observe that $\mathcal{C} \in \mathcal{A}_d(\mathcal{C})$, so we have to prove that $\mathcal{M}\mathcal{C}' \in \mathcal{A}_d(\mathcal{C})$. For this purpose, we first note that $\deg(\mathcal{M}\mathcal{C}') = d$. We analyze separately the required conditions on the singularities and on the neighboring points (see Definition 4.55).

(iii.i) Let $P \in \text{Sing}(\mathcal{C})$. Then, taking into account that $\mathcal{C}' \in \mathcal{A}_k(\mathcal{C})$, we have
$$\text{mult}_P(\mathcal{M}\mathcal{C}') = \text{mult}_P(\mathcal{M}) + \text{mult}_P(\mathcal{C}') \geq \text{mult}_P(\mathcal{C}') \geq \text{mult}_P(\mathcal{C}) - 1.$$

(iii.ii) Let $P \in \text{Ngr}(\mathcal{C})$, and let $\mathcal{Q}_P$ as above. Observe that $\mathcal{Q}_P(\mathcal{M}\mathcal{C}') = \mathcal{Q}_P(\mathcal{M})\mathcal{Q}_P(\mathcal{C}')$. Therefore,
$$\text{mult}_P(\mathcal{Q}_P(\mathcal{M}\mathcal{C}')) = \text{mult}_P(\mathcal{Q}_P(\mathcal{M})\mathcal{Q}_P(\mathcal{C}')) =$$
$$\text{mult}_P(\mathcal{Q}_P(\mathcal{M})) + \text{mult}_P(\mathcal{Q}_P(\mathcal{C}')) \geq \text{mult}_P(\mathcal{Q}_P(\mathcal{C}'))$$
$$\geq \text{mult}_P(\mathcal{Q}_P(\mathcal{C})) - 1.$$

Summarizing, we get that if $(\lambda, \mu) \neq (0, 0)$ then $\mu \mathcal{M}\mathcal{C}' + \lambda \mathcal{C} \in \mathcal{H}_d \cap \mathcal{H}(d, \sum_{P \in \mathcal{M} \cap \mathcal{C}} P)$. In order to prove that for almost all $(\lambda, \mu) \in K^2$ the curve $\mu \mathcal{M}\mathcal{C}' + \lambda \mathcal{C}$ does not have multiple components, one reasons analogously as in the proof of Statement (1). In this case, $A(\lambda, \mu, x, y, z) := \mu MG + \lambda F$. $\square$

The following lemma can be found in [Wal50] Chap. III, Theorem 7.6.

Lemma 4.60. *Let $\mathcal{C}_1$ and $\mathcal{C}_2$ be two projective curves of degrees d_1 and d_2 respectively, having no common components and neither $\mathcal{C}_1$ nor $\mathcal{C}_2$ having any multiple components. Then*
$$d_1 d_2 \geq \sum_{\substack{P \in \text{Ngr}_{P'}(\mathcal{C}_1) \\ P' \in \mathcal{C}_1 \cap \mathcal{C}_2}} \text{mult}_P(Q_p(\mathcal{C}_1)) \text{mult}_P(Q_p(\mathcal{C}_2)),$$
where $\text{Ngr}_{P'}(\mathcal{C}_1) = \{P'\}$ if $P' \in [\mathcal{C}_1 \cap \mathcal{C}_2] \setminus \text{Sing}(\mathcal{C}_1)$.

Now, we show how from linear systems of adjoint curves we may generate parametrizing linear systems.

Theorem 4.61. *Let $\mathcal{C}$ be a projective curve of degree d and genus 0, let $k \in \{d-1, d-2\}$, and let $\mathcal{S}_k \subset \mathcal{C} \setminus \mathrm{Sing}(\mathcal{C})$ be such that $\mathrm{card}(\mathcal{S}_k) = kd - (d-1)(d-2) - 1$. Then*

$$\mathcal{A}_k(\mathcal{C}) \cap \mathcal{H}(k, \sum_{P \in \mathcal{S}_k} P)$$

parametrizes $\mathcal{C}$.

Proof. Let $\mathcal{H} = \mathcal{A}_k(\mathcal{C}) \cap \mathcal{H}(k, \sum_{P \in \mathcal{S}_k} P)$. We check whether the conditions in Theorem 4.54 are satisfied. Note that Condition (3) holds trivially, because $\mathcal{C}$ is irreducible and $k < d$. Let us check Condition (1), i.e. $\dim(\mathcal{H}) = 1$. $\dim(\mathcal{H}) \geq \dim(\mathcal{A}_k(\mathcal{C})) - [kd - (d-1)(d-2) - 1]$, and by Theorem 4.58 we know that $\dim(\mathcal{H}) \geq 1$. Now, let us assume that $\dim(\mathcal{H}) > 1$. We take two different points $Q_1, Q_2 \in \mathcal{C} \setminus (\mathrm{Sing}(\mathcal{C}) \cup \mathcal{S}_k)$, and we consider the linear subsystem

$$\mathcal{H}' = \mathcal{H} \cap \mathcal{H}(k, Q_1 + Q_2).$$

Observe that $\dim(\mathcal{H}') \geq 0$. Thus, $\mathcal{H}' \neq \emptyset$. Let $\mathcal{C}' \in \mathcal{H}'$. Since $\deg(\mathcal{C}') < \deg(\mathcal{C})$ and $\mathcal{C}$ is irreducible, we know that $\mathcal{C}'$ and $\mathcal{C}$ do not have common components. Now, we distinguish two different cases:

(i) If $\mathcal{C}'$ does not have multiple components, then since $\mathcal{C}$ does not have common components either, by Lemma 4.60 and the fact that $\mathrm{genus}(\mathcal{C}) = 0$, we get that

$$kd \geq$$

$$\sum_{P \in \mathrm{Ngr}(\mathcal{C})} \mathrm{mult}_P(Q_P(\mathcal{C})) \, \mathrm{mult}_P(Q_P(\mathcal{C}')) + \sum_{P \in \mathcal{S}_k \cup \{Q_1, Q_2\}} \mathrm{mult}_P(\mathcal{C}) \, \mathrm{mult}_P(\mathcal{C}') \geq$$

$$\sum_{P \in \mathrm{Ngr}(\mathcal{C})} \mathrm{mult}_P(Q_P(\mathcal{C})) \, (\mathrm{mult}_P(Q_P(\mathcal{C})) - 1) + \sum_{P \in \mathcal{S}_k \cup \{Q_1, Q_2\}} \mathrm{mult}_P(\mathcal{C}) \, \mathrm{mult}_P(\mathcal{C}')$$

$$\geq (d-1)(d-2) + [kd - (d-1)(d-2) - 1] + 2 = kd + 1,$$

which is impossible.

(ii) Let us assume that $\mathcal{C}'$ has multiple components. Then, we consider $d - k$ different lines $\mathcal{L}_1, \ldots, \mathcal{L}_{d-k}$ such that $\mathcal{L}_i$ and $\mathcal{C}$ intersects in d different points and

$$(\mathcal{L}_i \cap \mathcal{C}) \cap (\mathrm{Sing}(\mathcal{C}) \cup \mathcal{S}_k \cup \{Q_1, Q_2\} \cup_{j \neq i} (\mathcal{L}_j \cap \mathcal{C})) = \emptyset.$$

Let L_i be the defining polynomial of $\mathcal{L}_i$ and let $\mathcal{M}$ be the curve of defining polynomial $L_1 \cdots L_{d-k}$. Now, applying Lemma 4.59 (2) to $\mathcal{C}$ and taking $\mathcal{F}$ as $\mathcal{S}_k \cup \{Q_1, Q_2\}$, we take $\lambda, \mu \in K$ such that

$$\mathcal{C}'' := \mu \mathcal{M} \mathcal{C}' + \lambda \mathcal{C} \in \mathcal{A}_d(\mathcal{C}) \cap \mathcal{H}(d, \sum_{P \in \mathcal{S}_k \cup \{Q_1, Q_2\}} P) \cap \mathcal{H}(d, \sum_{p \in \mathcal{M} \cap \mathcal{C}} P),$$

and $\mathcal{C}''$ does not have common components. In this situation we apply Lemma 4.60 to $\mathcal{C}''$ and $\mathcal{C}$; note that $\mathcal{C}''$ and $\mathcal{C}$ do not have common components because both curves have the same degree and $\mathcal{C}$ is irreducible. So

$$d^2 \geq$$

$$\sum_{P \in \mathrm{Ngr}(\mathcal{C})} \mathrm{mult}_P(Q_P(\mathcal{C}))\,\mathrm{mult}_P(Q_P(\mathcal{C}'')) + \sum_{P \in \mathcal{S}_k \cup \{Q_1, Q_2\}} \mathrm{mult}_P(\mathcal{C})\,\mathrm{mult}_P(\mathcal{C}'') +$$

$$\sum_{P \in \mathcal{M} \cap \mathcal{C}} \mathrm{mult}_P(\mathcal{C})\,\mathrm{mult}_P(\mathcal{C}'') \geq \sum_{P \in \mathrm{Ngr}(\mathcal{C})} \mathrm{mult}_P(Q_P(\mathcal{C}))\,(\mathrm{mult}_P(Q_P(\mathcal{C})) - 1)$$

$$+ \sum_{P \in \mathcal{S}_k \cup \{Q_1, Q_2\}} \mathrm{mult}_P(\mathcal{C})\,\mathrm{mult}_P(\mathcal{C}'') + d(d - k)$$

$$= (d - 1)(d - 2) + \sum_{P \in \mathcal{S}_k \cup \{Q_1, Q_2\}} \mathrm{mult}_P(\mathcal{C})\,\mathrm{mult}_P(\mathcal{C}'') + d(d - k)$$

$$\geq (d - 1)(d - 2) + [kd - (d - 1)(d - 2) - 1] + 2 + d(d - k)$$

$$= kd + 1 + d(d - k) = d^2 + 1,$$

which is impossible.

Now, let us check that Condition (2) holds in Theorem 4.54. For this purpose, we first prove that the set of base points $\mathcal{B}$ of $\mathcal{H}$ is $\mathrm{Sing}(\mathcal{C}) \cup \mathcal{S}_k$. It is clear that $\mathrm{Sing}(\mathcal{C}) \cup \mathcal{S}_k \subset \mathcal{B}$. Let us assume that $\mathcal{B} \neq \mathrm{Sing}(\mathcal{C}) \cup \mathcal{S}_k$, so there exists $Q \in \mathcal{B} \setminus (\mathrm{Sing}(\mathcal{C}) \cup \mathcal{S}_k)$. We choose a curve $\mathcal{C}' \in \mathcal{H}$ passing through a point $Q' \in \mathcal{C} \setminus \mathcal{B}$. This is possible because $\dim(\mathcal{H}) = 1$. Then, since $\mathcal{C}$ and $\mathcal{C}'$ do not have common components, we argue similarly as above distinguishing two different cases:

(i) Let $\mathcal{C}'$ be without multiple components. Since $\mathcal{C}$ does not have multiple components either, we can apply Lemma 4.60. Reasoning as in (i) above, we arrive at the contradiction $kd \geq kd + 1$. Thus, $\mathcal{B} = \mathrm{Sing}(\mathcal{C}) \cup \mathcal{S}_k$.

(ii) Let us assume that $\mathcal{C}'$ has multiple components. We consider $d - k$ different lines $\mathcal{L}_1, \ldots, \mathcal{L}_{d-k}$ such that $\mathcal{L}_i$ and $\mathcal{C}$ intersect in d different points and

$$(\mathcal{L}_i \cap \mathcal{C}) \cap (\mathrm{Sing}(\mathcal{C}) \cup \mathcal{S}_k \cup \{Q, Q'\} \cup_{j \neq i} (\mathcal{L}_j \cap \mathcal{C})) = \emptyset.$$

Let L_i be the defining polynomial of $\mathcal{L}_i$ and let $\mathcal{M}$ be the curve defined by $L_1 \cdots L_{d-k}$. Now, applying Lemma 4.59 (2) to $\mathcal{C}$ and taking $\mathcal{F}$ as $\mathcal{S}_k \cup \{Q, Q'\}$, we take $\lambda, \mu \in K$ such that

$$\mathcal{C}'' := \mu \mathcal{M} \mathcal{C}' + \lambda \mathcal{C} \in \mathcal{A}_d(\mathcal{C}) \cap \mathcal{H}(d, \sum_{P \in \mathcal{S}_k \cup \{Q, Q'\}} P) \cap \mathcal{H}(d, \sum_{p \in \mathcal{M} \cap \mathcal{C}} P),$$

and $\mathcal{C}''$ does not have multiple components. In this situation we apply Lemma 4.60 to $\mathcal{C}''$ and $\mathcal{C}$; note that $\mathcal{C}''$ and $\mathcal{C}$ do not have common components because both curves have the same degree and $\mathcal{C}$ is irreducible. So, reasoning as in (ii) above, we arrive at the contradiction $d^2 \geq d^2 + 1$. Thus, $\mathcal{B} = \mathrm{Sing}(\mathcal{C}) \cup \mathcal{S}_k$.

Now that we have proved that $\mathcal{B} = \text{Sing}(\mathcal{C}) \cup \mathcal{S}_k$, we show that Statement (2) in Theorem 4.54 holds. That is, we have to prove

$$\sum_{P \in \mathcal{B}} \text{mult}_P(\mathcal{C}, \mathcal{C}') = dk - 1$$

for almost all $\mathcal{C}' \in \mathcal{H}$. We structure the proof as follows: First, we prove that for all $\mathcal{C}' \in \mathcal{H}$ we have

$$\sum_{P \in \mathcal{B}} \text{mult}_P(\mathcal{C}, \mathcal{C}') \geq dk - 1.$$

Second, we show that there exists at least one curve $\mathcal{C}' \in \mathcal{H}$ such that

$$\sum_{P \in \mathcal{B}} \text{mult}_P(\mathcal{C}, \mathcal{C}') = dk - 1,$$

and finally we show that the equality holds for almost all curves in $\mathcal{H}$.

(a) Let us assume that there exists $\mathcal{C}' \in \mathcal{H}$ such that the sum of multiplicities of intersection at $\mathcal{B}$ is equal to $dk - \ell$, where $\ell > 1$. Then, since $\mathcal{C}$ and $\mathcal{C}'$ do not have common components, by Bézout Theorem we deduce that there exists a set of points $\mathcal{E} \subset (\mathcal{C} \cap \mathcal{C}') \setminus \mathcal{B}$ such that

$$\sum_{P \in \mathcal{E}} \text{mult}_P(\mathcal{C}, \mathcal{C}') = \ell.$$

Now we argue similarly as above distinguishing two different cases:

(a.1) Let $\mathcal{C}'$ be without multiple components. Since $\mathcal{C}$ does not have multiple components either, we can apply Lemma 4.60. Reasoning as in (i) above, and using the fact that $\mathcal{B} = \text{Sing}(\mathcal{C}) \cup \mathcal{S}_k$, we derive $kd \geq kd + \ell - 1$, which is impossible since $\ell > 1$.

(a.2) Let us assume that $\mathcal{C}'$ has multiple components. Then, we consider $d - k$ different lines $\mathcal{L}_1, \ldots, \mathcal{L}_{d-k}$ such that $\mathcal{L}_i$ and $\mathcal{C}$ intersect in d different points and

$$(\mathcal{L}_i \cap \mathcal{C}) \cap (\text{Sing}(\mathcal{C}) \cup \mathcal{S}_k \cup \mathcal{E} \cup_{j \neq i} (\mathcal{L}_j \cap \mathcal{C})) = \emptyset.$$

Let L_i be the defining polynomial of $\mathcal{L}_i$ and let $\mathcal{M}$ be the curve of defining polynomial $L_1 \cdots L_{d-k}$. Now, applying Lemma 4.59(2) to $\mathcal{C}$, and taking $\mathcal{F}$ as $\mathcal{S}_k \cup \mathcal{A}$, we take $\lambda, \mu \in K$ such that

$$\mathcal{C}'' := \mu \mathcal{M} \mathcal{C}' + \lambda \mathcal{C} \in \mathcal{A}_d(\mathcal{C}) \cap \mathcal{H}(d, \sum_{P \in \mathcal{S}_k \cup \mathcal{A}} P) \cap \mathcal{H}(d, \sum_{p \in \mathcal{M} \cap \mathcal{C}} P),$$

and $\mathcal{C}''$ does not have multiple components. In this situation we apply Lemma 4.60 to $\mathcal{C}''$ and $\mathcal{C}$; note that $\mathcal{C}''$ and $\mathcal{C}$ do not have common components because both curves have the same degree and $\mathcal{C}$ is irreducible. So, reasoning as in (ii), we derive $d^2 \geq d^2 + \ell - 1$, which is impossible since $\ell > 1$.

(b) Let us assume that for all curves in $\mathcal{H}$ the sum of multiplicities of intersection at $\mathcal{B}$ is dk. Then, since $\dim(\mathcal{H}) = 1$, we consider a point $Q \in \mathcal{C} \setminus \mathcal{B}$, and we take $\mathcal{C}' \in \mathcal{H}$ such that $Q \in \mathcal{C}'$. In this situation we have

$$\sum_{P \in \mathcal{C}' \cap \mathcal{C}} \operatorname{mult}_P(\mathcal{C}, \mathcal{C}') \geq \sum_{P \in \mathcal{B}} \operatorname{mult}_P(\mathcal{C}, \mathcal{C}') + \operatorname{mult}_Q(\mathcal{C}, \mathcal{C}') \geq dk + 1.$$

Therefore, by Bézout's theorem, the curves $\mathcal{C}$ and $\mathcal{C}'$ have a common component, which is impossible.

(c) Let $\mathcal{C}' \in \mathcal{H}$ be the curve whose existence ensures step (b) of our reasoning. Since the sum of multiplicities of intersection at $\mathcal{B}$ is $dk - 1$, and since $\mathcal{C}'$ and $\mathcal{C}$ do not have common components, by Bézout's theorem we know that $\mathcal{C} \cap \mathcal{C}' = \mathcal{B} \cup \{Q\}$, where $Q := (a : b : c) \notin \mathcal{B}$. Now, let $H(t_0, t_1, x, y, z) := t_0 G_0 + t_1 G_1$ be the defining polynomial of a generic element in $\mathcal{H}$, where we assume w.l.o.g. that G_0 is the defining polynomial of $\mathcal{C}'$. Furthermore we assume w.l.o.g., probably after performing a suitable linear change of coordinates, that $F(0, 0, 1) \neq 0$, $G_i(0, 0, 1) \neq 0$, and that $(0 : 0 : 1)$ is not on any line connecting two different points in $\mathcal{B} \cup \{Q\}$. Note that if $F(0, 0, 1) \neq 0$ then, in particular, one has that the leading coefficient of F w.r.t. z is constant and that $(0 : 0 : 1) \notin \mathcal{B} \cup \{Q\}$. Also, condition $G_i(0, 0, 1) \neq 0$ implies that $(0 : 0 : 1)$ is neither on $\mathcal{C}'$ nor on the curve defined by H over the algebraic closure of $K(t_0, t_1)$. Then let $R(t_0, t_1, x, y) := \operatorname{res}_z(H, F)$. Taking into account the previous steps (a) and (b), one has that R factors as

$$R(t_0, t_1, x, y) = (\alpha_2(t_0, t_1)x - \alpha_1(t_0, t_1)y) \prod_{(a_i : b_i : c_i) \in \mathcal{B}} (b_i x - a_i y)^{r_i},$$

where $\sum r_i = dk - 1$.

Now, for every i we introduce the polynomials $\delta_i(t_0, t_1) = \alpha_2 a_i - \alpha_1 b_i$. Let us see that none of these polynomials is identically zero. For this purpose, we first observe that, since the leading coefficient of F w.r.t. z is constant, the resultant specializes properly and therefore $(\alpha_2(1, 0)x - \alpha_1(1, 0)y) = \lambda(bx - ay)$ for some $\lambda \in K^*$. Therefore if δ_i is identically zero then $ba_i - ab_i = 0$, which is impossible because $(0 : 0 : 1)$ is not on any line connecting a point in $\mathcal{B}$ and Q. We consider the set

$$\Omega = \{(t_0, t_1) \in K^2 \setminus \{(0, 0)\} \mid \prod \delta_i(t_0, t_1) \neq 0\}.$$

Note that Ω is open and nonempty. Moreover, because of the construction, for every $(t_0, t_1) \in \Omega$, if $\mathcal{C}''$ is the curve defined by $H(t_0, t_1, x, y, z)$, then

$$\sum_{P \in \mathcal{B}} \operatorname{mult}_P(\mathcal{C}, \mathcal{C}'') = dk - 1. \qquad \square$$

Theorem 4.62. *Let $\mathcal{C}$ be a projective curve of degree d and genus 0, let $Q \notin \mathcal{C}$, and let $\mathcal{S}_d \subset \mathcal{C} \setminus \mathrm{Sing}(\mathcal{C})$ be such that $\mathrm{card}(\mathcal{S}_d) = 3(d-1)$. Then*

$$\mathcal{A}_d(\mathcal{C}) \cap \mathcal{H}(d, Q + \sum_{P \in \mathcal{S}_d} P)$$

parametrizes $\mathcal{C}$.

Proof. Let $\mathcal{H} = \mathcal{A}_d(\mathcal{C}) \cap \mathcal{H}(d, \sum_{P \in \mathcal{S}_d} P + Q)$. We prove that the conditions in Theorem 4.54 are satisfied. First, we observe that for every $\mathcal{C}' \in \mathcal{H}$, since $\deg(\mathcal{C}) = \deg(\mathcal{C}')$, since $Q \in \mathcal{C}'$ but $Q \notin \mathcal{C}$, and since $\mathcal{C}$ is irreducible, the curves $\mathcal{C}$ and $\mathcal{C}'$ do not have common components. Therefore, Condition (3) holds.

Let us now check Condition (1), i.e. $\dim(\mathcal{H}) = 1$. Since $\dim(\mathcal{H}) \geq \dim(\mathcal{A}_d(\mathcal{C})) - [3(d-1)] - 1$, by Theorem 4.58, we know that $\dim(\mathcal{H}) \geq 1$. Now, let us assume that $\dim(\mathcal{H}) > 1$. Then, we take two different points $Q_1, Q_2 \in \mathcal{C} \setminus (\mathrm{Sing}(\mathcal{C}) \cup \mathcal{S}_d)$, and we consider the linear subsystem

$$\mathcal{H}' = \mathcal{H} \cap \mathcal{H}(d, Q_1 + Q_2).$$

Observe that $\dim(\mathcal{H}') \geq 0$. Thus, $\mathcal{H}' \neq \emptyset$. Let $\mathcal{C}' \in \mathcal{H}'$. Note that, since $\mathcal{C}' \in \mathcal{H}' \subset \mathcal{H}$, reasoning as above one has that $\mathcal{C}'$ and $\mathcal{C}$ do not have common components. Now, we distinguish two different cases:

(i) If $\mathcal{C}'$ does not have multiple components, then since $\mathcal{C}$ does not have common components either, applying Lemma 4.60 and that genus$(\mathcal{C}) = 0$, one has that (note that $Q \notin \mathcal{C} \cap \mathcal{C}'$)

$$d^2 \geq$$

$$\sum_{P \in \mathrm{Ngr}(\mathcal{C})} \mathrm{mult}_P(Q_P(\mathcal{C})) \, \mathrm{mult}_P(Q_P(\mathcal{C}')) + \sum_{P \in \mathcal{S}_d \cup \{Q_1, Q_2\}} \mathrm{mult}_P(\mathcal{C}) \, \mathrm{mult}_P(\mathcal{C}') \geq$$

$$\sum_{P \in \mathrm{Ngr}(\mathcal{C})} \mathrm{mult}_P(Q_P(\mathcal{C})) \, (\mathrm{mult}_P(Q_P(\mathcal{C})) - 1) + \sum_{P \in \mathcal{S}_d \cup \{Q_1, Q_2\}} \mathrm{mult}_P(\mathcal{C}) \, \mathrm{mult}_P(\mathcal{C}')$$

$$= (d-1)(d-2) + \sum_{P \in \mathcal{S}_d \cup \{Q_1, Q_2\}} \mathrm{mult}_P(\mathcal{C}) \, \mathrm{mult}_P(\mathcal{C}')$$

$$\geq (d-1)(d-2) + 3(d-1) + 2 = d^2 + 1,$$

which is impossible.

(ii) Let us assume that $\mathcal{C}'$ has multiple components. Then, we consider the linear system $\mathcal{H}^* := \mathcal{A}_d(\mathcal{C}) \cap \mathcal{H}(d, \sum_{P \in \mathcal{S}_d} P) \cap \mathcal{H}(d, Q_1 + Q_2)$. We observe that, since $\mathcal{C}' \in \mathcal{H}'$, then $\mathcal{C}' \in \mathcal{H}^*$. Now, we apply Lemma 4.59(1) to $\mathcal{C}$ and $\mathcal{F} := \mathcal{S}_d \cup \{Q_1, Q_2\}$, and we take $\lambda, \mu \in K$ such that

$$\mathcal{C}'' := \mu \mathcal{C}' + \lambda \mathcal{C} \in \mathcal{H}^*,$$

and such that $\mathcal{C}''$ does not have common components. In this situation we apply Lemma 4.60 to $\mathcal{C}''$ and $\mathcal{C}$; note that $\mathcal{C}''$ and $\mathcal{C}$ do not have common components because otherwise this would imply that $\mathcal{C}'$ and $\mathcal{C}$ have a common component, which is a contradiction.

$$d^2 \geq$$

$$\sum_{P \in \mathrm{Ngr}(\mathcal{C})} \mathrm{mult}_P(Q_P(\mathcal{C}))\,\mathrm{mult}_P(Q_P(\mathcal{C}'')) + \sum_{P \in \mathcal{S}_d \cup \{Q_1, Q_2\}} \mathrm{mult}_P(\mathcal{C})\,\mathrm{mult}_P(\mathcal{C}'') \geq$$

$$\sum_{P \in \mathrm{Ngr}(\mathcal{C})} \mathrm{mult}_P(Q_P(\mathcal{C}))\,(\mathrm{mult}_P(Q_P(\mathcal{C})) - 1) + \sum_{P \in \mathcal{S}_d \cup \{Q_1, Q_2\}} \mathrm{mult}_P(\mathcal{C})\,\mathrm{mult}_P(\mathcal{C}'')$$

$$\geq (d-1)(d-2) + 3(d-1) + 2 = d^2 + 1,$$

which is impossible.

Now, let us check whether Condition (2) in Theorem 4.54 holds. For this purpose, we first prove that the set of base points $\mathcal{B}$ of $\mathcal{H}$ is $\mathrm{Sing}(\mathcal{C}) \cup \mathcal{S}_d \cup \{Q\}$. It is clear that $\mathrm{Sing}(\mathcal{C}) \cup \mathcal{S}_d \cup \{Q\} \subset \mathcal{B}$. Let us assume that $\mathcal{B} \neq \mathrm{Sing}(\mathcal{C}) \cup \mathcal{S}_d \cup \{Q\}$. Then there exists $R \in \mathcal{B} \setminus (\mathrm{Sing}(\mathcal{C}) \cup \mathcal{S}_d \cup \{Q\})$. We choose a curve $\mathcal{C}' \in \mathcal{H}$ passing through a point $R' \in \mathcal{C} \setminus \mathcal{B}$. This is possible because $\dim(\mathcal{H}) = 1$. Then, since $\mathcal{C}$ and $\mathcal{C}'$ do not have common components, we argue similarly as above distinguishing two different cases:

(i) Let $\mathcal{C}'$ be without multiple components. Since $\mathcal{C}$ does not have common components either, we can apply Lemma 4.60. Reasoning as in (i) above, we arrive at the contradiction $d^2 \geq d^2 + 1$. Thus, $\mathcal{B} = \mathrm{Sing}(\mathcal{C}) \cup \mathcal{S}_d \cup \{Q\}$.

(ii) Let us assume that $\mathcal{C}'$ has multiple components. Then, we consider the linear system $\mathcal{H}^* := \mathcal{A}_d(\mathcal{C}) \cap \mathcal{H}(d, \sum_{P \in \mathcal{S}_d} P) \cap \mathcal{H}(d, R + R')$. We observe that, since $\mathcal{C}' \in \mathcal{H} \cap \mathcal{H}(d, R + R')$, then $\mathcal{C}' \in \mathcal{H}^*$. Now, we apply Lemma 4.59(1) to $\mathcal{C}$ and $\mathcal{F} := \mathcal{S}_d \cup \{R, R'\}$, and we take $\lambda, \mu \in K$ such that

$$\mathcal{C}'' := \mu \mathcal{C}' + \lambda \mathcal{C} \in \mathcal{H}^*,$$

and $\mathcal{C}''$ does not have multiple components. In this situation we apply Lemma 4.60 to $\mathcal{C}''$ and $\mathcal{C}$; note that $\mathcal{C}''$ and $\mathcal{C}$ do not have common components because otherwise it would imply that $\mathcal{C}'$ and $\mathcal{C}$ have a common component, which is not the case. Now, reasoning as in (ii) above, we arrive at the contradiction $d^2 \geq d^2 + 1$. Thus, $\mathcal{B} = \mathrm{Sing}(\mathcal{C}) \cup \mathcal{S}_d \cup \{Q\}$.

Once we have proved that $\mathcal{B} = \mathrm{Sing}(\mathcal{C}) \cup \mathcal{S}_d \cup \{Q\}$, we show that Statement (2) in Theorem 4.54 holds. That is, we have to prove that for almost all $\mathcal{C}' \in \mathcal{H}$ one has that

$$\sum_{P \in \mathcal{B}} \mathrm{mult}_P(\mathcal{C}, \mathcal{C}') = d^2 - 1.$$

We structure the proof as follows: First, we prove that for all $\mathcal{C}' \in \mathcal{H}$ it holds that

$$\sum_{P \in \mathcal{B}} \operatorname{mult}_P(\mathcal{C}, \mathcal{C}') \geq d^2 - 1.$$

Second, we show that there exists at least one curve $\mathcal{C}' \in \mathcal{H}$ such that

$$\sum_{P \in \mathcal{B}} \operatorname{mult}_P(\mathcal{C}, \mathcal{C}') = d^2 - 1,$$

and finally we show that the equality holds for almost all curves in $\mathcal{H}$.

(a) Let us assume that there exists $\mathcal{C}' \in \mathcal{H}$ such that the sum of multiplicities of intersection at $\mathcal{B}$ is equal to $d^2 - \ell$, where $\ell > 1$. Then, since $\mathcal{C}$ and $\mathcal{C}'$ do not have common components, by Bézout Theorem we deduce that there exists a set of points $\mathcal{E} \subset (\mathcal{C} \cap \mathcal{C}') \setminus \mathcal{B}$ such that

$$\sum_{P \in \mathcal{E}} \operatorname{mult}_P(\mathcal{C}, \mathcal{C}') = \ell.$$

Now we argue similarly as above distinguishing two different cases:

(a.1) Let $\mathcal{C}'$ be without multiple components. Since $\mathcal{C}$ does not have multiple components either, we can apply Lemma 4.60. Reasoning as in (i) above, and using the fact that $\mathcal{B} = \operatorname{Sing}(\mathcal{C}) \cup \mathcal{S}_k$, we derive $d^2 \geq d^2 + \ell - 1$, which is impossible since $\ell > 1$.

(a.2) Let us assume that $\mathcal{C}'$ has multiple components. Then, we consider the linear system $\mathcal{H}^* := \mathcal{A}_d(\mathcal{C}) \cap \mathcal{H}(d, \sum_{P \in \mathcal{S}_d} P)$. We observe that, since $\mathcal{C}' \in \mathcal{H}$, then $\mathcal{C}' \in \mathcal{H}^*$. Now, we apply Lemma 4.59(1) to $\mathcal{C}$ and $\mathcal{F} := \mathcal{S}_d$, and we take $\lambda, \mu \in K$ such that

$$\mathcal{C}'' := \mu\mathcal{C}' + \lambda\mathcal{C} \in \mathcal{H}^*,$$

and $\mathcal{C}''$ does not have multiple components. In this situation we apply Lemma 4.60 to $\mathcal{C}''$ and $\mathcal{C}$; note that $\mathcal{C}''$ and $\mathcal{C}$ do not have common components because otherwise also $\mathcal{C}'$ and $\mathcal{C}$ would have common components, which we have excluded. Also, note that by assumption $\mathcal{E} \subset \mathcal{C}' \cap \mathcal{C}$, and therefore $\mathcal{E} \subset \mathcal{C}'' \cap \mathcal{C}$. So, reasoning as in (ii), we derive $d^2 \geq d^2 + \ell - 1$, which is impossible since $\ell > 1$.

(b) Let us assume that for all curves in $\mathcal{H}$ the sum of multiplicities of intersection is d^2. Then, since $\dim(\mathcal{H}) = 1$, we consider a point $R \in \mathcal{C} \setminus \mathcal{B}$, and we take $\mathcal{C}' \in \mathcal{H}$ such that $R \in \mathcal{C}'$. In this situation we have

$$\sum_{P \in \mathcal{C}' \cap \mathcal{C}} \operatorname{mult}_P(\mathcal{C}, \mathcal{C}') \geq \sum_{P \in \mathcal{B}} \operatorname{mult}_P(\mathcal{C}, \mathcal{C}') + \operatorname{mult}_R(\mathcal{C}, \mathcal{C}') \geq d^2 + 1.$$

Therefore, by Bézout's theorem, the curves $\mathcal{C}$ and $\mathcal{C}'$ have a common component, which is impossible.

(c) Let $C' \in \mathcal{H}$ be the curve whose existence ensures step (b) of our reasoning. Since the sum of multiplicities of intersection at $\mathcal{B}$ is $d^2 - 1$, since C' and C do not have common components, and since $Q \in \mathcal{B}$ but $Q \notin C \cap C'$, by Bézout's theorem we know that $C \cap C' = (\mathcal{B} \setminus \{Q\}) \cup \{R\}$, where $R :=$ $(a : b : c) \notin \mathcal{B}$. Now, let $H(t_0, t_1, x, y, z) := t_0 G_0 + t_1 G_1$ be the defining polynomial of a generic element in $\mathcal{H}$, where we assume w.l.o.g. that G_0 is the defining polynomial of C', and let F be the defining polynomial of C. We assume w.l.o.g., probably after performing a suitable linear change of coordinates, that $F(0, 0, 1) \neq 0$, $G_i(0, 0, 1) \neq 0$, and that $(0 : 0 : 1)$ is not on any line connecting two different points in $(\mathcal{B} \setminus \{Q\}) \cup \{R\}$. Note that if $F(0, 0, 1) \neq 0$ then, in particular, one has that the leading coefficient of F w.r.t. z is constant and that $(0 : 0 : 1) \notin (\mathcal{B} \setminus \{Q\}) \cup \{R\}$. Also, condition $G_i(0, 0, 1) \neq 0$ implies that $(0 : 0 : 1)$ is neither on C' nor on the curve defined by H over the algebraic closure of $K(t_0, t_1)$. Then let $R(t_0, t_1, x, y) := \operatorname{res}_z(H, F)$. Taking into account the previous steps (a) and (b) one has that R factors as

$$R(t_0, t_1, x, y) = (\alpha_2(t_0, t_1)x - \alpha_1(t_0, t_1)y) \prod_{(a_i : b_i : c_i) \in \mathcal{B} \setminus \{Q\}} (b_i x - a_i y)^{r_i} ,$$

where $\sum r_i = d^2 - 1$.

Now, for every i we introduce the polynomials $\delta_i(t_0, t_1) = \alpha_2 a_i - \alpha_1 b_i$. Let us see that none of these polynomials is identically zero. For this purpose, we first observe that, since the leading coefficient of F w.r.t. z is constant, the resultant specializes properly and therefore $(\alpha_2(1, 0)x - \alpha_1(1, 0)y) = \lambda(bx - ay)$ for some $\lambda \in K^*$. Therefore if δ_i is identically zero then $ba_i - ab_i = 0$ which is impossible because $(0 : 0 : 1)$ is not on any line connecting a point in $(\mathcal{B} \setminus \{Q\})$ and R. We consider the set

$$\Omega = \{(t_0, t_1) \in K^2 \setminus \{(0, 0)\} \mid \prod \delta_i(t_0, t_1) \neq 0\} .$$

Note that Ω is open and nonempty. Moreover, because of the construction, for every $(t_0, t_1) \in \Omega$, if C'' is the curve defined by $H(t_0, t_1, x, y, z)$, then

$$\sum_{P \in \mathcal{B}} \operatorname{mult}_P(C, C'') = d^2 - 1. \qquad \square$$

From these theorems, one deduces the following result:

Theorem 4.63. *An algebraic curve C is rational if and only if* genus$(C) = 0$.

Proof. One implication is already stated in Theorem 4.11. In this section we have developed an algorithm which can parametrize every curve of genus 0.

$\square$

The results proved in this section provide a family of algorithms for parametrizing any rational curve by means of adjoints.

Algorithm PARAMETRIZATION-BY-ADJOINTS.

Given the defining polynomial $F(x, y, z)$ of an irreducible projective curve C of degree d and genus 0, the algorithm computes a rational parametrization of C.

1. If $d \leq 3$ or $\text{Sing}(C)$ contains exactly one point of multiplicity $d - 1$, apply algorithm PARAMETRIZATION-BY-LINES.
2. Choose $k \in \{d - 2, d - 1, d\}$ and compute the defining polynomial of $\mathcal{A}_k(C)$.
3. Choose a set $\mathcal{S} \subset (C \setminus \text{Sing}(C))$ such that $\text{card}(\mathcal{S}) = kd - (d - 1)(d - 2) - 1$.
4. If $k < d$ then
   compute the defining polynomial H of $\mathcal{A}_k(C) \cap \mathcal{H}(k, \sum_{P \in \mathcal{S}} P)$;
   else (i.e. $k = d$)
   choose $Q \notin C$ and
   compute the defining polynomial H of $\mathcal{A}_k(C) \cap \mathcal{H}(k, Q + \sum_{P \in \mathcal{S}} P)$.
5. Set one of the parameters in H to 1 and let t be the remaining parameter in H. Return the solution in $\mathbb{P}^2(K(t))$ of $\{\text{pp}_t(\text{res}_y(F, H)) = 0, \text{pp}_t(\text{res}_x(F, H)) = 0\}$.

From the point of view of time efficiency one must choose $k = d - 2$ in Step 2, since then degrees of polynomials are the smallest. Nevertheless, the selection of $k = d$ can be also interesting in the sense that at most one algebraic number of degree d has to be introduced (see Theorem 4.72), and therefore it is a first approach to algebraic optimality of the output (see Chap. 5). In the next section, we will consider the algebraic extensions of the field of definition required by the parametrization algorithm. But first, we illustrate the algorithm by two examples.

Example 4.64. Let C be the quintic over $\mathbb{C}$ (see Figure 4.3) of defining polynomial (see Example 3.13)

$$F(x, y, z) = y^2 z^3 - x^5.$$

From the implicit equation it is clear that $(t^2, t^5, 1)$ is a parametrization of C. Nevertheless, let us see how the algorithm works. In Example 3.13 we have determined that

$$\text{Sing}(C) = \{(0 : 1 : 0), (0 : 0 : 1)\},$$

where $P_1 = (0 : 1 : 0)$ is a triple nonordinary point, and $P_2 = (0 : 0 : 1)$ is a nonordinary double point. Furthermore, in Example 3.13 the neighboring graph of C was computed. We have obtained that $P_{1,1} = (1 : 1 : 0)$ is an ordinary double point in the first neighborhood of P_1, $P_{2,1} = (1 : 1 : 0)$ is a nonordinary double point in the first neighborhood of P_2, and $P_{2,2} = (-2 : 1 : 0)$ is a simple point in the second neighborhood of P_2.

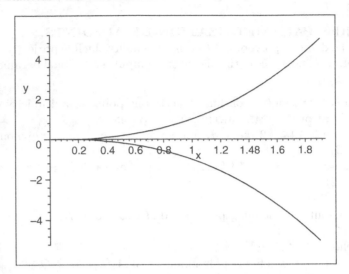

Fig. 4.3. $\mathcal{C}_{*,z}$

Therefore, genus($\mathcal{C}$) = 0, and hence $\mathcal{C}$ is rational (see Theorem 4.63). We proceed to parametrize the curve. In Step 2 we choose $k = d - 2 = 3$. In order to compute $\mathcal{A}_k(\mathcal{C})$, we consider a generic form in $\{x, y, z\}$ of degree 3:

$$H = a_{00}\, z^3 + a_{01}\, yz^2 + a_{02}\, y^2z + a_{03}\, y^3 + a_{10}\, xz^2 + a_{11}\, xyz + a_{12}\, xy^2$$
$$+a_{20}\, x^2z + a_{21}\, x^2y + a_{30}\, x^3.$$

First, we require P_1 to be a double point on $\mathcal{A}_3(\mathcal{C})$, and P_2 to be a simple point on $\mathcal{A}_3(\mathcal{C})$. That is, we consider the equations:

$$\left\{ \frac{\partial H}{\partial x}(P_1) = 0, \frac{\partial H}{\partial y}(P_1) = 0, \frac{\partial H}{\partial z}(P_1) = 0, H(P_2) = 0 \right\}.$$

Solving the linear system of equations in $a_{i,j}$ derived from the system above one gets:

$$H = a_{01}\, yz^2 + a_{10}\, xz^2 + a_{11}\, xyz + a_{20}\, x^2z + a_{21}\, x^2y + a_{30}\, x^3$$

Next, we consider the neighboring points. That is, we impose that

$$\{\mathcal{Q}_{P_1}(H)(P_{1,1}) = 0, \ \mathcal{Q}_{P_2}(H)(P_{2,1}) = 0\}.$$

This leads to
$$H = a_{01}yz^2 + a_{11}xyz + a_{20}x^2z + a_{30}x^3,$$

as the defining polynomial of $\mathcal{A}_3(\mathcal{C})$.

In Step 3 we choose a set $\mathcal{S} \subset (\mathcal{C} \setminus \mathrm{Sing}(\mathcal{C}))$ with 2 points, namely

$$\mathcal{S} = \{(1 : 1 : 1), (1 : -1 : 1)\}.$$

In Step 4 we compute the defining polynomial of $\mathcal{A}_3(\mathcal{C}) \cap \mathcal{H}(3, Q_1 + Q_2)$, where $Q_1 = (1 : 1 : 1)$ and $Q_2 = (1 : -1 : 1)$. That is, we solve the equations

$$H(1, 1, 1) = 0, \ H(1, -1, 1) = 0,$$

which leads to

$$H(x, y, z) = -a_{11}yz^2 + a_{11}xyz - x^2 z a_{30} + a_{30}x^3.$$

Setting $a_{11} = 1, a_{30} = t$, we get the defining polynomial

$$H(t, x, y, z) = -yz^2 + xyz - x^2 zt + tx^3$$

of the parametrizing system. Finally, in Step 5, the solution of the system

$$\begin{cases} -x + t^2 z = 0 \\ -y + t^5 z = 0 \end{cases}$$

provides the parametrization

$$\mathcal{P}(t) = (t^2, t^5, 1).$$

Example 4.65. Let $\mathcal{C}$ be the quartic over $\mathbb{C}$ (see Fig. 4.4) of

$$F(x, y, z) = -2xy^2 z - 48x^2 z^2 + 4xyz^2 - 2x^3 z + x^3 y - 6y^4 + 48y^2 z^2 + 6x^4.$$

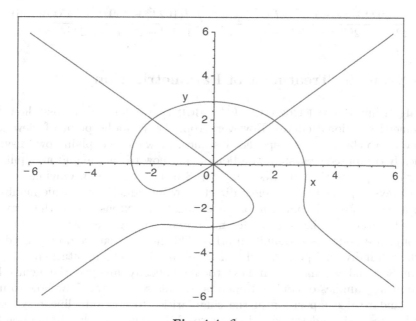

Fig. 4.4. $\mathcal{C}_{*,z}$

The singular locus of $\mathcal{C}$ is

$$\text{Sing}(\mathcal{C}) = \{(0:0:1), (2:2:1), (-2:2:1)\},$$

all three points being double points. Therefore, genus$(\mathcal{C}) = 0$, and hence $\mathcal{C}$ is rational (see Theorem 4.63). Note that no blowing up is required. We proceed to parametrize the curve. In Step 2 we choose $k = d - 2 = 2$. The defining polynomial of $\mathcal{A}_2(\mathcal{C})$ is

$$H(x, y, z) = (-2\,a_{02} - 2\,a_{20})\,yz + a_{02}y^2 - 2\,a_{11}xz + a_{1,1}xy + a_{20}x^2.$$

In Step 3 we choose a set $\mathcal{S} \subset (\mathcal{C} \setminus \text{Sing}(\mathcal{C}))$ with 1 point, namely $\mathcal{S} = \{(3:0:1)\}$. In Step 4, we compute the defining polynomial of $\mathcal{H} := \mathcal{A}_2(\mathcal{C}) \cap \mathcal{H}(2, Q)$, where $Q = (3:0:1)$. This leads to

$$H(x, y, z) = (-2\,a_{02} - 2\,a_{20})\,yz + a_{02}y^2 - 3\,a_{20}xz + \frac{3}{2}\,a_{20}xy + a_{20}x^2.$$

Setting $a_{02} = 1, a_{20} = t$, we get the defining polynomial

$$H(t, x, y, z) = (-2 - 2t)\,yz + y^2 - 3\,txz + \frac{3}{2}\,txy + tx^2$$

of the parametrizing system. Finally, in Step 5, the solution of the system defined by the resultants provides the following affine parametrization of $\mathcal{C}$:
$\mathcal{P}(t) =$

$$\left(12\,\frac{9\,t^4 + t^3 - 51\,t^2 + t + 8}{126\,t^4 - 297\,t^3 + 72\,t^2 + 8t - 36}, \; -2\,\frac{t(162\,t^3 - 459\,t^2 + 145\,t + 136)}{126\,t^4 - 297\,t^3 + 72\,t^2 + 8t - 36}\right).$$

4.8 Symbolic Treatment of Parametrization

In algorithm PARAMETRIZATION-BY-ADJOINTS we have described how to parametrize rational curves. However, from the symbolic point of view, we still want to clarify some steps. For instance, we want to explain how to symbolically compute the system of adjoints, and how to choose and manipulate the simple points that are taken in Step 3 of the algorithm. Obviously, one can always approach the problem directly, by introducing algebraic numbers and carrying out all computations over algebraic extensions of the ground field. However, we take here a different approach, using the notion of a family of conjugate points (see Definition 3.15). This means that we do not need to work with individual points, and hence we safe time in computation.

In Sect. 3.3 we have seen how to symbolically analyze the genus by introducing families of conjugate points. In this section we show how to use the standard decomposition of the singularities to compute linear systems of adjoints, and we describe a first approach for choosing the simple points.

In the next chapter, we will develop an optimal approach for the choice of the necessary simple points.

Throughout this section, we assume that $\mathcal{C}$ is a projective rational curve of degree d, and that its not necessarily algebraically closed ground field $\mathbb{K}$ (see Definition 3.14) is a computable field.

We start by proving that linear systems of adjoint curves can be computed without extending $\mathbb{K}$.

Theorem 4.66. *Let $\mathcal{C}$ be a rational projective curve of degree d and ground field $\mathbb{K}$. Then $\mathbb{K}$ is also the ground field of $\mathcal{A}_k(\mathcal{C})$, $k \geq d - 2$.*

Proof. The linear system of adjoints $\mathcal{A}_k(\mathcal{C})$ can be expressed as

$$\mathcal{A}_k(\mathcal{C}) = \mathcal{H}(k, \sum_{P \in \mathcal{D}(\mathrm{Ngr}(\mathcal{C}))} (\mathrm{mult}_P(Q_P(\mathcal{C})) - 1) \cdot P),$$

where $\mathcal{D}(\mathrm{Ngr}(\mathcal{C}))$ is the standard decomposition of the neighboring graph of $\mathcal{C}$ (see Definition 3.27). Observe that all the transformations for dealing with the neighboring graph of $\mathcal{C}$ are performed over the ground field. So Theorem 3.26 and Lemma 3.19 yield the result.   □

Combining the results in Sect. 4.6 and Theorem 4.66, we can guarantee that the output of Step 2 in algorithm PARAMETRIZATION-BY-ADJOINTS is defined over $\mathbb{K}$. In Step 3 we have to compute simple points on $\mathcal{C}$. In Chap. 5 we will see that we can find such points in a field extension of $\mathbb{K}$ of degree at most 2. Here we prove a more modest result, namely that we can always parametrize using a field extension of $\mathbb{K}$ of degree at most d. Note that if the simple points are taken randomly, and adjoints of degree $d - 2$ are considered, then in general a field extension of degree $(d - 3)^d$ has to be introduced. Moreover, this bound is even worse if adjoints of higher degree are used.

Lemma 4.67. *Let P be a simple point on an irreducible projective curve $\mathcal{C}$ of degree $d > 1$. There exist at most $d(d - 1)$ tangents to $\mathcal{C}$, at a simple point on $\mathcal{C}$, passing through P.*

Proof. We assume w.l.o.g. that $P = (0 : 0 : 1)$. If $Q \in \mathcal{C} \setminus \mathrm{Sing}(\mathcal{C})$, then the tangent to $\mathcal{C}$ at Q is given by

$$x \frac{\partial F}{\partial x}(Q) + y \frac{\partial F}{\partial y}(Q) + z \frac{\partial F}{\partial z}(Q) = 0 ,$$

where F is the defining polynomial of $\mathcal{C}$ (see Theorem 2.13). Thus, the simple points of $\mathcal{C}$ with tangent passing through P are solutions of $\{\frac{\partial F}{\partial z} = 0,\ F = 0\}$. Since F is irreducible and since $\partial F / \partial z$ has total degree $d - 1$, according to Bézout's Theorem there are at most $d(d - 1)$ different solutions.   □

In the following we show how the simple points in algorithm PARAMETRIZATION-BY-ADJOINTS can be taken in families of conjugate points for different options of the degree of the adjoint curves.

Theorem 4.68 (Parametrizing with adjoints of degree $d-2$**).** *If algorithm* PARAMETRIZATION-BY-ADJOINTS *is performed with adjoints of degree* $k = d - 2$, *the set* $\mathcal{S}$ *of simple points in Step 3 of the algorithm can be taken as a family of conjugate points over a field extension of* $\mathbb{K}$ *of degree at most* $d(d-1)(d-2)$.

Proof. Let $F(x, y, z) \in \mathbb{K}[x, y, z]$ be the defining polynomial of $\mathcal{C}$. The theorem obviously holds for curves of degree ≤ 4. So w.l.o.g. we may assume that $\deg(\mathcal{C}) > 4$. Note that $\operatorname{card}(\mathcal{S}) = d - 3$ in Step 3. Take $b_1, b_2 \in \mathbb{K}$ such that no singular point of $\mathcal{C}$ is of the form $(b_1 : b_2 : c)$. Now, compute an irreducible factor $p_1(t)$ of $F(b_1, b_2, t)$ over $\mathbb{K}$. Then,

$$P_1 = (b_1 : b_2 : \beta_1) \in \mathbb{P}^2(\mathbb{K}(\beta_1)),$$

where $p_1(\beta_1) = 0$, is a simple point of $\mathcal{C}$. Note that $\deg(p_1) \leq d$. In this situation, choose $\lambda, \mu \in \mathbb{K}$ such that:

(1) $b_1 \neq \lambda\beta_1$, $b_2 \neq \mu\beta_1$,

(2) $\operatorname{res}_t\left(\overline{q}(t), \frac{\partial \overline{q}}{\partial t}\right) \neq 0$, where $\overline{q}(t) = F(\lambda t + b_1, \mu t + b_2, t + \beta_1) \in \mathbb{K}(\beta_1)[t]$.

Condition (2) implies that the line $\mathcal{L} = \{(\lambda t + b_1 : \mu t + b_2 : t + \beta_1) \mid t \in K\}$ does not pass through the singularities and that $\mathcal{L}$ is not tangent to $\mathcal{C}$. The reason for Condition (1) will become clear later. Note that Lemma 4.67 implies that Condition (2) can always be achieved. Condition (1) is clearly reachable. Now, we consider the polynomial

$$q(t) = \frac{\overline{q}(t)}{t} \in \mathbb{K}(\beta_1)[t]$$

(note that $\overline{q}(0) = F(P_1) = 0$), and choose an irreducible factor $p_2(t)$ over $\mathbb{K}(\beta_1)$ of $q(t)$. Thus, from the above construction we deduce that

$$P_2 = (\lambda\beta_2 + b_1 : \mu\beta_2 + b_2 : \beta_2 + \beta_1) \in \mathbb{P}^2(\mathbb{K}(\beta_1, \beta_2)),$$

where $p_2(\beta_2) = 0$, is a simple point of $\mathcal{C}$ because of (2). Note that $\deg(p_2) \leq d - 1$. Then, we introduce the polynomial

$$q^\star(t) = \frac{q(t)}{t - \beta_2} \in \mathbb{K}(\beta_1, \beta_2)[t].$$

Take an irreducible factor $p_3(t)$ of $q^\star(t)$ over $\mathbb{K}(\beta_1, \beta_2)$, and consider the point

$$P_3 = (\lambda\beta_3 + b_1 : \mu\beta_3 + b_2 : \beta_3 + \beta_1) \in \mathbb{P}^2(\mathbb{K}(\beta_1, \beta_2, \beta_3)),$$

where $p_3(\beta_3) = 0$. Note that $\deg(p_3) \leq d - 2$. Observe that P_3 is a simple point on $\mathcal{C}$ because of (2). Finally, we introduce the polynomial

$$m(t) = \frac{q^\star(t)}{t - \beta_3} \in \mathbb{K}(\beta_1, \beta_2, \beta_3)[t].$$

In this situation, we claim that

$$\mathcal{F} = \{(\lambda t + b_1 : \mu t + b_2 : t + \beta_1)\}_{m(t)}$$

is a family of $(d-3)$ conjugate simple points on $\mathcal{C}$ over $\mathbb{K}(\beta_1, \beta_2, \beta_3)$. First, note that $m(t) \in \mathbb{K}(\beta_1, \beta_2, \beta_3)[t]$, thus $\mathcal{F}$ contains conjugate points over $\mathbb{K}(\beta_1, \beta_2, \beta_3)$. Moreover, because of Condition (1), the coordinate polynomials in $\mathcal{F}$ are coprime and with coefficients in $\mathbb{K}(\beta_1) \subset \mathbb{K}(\beta_1, \beta_2, \beta_3)$. Hence, Condition (1) in Definition 3.15 is satisfied. Furthermore, by construction $\overline{q}(t)$ is squarefree. Thus, since $m(t)$ is a factor of $\overline{q}(t)$, also $m(t)$ must be squarefree. So Condition (2) in Definition 3.15 is satisfied. Moreover $\deg(\overline{q}(t)) = d > 4$, hence $\deg(m(t)) = d - 3 > 1$ and the degree of the polynomials defining the coordinates of $\mathcal{F}$ is 1. Thus, Condition (3) in Definition 3.15 is also satisfied. Now, we check that $\operatorname{card}(\mathcal{F}) = d - 3$. Let α_1, α_2 be two different roots of $m(t)$, and let P_{α_i} be the point in $\mathcal{F}$ generated by the root α_i. If $\alpha_1 = -\beta_1$ then $\alpha_2 \neq -\beta_1$ and hence $P_{\alpha_1} \neq P_{\alpha_2}$ (similarly if $\alpha_2 = -\beta_1$). Let α_1, α_2 be different from $-\beta_1$. Then $P_{\alpha_1} = P_{\alpha_2}$ implies

$$\frac{\lambda \alpha_1 + b_1}{\alpha_1 + \beta_1} = \frac{\lambda \alpha_2 + b_1}{\alpha_2 + \beta_1}, \quad \frac{\mu \alpha_1 + b_2}{\alpha_1 + \beta_1} = \frac{\mu \alpha_2 + b_2}{\alpha_2 + \beta_1}.$$

Since $\alpha_1 \neq \alpha_2$, this implies $\lambda \beta_1 = b_1$ and $\mu \beta_1 = b_2$, which is impossible because of Condition (1) in the construction. Summarizing, $\mathcal{F}$ is a family of $(d-3)$ conjugate simple points over $\mathbb{K}(\beta_1, \beta_2, \beta_3)$, which is an extension of $\mathbb{K}$ of degree at most $d(d-1)(d-2)$. $\square$

Corollary 4.69. *If algorithm* PARAMETRIZATION-BY-ADJOINTS *is performed with adjoints of degree* $k = d-2$, *and* $\mathcal{S}$ *in Step 3 of the algorithm is taken as in Theorem 4.68, the algorithm outputs a parametrization over a field extension of* $\mathbb{K}$ *of degree at most* $d(d-1)(d-2)$.

Proof. In Theorem 4.66 we have seen that the defining polynomial of $\mathcal{A}_{d-2}(\mathcal{C})$ has coefficients over $\mathbb{K}$. By Theorem 4.68, points in $\mathcal{S}$ are in a family of conjugate points over a field extension $\mathbb{L}$ of $\mathbb{K}$ of degree at most $d(d-1)(d-2)$. Thus, by Lemma 3.19 the defining polynomial of

$$\mathcal{A}_{d-2}(\mathcal{C}) \cap \mathcal{H}\left(d-2, \sum_{P \in \mathcal{S}} P\right)$$

has coefficients in $\mathbb{L}$. Therefore, the resultant polynomials in Step 5 are over $\mathbb{L}$, and hence also the parametrization. $\square$

Theorem 4.70 (Parametrizing with adjoints of degree $d-1$). *If algorithm* PARAMETRIZATION-BY-ADJOINTS *is performed with adjoints of degree* $k = d - 1$, *the set* $\mathcal{S}$ *of simple points in Step 3 of the algorithm can be taken as the union of two families of conjugate points over a field extension of* $\mathbb{K}$ *of degree at most* $d(d-1)$.

Proof. Let $F(x, y, z) \in \mathbb{K}[x, y, z]$ be the defining polynomial of $\mathcal{C}$. The theorem obviously holds for curves of degree ≤ 4. So w.l.o.g. we may assume that $\deg(\mathcal{C}) > 4$. Note that $\text{card}(\mathcal{S}) = 2d - 3 = (d - 1) + (d - 2)$ in Step 3. Then, the idea is to express $\mathcal{S}$ as the union of two families of conjugate simple points, one with $(d - 2)$ points and the other with $(d - 1)$. More precisely, take $b_1, b_2 \in \mathbb{K}$ such that no singular point of $\mathcal{C}$ is of the form $(b_1 : b_2 : c)$. Now, compute an irreducible factor $p_1(t)$ of $F(b_1, b_2, t)$ over $\mathbb{K}$. Then,

$$P_1 = (b_1 : b_2 : \beta_1) \in \mathbb{P}^2(\mathbb{K}(\beta_1)),$$

where $p_1(\beta_1) = 0$, is a simple point of $\mathcal{C}$. Note that $\deg(p_1) \leq d$. In this situation, choose $\lambda_1, \lambda_2, \mu_1, \mu_2 \in \mathbb{K}$ such that:

(1) $\lambda_1 \mu_2 \neq \lambda_2 \mu_1$, $b_1 \neq \lambda_i \beta_1$, $b_2 \neq \mu_i \beta_1$ for $i = 1, 2$,
(2) $\text{res}_t \left(\overline{q_i}(t), \frac{\partial \overline{q_i}}{\partial t} \right) \neq 0$, for $i = 1, 2$, where $\overline{q_i}(t) = F(\lambda_i t + b_1, \mu_i t + b_2, t + \beta_1) \in \mathbb{K}(\beta_1)[t]$.

Condition (1) implies in particular that the lines $\mathcal{L}_i = \{(\lambda_i t + b_1 : \mu_i t + b_2 : t + \beta_1) \,|\, t \in K\}$ are different. Condition (2) guarantees that the lines $\mathcal{L}_i$ do not pass through any singularities and that $\mathcal{L}_i$ is not tangent to $\mathcal{C}$. Note that Lemma 4.67 implies that Condition (2) can always be achieved. Condition (1) is easily reachable. Now, we consider the polynomials

$$q_i(t) = \frac{\overline{q_i}(t)}{t} \in \mathbb{K}(\beta_1)[t], \quad i = 1, 2,$$

(note that $\overline{q_i}(0) = F(P_1) = 0$). We claim that

$$\mathcal{F}_1 = \{(\lambda_1 t + b_1 : \mu_1 t + b_2 : t + \beta_1)\}_{q_1(t)}$$

is a family of $(d - 1)$ conjugate simple points on $\mathcal{C}$ over $\mathbb{K}(\beta_1)$. The proof of this fact is similar to the proof of Theorem 4.68 and we leave it to the reader.

In order to generate the second family we use $q_2(t)$. More precisely, let $p_2(t)$ be an irreducible factor of $q_2(t)$ over $\mathbb{K}(\beta_1)$. Then, we introduce the point

$$P_2 = (\lambda_2 \beta_2 + b_1 : \mu_2 \beta_2 + b_2 : \beta_2 + \beta_1),$$

where $p_2(\beta_2) = 0$. Note that $\deg(p_2) \leq d - 1$. $P_2 \in \mathcal{L}_2 \cap \mathcal{C}$, and therefore it is a simple point on $\mathcal{C}$. Now, we take

$$m(t) = \frac{q_2(t)}{t - \beta_2} \in \mathbb{K}(\beta_1, \beta_2)[t].$$

Then, reasoning similarly as above, we deduce that

$$\mathcal{F}_2 = \{(\lambda_2 t + b_1 : \mu_2 t + b_2 : t + \beta_1)\}_{m(t)}$$

is a family of $(d - 2)$ conjugate simple points on $\mathcal{C}$ over $\mathbb{K}(\beta_1, \beta_2)$. Thus, we have expressed $\mathcal{S}$ as $\mathcal{F}_1 \cup \mathcal{F}_2$. The only thing that we still have to prove is that $\mathcal{F}_1 \cap \mathcal{F}_2 = \emptyset$. Indeed, $\mathcal{L}_1 \cap \mathcal{L}_2 = \{P_1\}$, and $\mathcal{F}_i \subset \mathcal{L}_i \cap \mathcal{C}$. Thus the only common point of $\mathcal{F}_1$ and $\mathcal{F}_2$ is P_1. But the root corresponding to P_1 has been crossed out in both polynomials defining the families. $\qquad\square$

Corollary 4.71. *If algorithm* PARAMETRIZATION-BY-ADJOINTS *is performed with adjoints of degree* $k = d-1$, *and* $\mathcal{S}$ *in Step 3 of the algorithm is taken as in Theorem 4.70, the algorithm outputs a parametrization over a field extension of* $\mathbb{K}$ *of degree at most* $d(d-1)$.

Proof. Similar to the proof of Corollary 4.69. □

Theorem 4.72 (Parametrizing with adjoints of degree d). *If algorithm* PARAMETRIZATION-BY-ADJOINTS *is performed with adjoints of degree* $k = d$, *the set* $\mathcal{S}$ *of simple points in Step 3 of the algorithm can be taken as the union of three families of conjugate points over a field extension of* $\mathbb{K}$ *of degree at most* d.

Proof. Let $F(x, y, z) \in \mathbb{K}[x, y, z]$ be the defining polynomial of $\mathcal{C}$. As in the previous proofs we may assume w.l.o.g. that $\deg(\mathcal{C}) > 4$. Note that $\operatorname{card}(\mathcal{S}) = 3(d - 1)$ in Step 3. The idea is to express $\mathcal{S}$ as the union of three families of $(d - 1)$ conjugate points. For this purpose, we proceed as in the previous theorems. We take a simple point on the curve. This implies, in general, an extension of degree d, and we consider three lines through this point. More precisely, take $b_1, b_2 \in \mathbb{K}$ such that no singular point of $\mathcal{C}$ is of the form $(b_1 : b_2 : c)$. Now, compute an irreducible factor $p_1(t)$ of $F(b_1, b_2, t)$ over $\mathbb{K}$. Therefore,

$$P_1 = (b_1 : b_2 : \beta_1) \in \mathbb{P}^2(\mathbb{K}(\beta_1)),$$

where $p_1(\beta_1) = 0$, is a simple point of $\mathcal{C}$. Note that $\deg(p_1) \leq d$. In this situation, choose $\lambda_1, \lambda_2, \lambda_3, \mu_1, \mu_2, \mu_3 \in \mathbb{K}$ such that:

(1) $\lambda_i \mu_j \neq \lambda_j \mu_i$, for $i \neq j$, and $b_1 \neq \lambda_i \beta_1$, $b_2 \neq \mu_i \beta_1$ for $i = 1, 2, 3$,

(2) $\operatorname{res}_t\left(\overline{q_i}(t), \frac{\partial \overline{q_i}}{\partial t}\right) \neq 0$, for $i = 1, 2, 3$, where $\overline{q_i}(t) = F(\lambda_i t + b_1, \mu_i t + b_2, t + \beta_1) \in \mathbb{K}(\beta_1)[t]$.

Condition (1) implies in particular that the lines $\mathcal{L}_i = \{(\lambda_i t + b_1 : \mu_i t + b_2 : t + \beta_1) \mid t \in K\}$ are pairwise different, i.e. $\mathcal{L}_i \neq \mathcal{L}_j$ for $i \neq j$. Condition (2) guarantees that the lines $\mathcal{L}_i$ do not pass through any singularities and that $\mathcal{L}_i$ is not tangent to $\mathcal{C}$. Note that Lemma 4.67 implies that Condition (2) can always be achieved. Condition (1) is easily reachable. Now, we consider the polynomials

$$q_i(t) = \frac{\overline{q_i}(t)}{t} \in \mathbb{K}(\beta_1)[t], \quad i = 1, 2, 3$$

(note that $\overline{q_i}(0) = F(P_1) = 0$). We claim that

$$\mathcal{F}_i = \{(\lambda_i t + b_1 : \mu_i t + b_2 : t + \beta_1)\}_{q_i(t)}, \quad i = 1, 2, 3$$

are families of $(d - 1)$ conjugate simple points on $\mathcal{C}$ over $\mathbb{K}(\beta_1)$. The proof of this fact is similar to the proof of Theorem 4.68 and we leave it to the reader. □

Corollary 4.73. *If algorithm* PARAMETRIZATION-BY-ADJOINTS *is performed with adjoints of degree* $k = d$, $\mathcal{S}$ *in Step 3 of the algorithm is taken as in Theorem 4.72, and Q in Step 4 is taken over* $\mathbb{K}$, *the algorithm outputs a parametrization over a field extension of* $\mathbb{K}$ *of degree at most d.*

Proof. Similar to the proof of Corollary 4.69. $\square$

From the constructive proof of Theorem 4.72 it is clear that the field extension of $\mathbb{K}$ introduced in the parametrization method is the one used to define the simple point P_1 through which the three families of $(d-1)$ conjugate simple points are taken. Therefore, if P_1 can be taken to be rational, i.e. with coordinates in $\mathbb{K}$, then the output parametrization is defined over $\mathbb{K}$. In addition, the following result can be also deduced from the proof of Theorem 4.72.

Theorem 4.74. *Let $\mathcal{C}$ be a rational projective curve with ground field* $\mathbb{K}$. *Then $\mathcal{C}$ is parametrizable over* $\mathbb{K}$ *if and only if there exists a simple point on $\mathcal{C}$ with coordinates over* $\mathbb{K}$.

Proof. If $\mathcal{C}$ is parametrizable over $\mathbb{K}$, giving values in $\mathbb{K}$ to the parameter, one generates infinitely many points on $\mathcal{C}$ over $\mathbb{K}$. Thus, since the curve has finitely many singularities, one generates simple points on the curve with coordinates in $\mathbb{K}$. Conversely, let $P \in \mathcal{C}$ be simple with coordinates over $\mathbb{K}$. Then, P can be taken as the point P_1 in the proof of Theorem 4.72 to generate the 3 families of $(d-1)$ conjugate simple points on $\mathcal{C}$. This implies that these families are over $\mathbb{K}$, and therefore the output parametrization of the algorithm PARAMETRIZATION-BY-ADJOINTS is over $\mathbb{K}$. $\square$

The proofs of the previous theorems are constructive and they provide algorithms. We will outline the algorithm corresponding to adjoint curves of degree $d = \deg(\mathcal{C})$ (see Theorem 4.72 and Corollary 4.73). Algorithms derived from corollaries to Theorems 4.68 and 4.70 are left as exercises.

Algorithm SYMBOLIC-PARAMETRIZATION-BY-DEGREE-d-ADJOINTS.

Given the defining polynomial $F(x, y, z) \in \mathbb{K}[x, y, z]$ of a rational irreducible projective curve $\mathcal{C}$ of degree d, and the standard decomposition $\mathcal{D}(\mathrm{Ngr}(\mathcal{C}))$ of $\mathrm{Ngr}(\mathcal{C})$, the algorithm computes a rational parametrization of $\mathcal{C}$.

1. If $d \leq 3$ or $\mathrm{Sing}(\mathcal{C})$ contains exactly one point of multiplicity $d - 1$, apply algorithm PARAMETRIZATION-BY-LINES.
2. Take $b_1, b_2 \in \mathbb{K}$ such that no singular point of $\mathcal{C}$ is of the form $(b_1 : b_2 : c)$.
3. Compute an irreducible factor $p(t)$ of $F(b_1, b_2, t)$ over $\mathbb{K}$. Let β be a root of $p(t)$.

4. Choose $\lambda_1, \lambda_2, \lambda_3, \mu_1, \mu_2, \mu_3 \in \mathbb{K}$ such that:
   (i) $\lambda_i \mu_j \neq \lambda_j \mu_i$, for $i \neq j$,
   (ii) $b_1 \neq \lambda_i \beta$, $b_2 \neq \mu_i \beta$ for $i = 1, 2, 3$,
   (iii) $\mathrm{res}_t \left(\overline{q_i}(t), \frac{\partial \overline{q_i}}{\partial t} \right) \neq 0$, for $i = 1, 2, 3$, where $\overline{q_i}(t) = F(\lambda_i t + b_1, \mu_i t + b_2, t + \beta)$.
5. Compute $q_i(t) = \frac{\overline{q_i}(t)}{t}$ for $i = 1, 2, 3$.
6. Set $\mathcal{F}_i := \{(\lambda_i t + b_1 : \mu_i t + b_2 : t + \beta)\}_{q_i(t)}$ for $i = 1, 2, 3$.
7. Choose a point $Q \in \mathbb{P}^2(\mathbb{K}) \setminus \mathcal{C}$.
8. Let H be the defining polynomial of $\mathcal{H} = \mathcal{A}_d(\mathcal{C}) \cap \mathcal{H}(d, Q + \sum_{i=1}^3 \sum_{P \in \mathcal{F}_i} P)$; use $\mathcal{D}(\mathrm{Ngr}(\mathcal{C}))$ to compute symbolically $\mathcal{H}$ (see Theorem 4.72).
9. Set one of the parameters in H to 1 and let t be the remaining parameter in H. Return the solution in $\mathbb{P}^2(\mathbb{K}(\beta)(t))$ of $\{\mathrm{pp}_t(\mathrm{res}_y(F, H)) = 0, \mathrm{pp}_t(\mathrm{res}_x(F, H)) = 0\}$.

Remarks. Note that in Step 4 and also in the next step, we do not need to isolate an individual root of $p(t)$, but we can simply work modulo $p(t)$.

Example 4.75. We consider the quintic curve $\mathcal{C}$ over $\mathbb{C}$ defined by the polynomial

$$F(x, y, z) = 3y^3 z^2 - 3xy^2 z^2 - 2xy^3 z + y^3 x^2 + x^3 z^2.$$

The ground field of $\mathcal{C}$ is $\mathbb{Q}$. In Step 2, we take $b_1 = -1, b_2 = 1$. Thus,

$$F(-1, 1, t) = 5t^2 + 2t + 1.$$

In Step 3, we consider $p(t) = 5t^2 + 2t + 1$ and β with minimal polynomial $p(t)$. In Step 4, we take $\lambda_1 = 1, \lambda_2 = 0, \lambda_3 = 1$ and $\mu_1 = 0, \mu_2 = 1, \mu_3 = 2$. It is easy to check that conditions (i),(ii),(iii) are satisfied. In this situation, the polynomials $q_i(t)$ in Step 5 are

$$q_1(t) = \frac{23}{5} t + t^4 - 3 t^3 - \frac{1}{5} t^2 + 8 \beta + 2 \beta t^3 - \frac{32}{5} \beta t^2 + \frac{6}{5} \beta t,$$

$$q_2(t) = 3 t^4 + 14 t^3 + \frac{107}{5} t^2 + \frac{58}{5} t + 2 + 6 \beta t^3 + \frac{124}{5} \beta t^2 + \frac{156}{5} \beta t + 10 \beta,$$

$$q_3(t) = 5 t^4 + 21 t^3 + \frac{147}{5} t^2 + \frac{47}{5} t + 10 \beta t^3 + \frac{264}{5} \beta t^2 + \frac{294}{5} \beta t + 8 \beta.$$

Therefore, the families in Step 6 are

$$\mathcal{F}_1 = \{(t - 1 : 1 : t + \beta)\}_{q_1(t)}, \quad \mathcal{F}_2 = \{(-1 : t + 1 : t + \beta)\}_{q_2(t)},$$

$$\mathcal{F}_3 = \{(t - 1 : 2t + 1 : t + \beta)\}_{q_3(t)}.$$

In Step 7, we consider $Q = (1 : -1 : 1)$. In Step 8, we compute $\mathcal{H}$. For this purpose, first we apply the results on the symbolic computation of the genus (see Sect. 3.3), and we determine the standard decomposition of the singular locus:

$$\mathcal{D}(\mathrm{Sing}(\mathcal{C})) = \mathcal{D}(\mathrm{Ngr}(\mathcal{C})) =$$

$$\overbrace{\{(0:0:1)\}}^{\text{triple}} \cup \overbrace{\{(1:1:1)\} \cup \{(1:0:0)\} \cup \{(0:1:0)\}}^{\text{double}},$$

where the first singularity is a triple point and the others are double points. Let H be the defining polynomial of a generic form in x, y, z of degree 5:

$$H = a_{00}z^5 + a_{01}yz^4 + a_{02}y^2z^3 + a_{03}y^3z^2 + a_{04}y^4z + a_{05}y^5 + a_{10}xz^4 + a_{11}xyz^3 + a_{12}xy^2z^2 + a_{13}xy^3z + a_{14}xy^4 + a_{20}x^2z^3 + a_{21}x^2yz^2 + a_{22}x^2y^2z + a_{23}x^2y^3 + a_{30}x^3z^2 + a_{31}x^3yz + a_{32}x^3y^2 + a_{40}x^4z + a_{41}x^4y + a_{50}x^5.$$

Next, we compute the defining polynomial of $\mathcal{A}_5(\mathcal{C})$. That is, we consider the equations

$$\frac{\partial H}{\partial x}(0,0,1) = 0, \quad \frac{\partial H}{\partial y}(0,0,1) = 0, \quad \frac{\partial H}{\partial z}(0,0,1) = 0,$$

$$H(1,1,1) = 0, \quad H(1,0,0) = 0, \quad H(0,1,0) = 0.$$

Solving them and substituting in H we get the defining polynomial of the linear system of adjoints, which we denote again by H:

$$H = (-a_{03} - a_{04} - a_{11} - a_{12} - a_{13} - a_{14} - a_{20} - a_{21} - a_{22} - a_{23} - a_{30} - a_{31} - a_{32} - a_{40} - a_{41})y^2z^3 + a_{03}y^3z^2 + a_{04}y^4z + a_{11}xyz^3 + a_{12}xy^2z^2 + a_{13}xy^3z + a_{14}xy^4 + a_{20}x^2z^3 + a_{21}x^2yz^2 + a_{22}x^2y^2z + a_{23}x^2y^3 + a_{30}x^3z^2 + a_{31}x^3yz + a_{32}x^3y^2 + a_{40}x^4z + a_{41}x^4y.$$

Now, we introduce the new conditions

$$\begin{aligned}
H(1,-1,1) &= 0, \\
H(t-1,1,t+\beta) &= 0 \qquad \mathrm{mod}\ q_1(t), \\
H(-1,t+1,t+\beta) &= 0 \quad \mathrm{mod}\ q_2(t), \\
H(t-1,2t+1,t+\beta) &= 0 \ \mathrm{mod}\ q_3(t).
\end{aligned}$$

Solving these equations, and substituting in H we get the new linear subsystem of dimension 1 corresponding to Step 8 (we denote it again by H):

$$H = a_{41} - 12340xy^3za_{30} + 47562xy^3za_{41} - 4670xy^2z^2a_{30} - 3024xy^2z^2a_{41} - 1275xyz^3a_{30} - 3435xyz^3a_{41} + 4500y^4za_{30}\beta - 47100y^4z\beta a_{41} - 11280y^3z^2a_{30}\beta + 7425y^2z^3a_{30}\beta - 2900x^4za_{30}\beta - 9130x^4z\beta a_{41} + 600x^3y^2a_{30}\beta - 16505x^3y^2\beta a_{41} - 595x^3yza_{30} - 16824x^3yza_{41} + 3160x^2y^3a_{30}\beta - 39113x^2y^3\beta a_{41} + 7830y^2z^3a_{41} + 675y^2z^3a_{30} - 5940y^2z^3\beta a_{41} + 362x^3yz\beta a_{41} - 10965x^3yza_{30}\beta + 10565x^2y^2za_{30}\beta + 48008x^2y^2z\beta a_{41} - 36677x^2yz^2\beta a_{41} + 13890x^2yz^2a_{30}\beta + 7729x^2yz^2a_{41} -$$

$915x^2z^3a_{41} - 300x^2z^3a_{30} + 94094xy^3z\beta a_{41} + 11420xy^3za_{30}\beta - 3200xy^4a_{30} - 6120xy^4a_{41} - 11590xy^2z^2a_{30}\beta - 6225xyz^3a_{30}\beta + 22712xy^2z^2\beta a_{41} + 19005xyz^3\beta a_{41} - 27633y^3z^2a_{41} + 4860y^3z^2a_{30} - 8700y^4za_{41} + 7500y^4za_{30} - 37671y^3z^2\beta a_{41} + 3600x^3y^2a_{30} - 4990x^4za_{41} - 500x^4za_{30} - 3115x^3y^2a_{41} - 9999x^2y^3a_{41} + 5180x^2y^3a_{30}.$

Normalizing to $a_{30} = 1$ and $a_{41} = t$ and performing Step 9 we get the output parametrization

$$\mathcal{P}(t) = \left(\frac{\chi_{11}(t)}{\chi_{12}(t)}, \frac{\chi_{21}(t)}{\chi_{22}(t)} \right),$$

where,

$\chi_{11}(t) = -293257020t^2 - 37389240\beta t^2 + 1396500 + 12020500\beta + 23655150t - 116431950\beta t + 480367237t^3 + 1072866719\beta t^3,$

$\chi_{12}(t) = 54925000t^3,$

$\chi_{21}(t) = -5618255790\beta t^2 - 1542285990t^2 + 2931800\beta - 38638880 + 693472350\beta t + 588212970t + 7167937919\beta t^3 - 1401717583t^3,$

$\chi_{22}(t) = -260t \left(-69001t + 6701585\beta t - 201199 - 839635\beta + 6739937t^2 \right)$

which requires a field extension of degree 2. However, if in Step 3 we consider $b_1 = -3, b_2 = 1$, then $\beta = 1$, and the algorithm leads to the parametrization

$$\mathcal{P}(t) = \left(\frac{21\,t + 343\,t^3 + 1 + 1470\,t^2}{-9261t^3}, \frac{21\,t + 343\,t^3 + 1 + 1470\,t^2}{21t\,(931\,t^2 + 14\,t + 1)} \right)$$

which is over the ground field. In the next chapter, we will see how to parametrize over the smallest possible field extension.

Exercises

4.1. Let $R(t) \in K(t)$ be nonconstant. Prove that the following statements are equivalent:

(i) $R(t)$ is invertible.
(ii) $R(t)$ is linear.
(iii) $R(t) = \frac{at+b}{ct+d}$, where $a, b, c, d \in K$ and $ad - bc \neq 0$.

4.2. Consider a rational curve $\mathcal{C}$ and a parametrization $\mathcal{P}$ of $\mathcal{C}$. Is it true that if the degree of $\mathcal{P}$ is prime then $\mathcal{P}$ is proper? If not, what are the exceptions?.

4.3. Compute the tracing index of the parametrization

$$\mathcal{P}(t) = \left(\frac{t^4 + 3\,t^2 + 3}{t^4 + 3\,t^2 + 1}, \frac{t^4 + 2\,t^2 + 3}{t^2 + 2} \right).$$

4.4. May it happen that a proper parametrization is not injective for finitely many parameter values?. If so, give an example.

4.5. Let $R(t) = \frac{p(t)}{q(t)} \in K(t)$ be a nonconstant rational function in reduced form, let $U = \{\alpha \in K \,|\, q(\alpha) \neq 0\}$, and let $R : K \longrightarrow K$ be the rational mapping induced by $R(t)$. Prove that $\operatorname{card}(K \setminus R(U)) \leq 1$.

4.6. Apply Exercise 4.5 to show that the number of exceptions in Lemma 4.32 is bounded by $2 \deg(R(t)) + 1$.

4.7. Carry out the computations in Example 4.38 without using the implicit equation of the curve $\mathcal{C}$.

4.8. Let $\mathcal{C}$ be the plane curve defined by the irreducible polynomial

$$f = -2+5y-2yx+5y^2x-4y^2+9yx^2+y^3-2x^2-12y^2x^2+4y^3x^2-2y^3x \in \mathbb{C}[x,y],$$

and consider the rational parametrization

$$\mathcal{P}(t) = \left(\frac{t+1}{t^3+1}, \frac{t^2+1}{t^2+t+1} \right),$$

of $\mathcal{C}$. Determine whether $\mathcal{P}$ is proper, and in the affirmative case compute its inverse.

4.9. Compute the defining polynomial of the curve defined by the rational parametrization

$$\mathcal{P}(t) = \left(\frac{t^5+1}{t^2+3}, \frac{t^3+t+1}{t^2+1} \right)$$

and the inverse of $\mathcal{P}(t)$.

4.10. Prove that the curve $\mathcal{C}$ defining by the polynomial

$$f(x,y) = y^4 + x - \frac{75}{8}x^2y^2 + \frac{125}{8}x^3y - \frac{1875}{256}x^4$$

is parametrizable by lines. Compute a proper parametrization of $\mathcal{C}$ and its inverse.

4.11. Let $\mathcal{C}$ be the affine quintic curve defined by the polynomial

$$-\frac{75}{8}x^2y^2 + \frac{125}{8}x^3y - \frac{1875}{256}x^4 + x + y^4 + \frac{625}{16}x^3y^2 - \frac{9375}{256}x^4y$$

$$-\frac{125}{8}x^2y^3 + \frac{3125}{256}x^5 + y^5.$$

Apply algorithm PARAMETRIZATION-BY-LINES to parametrize $\mathcal{C}$.

4.12. Prove that any line can by parametrized by lines.

4.13. Give an example of a nonrational curve for which there exists a pencil of lines with the property required in Definition 4.48.

4.14. Prove that for a curve with no rational component, there does not exist a pencil of lines with the property required in Definition 4.48.

4.15. Let C be an affine curve such that its associated projective curve C^* is parametrizable by the pencil of lines $\mathcal{H}(t)$ of equation $L_1(x, y, z) - tL_2(x, y, z)$. Then, the affine parametrization of C, generated by $\mathcal{H}(t)$, is proper and $\frac{L_1(x,y,1)}{L_2(x,y,1)}$ is its inverse.

4.16. Extend the notion of proper rational parametrization to hypersurfaces over algebraically closed fields of characteristic zero.

4.17. Construct an algorithm that, given the defining polynomial of a plane rational curve and the inverse φ of a proper rational parametrization, computes the parametrization φ^{-1}. Apply the algorithm to the inverse mapping computed in Example 4.38.

4.18. Prove that irreducible nonrational curves of degree d may have adjoints of degree $d - 3$.

4.19. Let C be the affine curve defined by $f(x, y) = (x^2 + 4y + y^2)^2 - 16(x^2 + y^2) = 0$. Compute a rational parametrization of C.

4.20. Let C be the affine curve defined by $f(x, y) = x^4 + 5xy^3 + y^4 - 20y^3 + 23y^2 - 9x^2y - 6x^3y + 16xy^2 - 11xy$. Compute a rational parametrization of C.

4.21. Describe an algorithm for parametrizing curves based on Theorem 4.68.

4.22. Describe an algorithm for parametrizing curves based on Theorem 4.70.

5

Algebraically Optimal Parametrization

Summary. In Chap. 4 we have analyzed the parametrization problem for rational curves, and we have presented algorithms for this purpose. Furthermore, we have proved that these algorithms determine proper parametrizations. Therefore, we can ensure that the parametrizations generated by these algorithms are optimal w.r.t. the degree of the components (see Theorem 4.21 and Corollary 4.22). In this chapter we analyze a different optimality criterion for parametrizations, namely the degree of the field extension necessary for representing coefficients of the parametrization. For instance, the parametrization $(\sqrt{2}t, 2t^2)$ of the parabola is optimal w.r.t. the degree (i.e., is proper) but it is expressed over $\mathbb{Q}(\sqrt{2})$, while the alternative parametrization (t, t^2) is expressed over $\mathbb{Q}$ and is also optimal w.r.t. the degree. Thus, we are interested in computing proper parametrizations that require the smallest possible field extension of the ground field. After introducing the notion of the field of parametrization in Section 5.1 and describing the Legendre method for finding rational points on conics in Section 5.2, we present in Section 5.3 an algebraically optimal parametrization of algebraic curves.

Different tools may be applied for attacking this problem, such as anticanonical divisors or adjoint curves. The approach based on anticanonical divisors consists of computing a basis of the vector space $\mathcal{L}(D)$, where D is the anticanonical divisor, that defines a bijective morphism from the original curve to a conic. An optimal parametrization of this conic can then be transformed, via the bijective morphism, to an optimal parametrization of the curve. The approach based on adjoints consists of using adjoint curves to birationally transform the original curve into a conic or a line. Then, as in the previous method, such an optimal parametrization of the conic or line is transformed to an optimal parametrization of the curve via the birational morphism. In this chapter we describe the adjoint approach which is based on [SeW97], and we refer to [VaH97] for the method using anticanonical divisors.

Throughout this chapter we assume that $\mathbb{K}$ is a computable field of characteristic zero, K is its algebraic closure, and $\mathcal{C}$ is a rational projective curve over K of degree d, whose defining polynomial is $F(x, y, z)$ and whose ground field is $\mathbb{K}$. In addition, we will always assume that $\mathcal{C}$ is not a line. Observe that for lines the problem is trivial.

5.1 Fields of Parametrization

We want to compute algebraically optimal parametrizations, that is parametrizations defined over the "smallest" possible field extension of the ground field. Before we can present a solution to the problem of the optimal field of parametrization, we first need to clarify what we mean by the term "smallest" extension field.

Definition 5.1. *We say that a subfield* $\mathbb{L} \subset K$ *is a* field of parametrization *of* $\mathcal{C}$ *(or a* parametrizing field *of* $\mathcal{C}$*) if* $\mathcal{C}$ *can be parametrized over* $\mathbb{L}$*; that is, if there exists a parametrization* $\mathcal{P}(t)$ *of* $\mathcal{C}$ *with coefficients in* $\mathbb{L}$*. In this case we say that* $\mathbb{L}$ *is the* field of definition *of* $\mathcal{P}(t)$*.*

The following result limits the possibilities for being a field of parametrization.

Theorem 5.2. (1) *If* $\mathbb{L}$ *is a field of parametrization of* $\mathcal{C}$ *then* $\mathbb{L}$ *is a field extension of the ground field.*
(2) *Every rational plane curve over* K *has fields of parametrization that are finite extensions of its ground field.*

Proof. (1) If $\mathbb{L}$ is a field of parametrization of $\mathcal{C}$, then there exists a parametrization of $\mathcal{C}$ with coefficients in $\mathbb{L}$. Now, applying elimination techniques, one gets that the implicit equation of $\mathcal{C}$ is in $\mathbb{L}$, and then $\mathbb{K} \subset \mathbb{L}$.
(2) This follows from Sect. 4.8.   □

This theorem motivates Definition 5.3.

Definition 5.3. *A point of* $\mathcal{C}$ *with coordinates in a field extension* $\mathbb{L}$ *of* $\mathbb{K}$ *is called an* $\mathbb{L}$-rational point *of* $\mathcal{C}$*. Moreover, we say that a point of* $\mathcal{C}$ *is* rational *if it is* $\mathbb{K}$-rational.

With this new terminology we can rewrite Theorem 4.74.

Theorem 5.4. *An algebraic extension field* $\mathbb{L}$ *of* $\mathbb{K}$ *is a field of parametrization of* $\mathcal{C}$ *if and only if* $\mathcal{C}$ *has at least one simple* $\mathbb{L}$-rational point.

Now we are ready to introduce the precise notion of algebraic optimality.

Definition 5.5. *A field of parametrization* $\mathbb{L}$ *of* $\mathcal{C}$ *is optimal if it has minimal degree among all the finite algebraic fields of parametrization of* $\mathcal{C}$, *i.e., if*

$$[\mathbb{L} : \mathbb{K}] = \min \left\{ [\mathbb{F} : \mathbb{K}] \, \middle| \, \begin{array}{l} \mathbb{F} \text{ is a field of parametrization of } \mathcal{C}, \\ \text{and } \mathbb{F} \text{ is a finite algebraic extension of } \mathbb{K}. \end{array} \right\}$$

Also, a rational parametrization of $\mathcal{C}$ *is* (algebraically) *optimal if its field of definition is optimal.*

In general, optimal fields of parametrization of a rational curve are not unique. For instance, let us consider the ellipse over $\mathbb{C}$ of equation $2x^2 + 3y^2 = 1$. Its ground field is $\mathbb{Q}$. In Exercise 5.9, we will see that this curve does not have any point with coordinates in $\mathbb{Q}$. Therefore, by Theorem 5.4, $\mathbb{Q}$ is not a field of parametrization. Now, applying Algorithm CONIC-PARAMETRIZATION (see Sect. 4.6) first with the simple point $(\frac{\sqrt{2}}{2}, 0)$, and second with $(0, \frac{\sqrt{3}}{3})$ one gets the following parametrizations of the ellipse

$$\left(\frac{1}{2} \frac{\sqrt{2}\,(3\,t^2 - 2)}{2 + 3\,t^2}, -2 \frac{t\sqrt{2}}{2 + 3\,t^2} \right), \quad \left(-2 \frac{t\sqrt{3}}{2\,t^2 + 3}, \frac{1}{3} \frac{\sqrt{3}\,(2\,t^2 - 3)}{2\,t^2 + 3} \right).$$

Therefore, $\mathbb{Q}(\sqrt{2})$ and $\mathbb{Q}(\sqrt{3})$ are optimal fields of parametrization for the ellipse.

Now we can present a proper statement of the problem as follows: given the rational curve $\mathcal{C}$, whose ground field is $\mathbb{K}$, compute an (algebraically) optimal parametrization of $\mathcal{C}$. Note that, taking into account Theorem 5.4, in general this means finding optimal simple points on $\mathcal{C}$; that is simple points with coordinates in an optimal field of $\mathcal{C}$.

Let us first consider the case of irreducible conics. Using Algorithm CONIC-PARAMETRIZATION and taking into account that by intersecting the conic with lines we generate points on the conic with coordinates in a field extension of $\mathbb{K}$ of degree at most 2, we get Theorem 5.6.

Theorem 5.6. (1) *The optimal fields of parametrization of an irreducible conic have degree at most 2.*

(2) *The optimal field of parametrization of an irreducible conic is the ground field if and only if the conic has at least one rational point.*

We have seen that optimal fields for the conic $2x^2 + 3y^2 = 1$ are quadratic over $\mathbb{Q}$, and therefore the bound in Theorem 5.6 is sharp. In Sect. 5.2 we will see how the existence of rational points on a conic can be checked for certain ground fields, and how such rational points can be computed if they exist. This will lead to an algorithm for computing optimal parametrizations of conics.

Now, we consider the case where $\deg(\mathcal{C}) \geq 3$ and $\mathcal{C}$ can be parametrized by lines. This case includes for instance the rational cubics. So $\mathcal{C}$ has exactly one singularity and its coefficients are in $\mathbb{K}$. Therefore, the output of Algorithm

PARAMETRIZATION-BY-LINES (see Sect. 4.6) has coefficients in $\mathbb{K}$, and hence it is algebraically optimal. Furthermore, Algorithm PARAMETRIZATION-BY-LINES always provides optimal parametrizations. Thus, we have Theorem 5.7.

Theorem 5.7. *If C is parametrizable by lines and $d \neq 2$, then its optimal field of parametrization is the ground field.*

To complete this analysis we consider now the general case. In 1890, Hilbert and Hurwitz introduced a method, based on adjoints, for dealing with this problem (see [HiH90]). In fact, from the proof of their result one may derive a first approach to compute optimal parametrizations.

Theorem 5.8 (Hilbert–Hurwitz). *Let C be a rational plane curve of degree d. Then, for almost all adjoint curves $\Phi_1, \Phi_2, \Phi_3 \in \mathcal{A}_{d-2}(C)$, the mapping*

$$\mathcal{T} = \{y_1 : y_2 : y_3 = \Phi_1 : \Phi_2 : \Phi_3\}$$

transforms C birationally to an irreducible curve of degree $d - 2$.

The main consequence of the theorem by Hilbert–Hurwitz is Corollary 5.9.

Corollary 5.9. *If $\mathbb{K}$ is the ground field of C, and $\deg(C) = d$, the following holds:*

(1) If d is odd then $\mathbb{K}$ is the optimal field of parametrization of C.
(2) If d is even then the optimal fields of parametrization of C are field extensions of $\mathbb{K}$ of degree at most 2. Furthermore, the optimal field is $\mathbb{K}$ if and only if C has at least one rational simple point.

Proof. First we observe that the ground field of all curves in $\mathcal{A}_{d-2}(C)$ is $\mathbb{K}$ (see Theorem 4.66). Let $\mathcal{T}$ be a birational transformation as introduced in Theorem 5.8. Then $\mathcal{T}(C)$ is rational, its degree is $d - 2$ and because rational maps do not extend the ground field one has that the ground field of $\mathcal{T}(C)$ is again $\mathbb{K}$. Now we repeat the process until we arrive at a curve of degree 1 or 2, depending on whether the degree of the original curve is odd or even, respectively. Therefore, we have a birational map defined over $\mathbb{K}$ from C onto a rational curve $\mathcal{D}$ which is either a line or a conic depending on whether d is odd or even. Now the results follow from Theorems 5.6 and 5.7. $\qquad\square$

From the argument used in the proof of the Corollary 5.9 we can derive an algorithm for computing optimal parametrizations if a method for optimally parametrizing conics is provided. Although the problem for conics will be discussed in Sect. 5.2, for the sake of completeness we outline here the algorithm derived from the theorem of Hilbert–Hurwitz, referring to Sect. 5.2 when necessary.

Algorithm HILBERT–HURWITZ

Given a rational plane curve $\mathcal{C}$, the algorithm computes an (algebraically) optimal parametrization of the curve $\mathcal{C}$.

1. If $\mathcal{C}$ is parametrizable by lines and $d \neq 2$ apply Algorithm PARAMETRIZATION-BY-LINES.
2. Consider $\mathcal{D} := \mathcal{C}$ and $\mathcal{G} := \{x : y : z = x : y : z\}$.
3. While $\deg(\mathcal{D}) \geq 3$ do
   3.1. Compute $\mathcal{A}_{d-2}(\mathcal{D})$.
   3.2. Take three adjoints $\Phi_1, \Phi_2, \Phi_3 \in \mathcal{A}_{d-2}(\mathcal{D})$ and consider the mapping
   $$\mathcal{T} = \{y_1 : y_2 : y_3 = \Phi_1 : \Phi_2 : \Phi_3\}.$$
   If $\mathcal{T}$ is not birational try with three new adjoints.
   3.3. $\mathcal{G} := \mathcal{T} \circ \mathcal{G}$ and $\mathcal{D} := \mathcal{T}(\mathcal{D})$.
4. Determine an optimal parametrization $\mathcal{P}(t)$ of $\mathcal{D}$ (this is obvious for a line, and in the case of a conic we use Sect. 5.2).
5. Return $\mathcal{G}^{-1}(\mathcal{P}(t))$.

In Step 5 of Algorithm HILBERT–HURWITZ, instead of inverting the parametrization $\mathcal{P}(t)$, we might as well compute $d - 3$ simple points in $\mathcal{D}$ using $\mathcal{P}(t)$, and giving values to t in $\mathbb{K}$. Then, we might transform them onto points in $\mathcal{C}$ using $\mathcal{G}^{-1}$. If any of these transformed points in not simple on $\mathcal{C}$ replace it by a new one. With these $d - 3$ simple points on $\mathcal{C}$, we can parametrize $\mathcal{C}$ as described in Sect. 4.7. In Sect. 5.3 we will return to this problem and we will present a computationally more efficient solution. We finish this section with a simple example where we illustrate the method described above.

Example 5.10. We consider the projective curve $\mathcal{C}$ defined over $\mathbb{C}$ by the poynomial

$$F(x, y, z) = \frac{3}{2} y^3 z^2 + x y^2 z^2 - 2 x y^3 z - \frac{5}{2} x^2 y z^2 + x^3 z^2 + x^3 y^2$$

The ground field of $\mathcal{C}$ is $\mathbb{Q}$. $\mathcal{C}$ has a triple point at the $(0 : 0 : 1)$ and three double points at $(0 : 1 : 0)$, $(1 : 0 : 0)$, and $(1 : 1 : 1)$. So the genus of $\mathcal{C}$ is 0, i.e., $\mathcal{C}$ is rational. Furthermore, since the degree of the curve is odd, by Corollary 5.9, we know that $\mathbb{Q}$ is in fact an optimal field of parametrization of $\mathcal{C}$. We now proceed to compute an optimal parametrization applying the previous algorithm. Note that only one adjoint reduction suffices. We compute $\mathcal{A}_3(\mathcal{C})$. Its defining polynomial is

$$H(x, y, z, \lambda_1, \lambda_2, \lambda_3, \lambda_4) = -y^2 z \lambda_1 - \lambda_2 y^2 z - y^2 z \lambda_3 - y^2 z \lambda_4 + \lambda_1 x y z$$
$$+ \lambda_2 x y^2 + \lambda_3 x^2 z + \lambda_4 x^2 y \,.$$

Now we choose three adjoints in $\mathcal{A}_3(\mathcal{C})$:

$$\Phi_1(x,y,z) = H(x,y,z,0,0,0,1) = -y^2z + x^2y$$
$$\Phi_2(x,y,z) = H(x,y,z,0,1,0,0) = -y^2z + xy^2$$
$$\Phi_3(x,y,z) = H(x,y,z,0,0,1,0) = -y^2z + x^2z,$$

and we consider the rational map $\mathcal{T} = \{y_1 : y_2 : y_3 = \Phi_1 : \Phi_2 : \Phi_3\}$. By the method of Gröbner bases one deduces that $\mathcal{T}$ is birational (see [Sch98b]), that the implicit equation of $\mathcal{D} = \mathcal{T}(\mathcal{C})$ is

$$G(x,y,z) = -10\,x\,z\,y + 2\,y^2z - 11\,z^2y + 2\,x\,y^2 + 6\,x^3 + 4\,x^2y + 6\,z^2x - 4\,z\,x^2 - 4\,z^3\,,$$

and that $\mathcal{T}^{-1}$ is given by

$$(x : y : z) = (T_1(x,y,z) : T_2(x,y,z) : T_3(x,y,z))\,,$$

where

$$
\begin{aligned}
T_1 &= (-x^2 + y^2 + 5\,z\,x - 5\,z\,y)z\,(4\,y - z),\\
T_2 &= -18\,y^2z^2 + 2\,x^2z\,y + 2\,z^3y + 14\,y^3z + 24\,x\,z^2y - 2\,z\,x^3 - 2\,z^3x\\
&\quad - 14\,x\,y^2z + 8\,x^3y - 2\,y^4 - 6\,z^2x^2 - 6\,x^4,\\
T_3 &= z\,(4\,y - z)(-z\,y + 2\,x^2 + 2\,z^2).
\end{aligned}
$$

Since $\mathcal{D}$ is a rational cubic (it has a double point at $(-3 : 13 : 2)$) we apply Algorithm PARAMETRIZATION-BY-LINES (see Sect. 4.6) and we get the affine parametrization

$$\mathcal{Q}(t) = \left(-\frac{5\,t + 2 + t^2}{t^2 + 3 + 2\,t},\ \frac{39 + 31\,t + 9\,t^2 + t^3}{2(t^2 + 3 + 2\,t)}\right).$$

Finally, applying $\mathcal{T}^{-1}$ to $\mathcal{Q}(t)$, we get the optimal parametrization of $\mathcal{C}$

$$\mathcal{T}^{-1}(\mathcal{Q}(t)) = \left(-\frac{t^3 + 11\,t^2 + 41\,t + 43}{2(t^2 + 6\,t + 13)},\ \frac{t^3 + 11\,t^2 + 41\,t + 43}{4(2\,t + 5)}\right).$$

In Exercise 5.1 we propose the computation of $\mathcal{D}$ using interpolation techniques.

5.2 Rational Points on Conics

In Sect. 5.1 we have analyzed optimal fields of parametrization of a curve $\mathcal{C}$ over K with ground field $\mathbb{K}$. Corollary 5.9 tells us that if the degree of $\mathcal{C}$ is odd then the optimal field of parametrization is $\mathbb{K}$, otherwise it is a field extension of $\mathbb{K}$ of degree at most 2. Furthermore, the algorithm HILBERT–HURWITZ shows how the problem of checking the precise degree of this field extension can be reduced to the case of conics.

In this section, we focus on the case of conics. For certain fields we can decide the existence of rational points and, in the positive case, actually compute such points. Therefore, over such fields, we can derive an optimal parametrization algorithm. We present here the classical approach, based on the Legendre theory, for the case $\mathbb{K} = \mathbb{Q}$. For a description of the Legendre theory we refer to [CrR03], [IrR82], [Krä81], [LiN94], or [Ros88]. The relation of rational points and optimal parametrization has been investigated in [HiW98].

Throughout this section, $\mathcal{C}$ is a projective conic with ground field $\mathbb{Q}$, defined by

$$F(x, y, z) = a_1 x^2 + a_2 xy + a_3 y^2 + a_4 xz + a_5 yz + a_6 z^2,$$

where $a_i \in \mathbb{Q}$.

Our goal is to decide whether there is a rational point on $\mathcal{C}$, and if so, to compute one. By Theorem 5.6 we know that $\mathcal{C}$ has a rational point if and only if $\mathbb{Q}$ is an optimal field of parametrization; hence, if and only if, $\mathcal{C}$ has infinitely many rational points. In the following study, we distinguish between parabolas, ellipses (including circles), and hyperbolas. The case of parabolas is the easy one, and one may always, in fact, give an explicit formula for a rational point on it. However, the case of ellipses and hyperbolas is not so straight-forward. We need to manipulate the equation to reach a Legendre equation. The Legendre equation lets us decide the existence of rational points, and actually allows to compute such a point if it exists. In the sequel we denote by $\mathbb{Z}^*$ the set of nonzero integers, and by $\mathbb{Z}^+$ the set of positive integers.

5.2.1 The Parabolic Case

We start by observing that $\mathcal{C}$ is a parabola if and only if the coefficients of $F(x, y, z)$ satisfy one of the following relations (see Exercise 2.3):

$$a_2^2 = 4a_1 a_3 \quad \text{or} \quad a_4^2 = 4a_1 a_6 \quad \text{or} \quad a_5^2 = 4a_3 a_6.$$

Let us assume w.l.o.g that $a_2^2 = 4a_1 a_3$, i.e., we consider a parabola with respect to x and y, where z is the homogenizing variable. Furthermore, let us assume that $a_3 \neq 0$ (otherwise, we may reason similarly by interchanging x and y). Then we have the relation

$$4a_3 F(x, y, z) = (a_2 x + 2a_3 y + a_5 z)^2 + (4a_3 a_4 - 2a_2 a_5)xz + (4a_3 a_6 - a_5^2)z^2.$$

Since $\mathcal{C}$ is irreducible, this implies $4a_3 a_4 - 2a_2 a_5 \neq 0$. Thus,

$$(-2a_3(4a_3 a_6 - a_5^2), -4a_5 a_3 a_4 + a_2 a_5^2 + 4a_2 a_3 a_6, 4a_3(a_3 a_4 - a_2 a_5))$$

is a rational point on $\mathcal{C}$.

Example 5.11. Consider the affine parabola defined by

$$f(x, y) = x^2 + 2xy + y^2 + x + 2y - 2.$$

Since $a_3 \neq 0$, we can use the formula, and we get the rational point $(-3, 2)$ on the parabola.

5.2.2 The Hyperbolic and the Elliptic Case

The hyperbolic/elliptic case is characterized by the conditions

$$a_2^2 \neq 4a_1a_3 \quad \text{and} \quad a_4^2 \neq 4a_1a_6 \quad \text{and} \quad a_5^2 \neq 4a_3a_6$$

on the coefficients of $F(x, y, z)$. By well-known techniques in linear algebra (see [Krä81]) we can find a linear change of coordinates over $\mathbb{Q}$ transforming the conic C onto a conic of the form

$$x^2 + ky^2 = \ell z^2 \tag{5.1}$$

where $k, \ell \in \mathbb{Q}$, and where either $k < 0$ or $\ell > 0$. This implies the existence of real points. Now, expressing k and ℓ as $k = \dfrac{k_1}{k_2}$, $\ell = \dfrac{\ell_1}{\ell_2}$ with $k_i, \ell_j \in \mathbb{Z}^*$, and cleaning up denominators in (5.1) we get the following equation over $\mathbb{Z}$

$$a'x^2 + b'y^2 + c'z^2 = 0 \tag{5.2}$$

where $a' = \text{lcm}(k_2, \ell_2)$, $b' = \dfrac{k_1\,\ell_2}{\gcd(k_2, \ell_2)}$, and $c' = -\dfrac{\ell_1\,k_2}{\gcd(k_2, \ell_2)}$.

Clearly, a', b', c' are nonzero and do not have the same sign. Now, we want to reduce (5.2) to an equation of similar form whose coefficients are squarefree and pairwise relatively prime.

First, we express a', b', c' as

$$a' = a_1'r_1^2, \quad b' = b_1'r_2^2, \quad c' = c_1'r_3^2,$$

where a_1', b_1', c_1' are squarefree (see Exercise 5.5). We get the equation

$$a_1'x^2 + b_1'y^2 + c_1'z^2 = 0 \tag{5.3}$$

Note that (5.3) has an integral solution if and only if (5.2) has one.
Next, we divide (5.3) by $\gcd(a_1', b_1', c_1')$, getting

$$a''x^2 + b''y^2 + c''z^2 = 0 \tag{5.4}$$

Now, we make the coefficients pairwise relatively prime. For this purpose, let $g_1 = \gcd(a'', b'')$, $a''' = a''/g_1$, $b''' = b''/g_1$, and let $(\bar{x}, \bar{y}, \bar{z})$ be an integral solution of (5.4). Then $g_1 \mid c''\bar{z}^2$, and hence, since $\gcd(a'', b'', c'') = 1$, we have $g_1 \mid \bar{z}^2$. Furthermore, since g_1 is squarefree (since a'', b'' are), we have $g_1 \mid \bar{z}$. So, letting $\bar{z} = g_1z'$ and dividing (5.4) by g_1, we arrive at the equation

$$a'''\bar{x}^2 + b'''\bar{y}^2 + \underbrace{c''g_1}_{c'''}(z')^2 = 0.$$

At this point $\gcd(a''', b''') = 1$ and c''' is squarefree since g_1 and c'' are relatively prime. Repeating this process with $g_2 = \gcd(a''', c''')$ and $g_3 = \gcd(b'''', c'''')$ we finally arrive at an equation

$$a(x')^2 + b(y')^2 + c(z')^2 = 0 ,$$

where a, b, c satisfy the requirements in Definition 5.12.

Definition 5.12. *Let* $a, b, c \in \mathbb{Z}$ *be such that* $abc \neq 0$, *and they satisfy the following conditions:*

$$(i) \quad a > 0, \ b < 0, \ and \ c < 0$$

$$(ii) \quad a, \ b, \ and \ c \ are \ squarefree \tag{5.5}$$

$$(iii) \quad \gcd(a, b) = \gcd(a, c) = \gcd(b, c) = 1.$$

Then, the equation

$$ax^2 + by^2 + cz^2 = 0 \tag{5.6}$$

is called a Legendre equation.

5.2.3 Solving the Legendre Equation

The problem of finding a rational point on an ellipse or hyperbola reduces to the problem of finding a nontrivial integral solution of the so called Legendre Equation. Let us investigate necessary and sufficient conditions for the Legendre equation to have nontrivial integral solutions. By a nontrivial integral solution we mean a solution $(\overline{x}, \overline{y}, \overline{z}) \in \mathbb{Z}^3$ with $(\overline{x}, \overline{y}, \overline{z}) \neq (0, 0, 0)$ and $\gcd(\overline{x}, \overline{y}, \overline{z}) = 1$. Such conditions are given by Legendre's Theorem (Theorem 5.18). Based on the description in [IrR82] we develop here a constructive proof from which we can extract an algorithm to compute integral solutions. For the formulation of Legendre's Theorem we need to introduce the notion of quadratic residues.

Definition 5.13. *Let* $m, n \in \mathbb{Z}^\star$. *Then we say that* m *is a* quadratic residue *modulo* n, *and we denote this by* $m \mathcal{R} n$, *if there exists* $x \in \mathbb{Z}$ *such that* $x^2 \equiv_n m$.

The problem of deciding whether $m \mathcal{R} n$ can be solved directly by checking all the elements in $\mathbb{Z}_n$. Alternatively, one may approach the problem by using Legendre's symbol (see [Coh00] or [IrR82] for the notion of Legendre's symbol). We outline here a method based on the notion of quadratic reciprocity to solve this question. If $n = 1$ or $n = 2$ the problem is trivial, and we may always assume w.l.o.g that $n > 0$ (see Exercise 5.7). In this situation, if n is an odd prime number one can prove from the Law of quadratic reprocity (see Sect. 18.5 in [vGG99]) that $m \mathcal{R} n$ if and only if

$$m^{\frac{n-1}{2}} \equiv_n 1.$$

So we will have to deal with the case where $n > 2$ and n is not a prime number.

Lemma 5.14. *Let* $n, m \in \mathbb{Z}^\star$ *such that* $\gcd(m, n) = 1$. *If* $a \in \mathbb{Z}^\star$ *satisfies* $a \mathcal{R} n$ *and* $a \mathcal{R} m$, *then* a *also satisfies* $a \mathcal{R} nm$.

Proof. Since $a \mathcal{R} n$ and $a \mathcal{R} m$, there exist x_1, $x_2 \in \mathbb{Z}$ such that

$$x_1^2 \equiv_n a, \ x_2^2 \equiv_m a.$$

In addition, $\gcd(n, m) = 1$ implies that there exist ℓ_1', $\ell_2' \in \mathbb{Z}$ such that $\ell_1' n - \ell_2' m = 1$. Thus, there exist ℓ_1, $\ell_2 \in \mathbb{Z}$, namely $\ell_i = \ell_i'(x_2 - x_1)$, $i = 1, 2$, such that

$$x_1 + \ell_1 n = x_2 + \ell_2 m .$$

In this situation we prove that for $x_3 = x_1 + \ell_1 n$ we get $x_3^2 \equiv_{nm} a$, from where we deduce that $a \mathcal{R} nm$. Indeed,

$$x_3^2 = (x_1 + \ell_1 n)^2 \equiv_n x_1^2 \equiv_n a, \quad x_3^2 = (x_2 + \ell_2 m)^2 \equiv_m x_2^2 \equiv_m a .$$

Thus, there exist $k_1, k_2 \in \mathbb{Z}$ such that

$$x_3^2 = a + k_1 n = a + k_2 m .$$

This implies that $k_1 n = k_2 m$. But $\gcd(n, m) = 1$, and therefore n divides k_2. Hence, there exists $k_3 \in \mathbb{Z}$ such that $k_2 = k_3 n$, and then

$$x_3^2 = a + k_3 nm \equiv_{nm} a . \qquad \square$$

Now let us return to consider the case where $n > 2$ and n is not a prime number. Let

$$n = \prod_{i=1}^{r} n_i^{e_i}$$

be the irreducible factorization of n. Then, $m \mathcal{R} n$ implies that $m \mathcal{R} n_i$ for $i = 1, \dots, r$ (see Exercise 5.7). Now, we distinguish two cases depending on whether n is squarefree or not.

Suppose n is squarefree. Then, $m \mathcal{R} n$ if and only if $m \mathcal{R} n_i$ for $i = 1, \dots, r$ (note that the left–right implication always holds and for the right–left implication see Lemma 5.14). Thus, in this case, one may check whether $m \mathcal{R} n$ by checking whether $m \mathcal{R} n_i$, $\forall i = 1, \dots, r$. We have seen above how this can be done for any prime number.

Now assume that n is not squarefree. Let $m \mathcal{R} n$, and let $x \in \mathbb{Z}$ be such that $x^2 \equiv_n m$. Then, we know that $x^2 \equiv_{n_i} m$ for $i = 1, \dots, r$. Thus, one may check the existence of x as follows: if for some $i \in \{1, \dots, r\}$ we have that $m \mathcal{\bar{R}} n_i$, then $m \mathcal{\bar{R}} n$. On the other hand, assume that for every $i \in \{1, \dots, r\}$ we have $m \mathcal{R} n_i$ and $x_i \in \mathbb{Z}$ is such that $x_i^2 \equiv_{n_i} m$ (these x_i are usually not unique). For the possible $x \in \mathbb{Z}$ such that $x^2 \equiv_n m$ we must have that $x \equiv_{n_i} x_i$ for some candidates x_i. Thus, applying the Chinese Remainder Algorithm to these congruences one determines the possible candidates for x (observe that these candidates are $x + k \prod_{i=1}^{r} n_i$, for some k). Finally the problem is solved by checking whether any of these candidates satisfies $x^2 \equiv_n m$.

So now let us return to our original problem of solving the Legendre equation. First, we state some preliminary technical lemmas.

Lemma 5.15. *Let $n \in \mathbb{Z}^+$, and let α, β, and γ be positive nonintegral real numbers such that $\alpha\beta\gamma = n$. Then, for every triple $(a_1, a_2, a_3) \in \mathbb{Z}^3 \setminus \{(0,0,0)\}$, the congruence*

$$a_1 x + a_2 y + a_3 z \equiv_n 0$$

has a solution $(\overline{x}, \overline{y}, \overline{z}) \neq (0,0,0)$ such that

$$|\overline{x}| < \alpha \ , |\overline{y}| < \beta, \ and \ |\overline{z}| < \gamma.$$

Proof. Consider the set

$$S = \{(x, y, z) \in \mathbb{N}^3 \mid x \leq \lfloor \alpha \rfloor \ \text{and} \ y \leq \lfloor \beta \rfloor \ \text{and} \ z \leq \lfloor \gamma \rfloor \} \ .$$

Note that $\text{card}(S) = (1 + \lfloor \alpha \rfloor)(1 + \lfloor \beta \rfloor)(1 + \lfloor \gamma \rfloor) > \alpha\beta\gamma = n$. Now we consider the set $\mathcal{A} = \{a_1 x + a_2 y + a_3 z \mid (x, y, z) \in S\}$. If $\text{card}(\mathcal{A}) < \text{card}(S)$, this means that there exist at least two distinct elements (x_1, y_1, z_1), $(x_2, y_2, z_2) \in S$ such that

$$a_1 x_1 + a_2 y_1 + a_3 z_1 = a_1 x_2 + a_2 y_2 + a_3 z_2 \ .$$

Now, let us assume that $\text{card}(\mathcal{A}) = \text{card}(S)$. Then, since $\text{card}(S) > n$ and there are n residue classes modulo n, one deduces that there exist at least two distinct elements (x_1, y_1, z_1), $(x_2, y_2, z_2) \in S$ such that

$$a_1 x_1 + a_2 y_1 + a_3 z_1 \equiv_n a_1 x_2 + a_2 y_2 + a_3 z_2 \ .$$

In any case, $(\overline{x}, \overline{y}, \overline{z}) = (x_1 - x_2, y_1 - y_2, z_1 - z_2) \neq (0,0,0)$ is a solution of the congruence

$$a_1 x + a_2 y + a_3 z \equiv_n 0 \ .$$

In addition, since α, β, and γ are positive nonintegral real numbers, and $x_i, y_i, z_i \in \mathbb{N}$, we know that $0 \leq x_i < \alpha$, $0 \leq y_i < \beta$, $0 \leq z_i < \gamma$, for $i = 1, 2$. Thus,

$$|\overline{x}| = |x_1 - x_2| \leq \max\{x_1, x_2\} < \alpha \ ,$$

and similarly

$$|\overline{y}| < \beta, \quad \text{and} \quad |\overline{z}| < \gamma \ . \qquad \square$$

Lemma 5.16. *Let $m, n \in \mathbb{N}$ such that $\gcd(m, n) = 1$, and let $ax^2 + by^2 + cz^2$, with $a, b, c \in \mathbb{Z}$, be a form that factors modulo m and modulo n. Then, $ax^2 + by^2 + cz^2$ also factors modulo mn.*

Proof. See Exercise 5.6. $\qquad \square$

Lemma 5.17. *Let $r \in \mathbb{Z}^+$ such that $-1 \mathcal{R} r$. Then, the equation*

$$x^2 + y^2 = r$$

has an integral solution.

Proof. First, since $-1 \mathcal{R} r$, there exists $x_0 \in \mathbb{Z}$ and $k \in \mathbb{Z}$ such that

$$x_0^2 + 1 = kr .$$

Moreover, $k \in \mathbb{N}^\star$ because $r > 0$. Let us assume that $k = 1$. Then, $(x_0, 1)$ is a nontrivial integral solution of the equation $x^2 + y^2 = r$; hence the statement holds. Now, let $k > 1$. Then, taking $y_0 = 1$, one has that (x_0, y_0) is a nontrivial integral solution of the equation $x^2 + y^2 = kr$. Then, let x_1, y_1 be integers of least absolute value such that $x_1 \equiv_k x_0$, and $y_1 \equiv_k y_0$. Note that there exist $c, d \in \mathbb{Z}$ such that $x_1 = x_0 + ck$, and $y_1 = y_0 + dk$. Thus,

$$x_1^2 + y_1^2 = (x_0 + ck)^2 + (y_0 + dk)^2 \equiv_k x_0^2 + y_0^2 \equiv_k 0 .$$

Therefore, there exists k' such that $x_1^2 + y_1^2 = k'k$. Moreover, because of $|x_1|, |y_1| \le k/2$, we get

$$x_1^2 + y_1^2 \le \left(\frac{k}{2}\right)^2 + \left(\frac{k}{2}\right)^2 = \frac{1}{2}k^2 ,$$

and hence $0 < k' \le \frac{k}{2}$. Additionally

$$k'k^2 r = (k'k)(kr) = (x_1^2 + y_1^2)(x_0^2 + y_0^2) = (x_0 x_1 + y_0 y_1)^2 + (x_0 y_1 - x_1 y_0)^2 ,$$

and therefore

$$k'r = \left(\frac{x_0 x_1 + y_0 y_1}{k}\right)^2 + \left(\frac{x_0 y_1 - x_1 y_0}{k}\right)^2 .$$

Now, let $x_2 = (x_0 x_1 + y_0 y_1)/k$, and $y_2 = (x_0 y_1 - x_1 y_0)/k$. We observe that

$$x_0 x_1 + y_0 y_1 = x_0(x_0 + ck) + y_0(y_0 + dk) \equiv_k x_0^2 + y_0^2 \equiv_k 0 ,$$

and

$$x_0 y_1 - x_1 y_0 = x_0(y_0 + dk) - y_0(x_0 + ck) \equiv_k 0 .$$

Thus, $x_2, y_2 \in \mathbb{Z}$, and (x_2, y_2) is a nontrivial integral solution of $x^2 + y^2 = k'r$. In this situation, since $k' \le k$, we either have a solution of $x^2 + y^2 = r$ (i.e., if $k' = 1$) or we may apply the previous reasoning again. Proceeding inductively we finally finish the proof. $\qquad\square$

Remarks. The proof of Lemma 5.17 is constructive. In the following we outline the corresponding algorithmic process. For this purpose, we will denote by "qr" an algorithmic procedure that decides whether $m\mathcal{R}n$, and that in the affirmative case outputs $x \in \mathbb{Z}$ such that $x^2 \equiv_n m$. Then, given $r \in \mathbb{Z}^+$ such that $-1 \mathcal{R} r$, the computation of $\alpha, \beta \in \mathbb{Z}$ such that $r = \alpha^2 + \beta^2$, can be performed as follows.

1. Determine

$$\alpha := \text{qr}(-1, r), \quad \beta := 1, \quad k := \frac{\alpha^2 + \beta^2}{r} .$$

2. While $k > 1$ do

$$\alpha_1 := \alpha \bmod k, \quad \beta_1 := \beta \bmod k, \quad \alpha_2 := \frac{\alpha_1 \alpha + \beta_1 \beta}{k},$$

$$\beta_2 := \frac{\alpha \beta_1 - \alpha_1 \beta}{k}, \quad \alpha := \alpha_2, \quad \beta := \beta_2, \quad k := \frac{\alpha^2 + \beta^2}{r}.$$

3. Return (α, β).

Now we are ready for stating Legendre's Theorem.

Theorem 5.18 (Legendre's Theorem). *The Legendre equation $ax^2 + by^2 + cz^2 = 0$ has a nontrivial integral solution if and only if $(-ab)\mathcal{R}c$, $(-bc)\mathcal{R}a$, and $(-ac)\mathcal{R}b$.*

Proof. Let $(\overline{x}, \overline{y}, \overline{z})$ be a nontrivial integral solution of $ax^2 + by^2 + cz^2 = 0$. Note that we can assume w.l.o.g that $\gcd(\overline{x}, \overline{y}, \overline{z}) = 1$. First we prove that $\gcd(c, \overline{x}) = 1$. Indeed, if any prime p divides $\gcd(c, \overline{x})$, then p divides $b\overline{y}^2$. Because of (5.5), $\gcd(b, c) = 1$, so p does not divide b. Thus, p divides $\overline{y}$. Consequently, p^2 divides $a\overline{x}^2 + b\overline{y}^2$, and hence p^2 divides $c\overline{z}^2$. Because of (5.5) c is squarefree, which implies that p divides $\overline{z}$. Therefore, p divides $\gcd(\overline{x}, \overline{y}, \overline{z})$, which is impossible. So, we have proved that $\gcd(c, \overline{x}) = 1$.
Now, since $\gcd(c, \overline{x}) = 1$ there exist $\lambda, \mu \in \mathbb{Z}$ such that $\lambda c + \mu \overline{x} = 1$. This implies that $\mu \overline{x} \equiv_c 1$. Furthermore, from the equality $a\overline{x}^2 + b\overline{y}^2 + c\overline{z}^2 = 0$ we get that $a\overline{x}^2 \equiv_c -b\overline{y}^2$. Thus,

$$b^2 \mu^2 \overline{y}^2 \equiv_c -ab(\overline{x}\mu)^2 \equiv_c -ab ,$$

and consequently we have $(-ab)\mathcal{R}c$. The remaining conditions can be derived analogously.

In order to prove the reverse implication we first deal with three special cases.

1. Case $b = c = -1$. In this case, the hypothesis $(-bc)\mathcal{R}a$ implies that $-1\mathcal{R}a$. So, by Lemma 5.17 there exist $r, s \in \mathbb{Z}$ such that $r^2 + s^2 = a$. Hence, in this case, $(1, r, s)$ is a nontrivial integral solution of the Legendre equation.
2. Case $a = 1$, $b = -1$. In this case $(1, 1, 0)$ is a nontrivial integral solution of the Legendre equation.
3. Case $a = 1$, $c = -1$. In this case $(1, 0, 1)$ is a nontrivial integral solution of the Legendre equation.

Now, we treat the general case. Since $(-ab)\mathcal{R}c$, there exists $t \in \mathbb{Z}$ such that

$$t^2 \equiv_c -ab .$$

On the other hand, because of (5.5) we have $\gcd(a, c) = 1$. Thus, there exists $a^* \in \mathbb{Z}$ such that $aa^* \equiv_c 1$, and therefore

$$
\begin{aligned}
ax^2 + by^2 + cz^2 &\equiv_c aa^*(ax^2 + by^2) \equiv_c a^*(a^2x^2 + aby^2) \\
&\equiv_c a^*(a^2x^2 - t^2y^2) = a^*(ax - ty)(ax + ty) \\
&\equiv_c (x - a^*ty)(ax + ty) \ .
\end{aligned}
$$

Using the remaining hypotheses (i.e., $(-bc)\,\mathcal{R}\,a$, and $(-ac)\,\mathcal{R}\,b$) and reasoning similarly as above we see that $ax^2 + by^2 + cz^2$ can also be expressed as a product of linear factors modulo b and modulo a. Then, taking into account Lemma 5.16 and (5.5), we deduce that there exist $a_1, \ldots, a_6 \in \mathbb{Z}$ such that

$$
ax^2 + by^2 + cz^2 \equiv_{abc} (a_1x + a_2y + a_3z)(a_4x + a_5y + a_6z) \ .
$$

Now, we consider the congruence

$$
(a_1x + a_2y + a_3z) \equiv_{abc} 0 \ .
$$

Since we are not in any of the special cases, and since a, b, and c satisfy (5.5), we have that $\sqrt{bc}$, $\sqrt{-ac}$, and $\sqrt{-ab}$ are nonintegral real numbers, and their product is abc. Applying Lemma 5.15 to the previous congruence, with $\alpha = \sqrt{bc}$, $\beta = \sqrt{-ac}$, and $\gamma = \sqrt{-ab}$, we deduce that there exists a nontrivial integral solution, say (x_1, y_1, z_1) of $a_1x + a_2y + a_3z \equiv_{abc} 0$, where

$$
|x_1| < \sqrt{bc}, \quad |y_1| < \sqrt{-ac}, \quad \text{and} \quad |z_1| < \sqrt{-ab} \ .
$$

Thus, taking into account that

$$
ax^2 + by^2 + cz^2 \equiv_{abc} (a_1x + a_2y + a_3z)(a_4x + a_5y + a_6z) \ ,
$$

we deduce that

$$
ax_1^2 + by_1^2 + cz_1^2 \equiv_{abc} 0 \ .
$$

Furthermore, since b and c are negative and $|x_1| < \sqrt{bc}$, the above inequalities imply that

$$
ax_1^2 + by_1^2 + cz_1^2 \leq ax_1^2 < abc \ .
$$

Moreover, since $a > 0$, $|y_1| < \sqrt{-Ac}$, $|z_1| < \sqrt{-ab}$ and b, c are negative, we have

$$
ax_1^2 + by_1^2 + cz_1^2 \geq by_1^2 + cz_1^2 > b(-ac) + c(-ab) = -2abc \ .
$$

Thus, $ax_1^2 + by_1^2 + cz_1^2$ is a multiple of abc, and $-2abc < ax_1^2 + by_1^2 + cz_1^2 < abc$. Hence, we are in one of the following cases:

$$
ax_1^2 + by_1^2 + cz_1^2 = 0, \quad \text{or} \quad ax_1^2 + by_1^2 + cz_1^2 = -abc \ .
$$

In the first case the result follows immediately. So let us assume that $ax_1^2 + by_1^2 + cz_1^2 = -abc$. In this situation, we introduce the integers

$$x_2 = x_1 z_1 - b y_1, \quad y_2 = y_1 z_1 + a x_1, \quad z_2 = z_1^2 + ab .$$

For these numbers we get the relation

$$\begin{aligned}
ax_2^2 + by_2^2 + cz_2^2 &= a(x_1 z_1 - b y_1)^2 + b(y_1 z_1 + a x_1)^2 + c(z_1^2 + ab)^2 \\
&= (ax_1^2 + by_1^2 + cz_1^2)z_1^2 - 2abx_1 y_1 z_1 + 2abx_1 y_1 z_1 \\
&\quad + ab(by_1^2 + ax_1^2 + cz_1^2) + abcz_1^2 + a^2 b^2 c \\
&= (-abc)z_1^2 + ab(-abc) + abcz_1^2 + a^2 b^2 c = 0 .
\end{aligned}$$

Thus, (x_2, y_2, z_2) is a solution. Furthermore, it is a nontrivial solution. Indeed, if $z_1^2 + ab = 0$, the coprimality and squarefreeness of a and b imply that $a = 1$ and $b = -1$. But this case has been treated above.

This completes the proof. Nontrivial solutions have been found in all cases. $\square$

Theorem 5.18 characterizes the existence of nontrivial solutions of the Legendre equation by means of quadratic residues. However, from the proof it is not clear how to compute a solution if it exists. In the following, we see how to approach the problem algorithmically. For this purpose, we first introduce the following notion.

Definition 5.19. *Let* $ax^2 + by^2 + cz^2 = 0$ *be a Legendre equation. The equation*

$$-x^2 + (-ba)y^2 + (-ca)z^2 = 0$$

is called the associated equation *to the Legendre equation.*

Remarks. Consider the equation of the form $-x^2 + Ay^2 + Bz^2 = 0$, where A, B are positive squarefree integers. This equation is associated to the Legendre equation $\gcd(A, B)x^2 - \frac{A}{\gcd(A,B)} y^2 - \frac{B}{\gcd(A,B)} z^2 = 0$.

Theorem 5.20. *The Legendre equation has a nontrivial integral solution if and only if its associated equation has a nontrivial integral solution.*

Proof. Let (λ, μ, γ) be a nontrivial integral solution of the Legendre equation $ax^2 + by^2 + cz^2 = 0$, i.e.,

$$a\lambda^2 + b\mu^2 + c\gamma^2 = 0 .$$

Multiplying by $-a$, we get

$$-(a\lambda)^2 + (-ab)\mu^2 + (-ac)\gamma^2 = 0 .$$

Thus, $(-a\lambda, \mu, \gamma)$ is a nontrivial integral solution of the associated equation to the Legendre equation (note that $a > 0$).

Conversely, if (λ, μ, γ) is a nontrivial integral solution of the associated equation to the Legendre equation, then

$$-\lambda^2 + (-Ba)\mu^2 + (-ca)\gamma^2 = 0 .$$

Multiplying by $-a$, we get

$$a\lambda^2 + b(a\mu)^2 + c(a\gamma)^2 = 0 ,$$

so $(\lambda, a\mu, a\gamma)$ is a nontrivial integral solution of the Legendre equation (note that $a > 0$). $\qquad\square$

Remarks. The proof of Theorem 5.20 provides an explicit transformation of solutions of the Legendre equation and solutions of the associated equation. More precisely,

(i) if (λ, μ, γ) is a nontrivial integral solution of the Legendre equation, then $(-a\lambda, \mu, \gamma)$ is a nontrivial integral solution of the associated equation, and

(ii) if (λ, μ, γ) is a nontrivial integral solution of the associated equation, then $(\lambda, a\mu, a\gamma)$ is a nontrivial integral solution of the Legendre equation.

Applying Legendre's Theorem (Theorem 5.18) and Theorem 5.20, we may also characterize the existence of solutions of the associated equation by means of quadratic residues of its coefficients. More precisely, we get Theorem 5.21.

Theorem 5.21. *The associated equation to the Legendre equation has a nontrivial solution if and only if $(-ab)\,\mathcal{R}\,(-ac)$, $(-ac)\,\mathcal{R}\,(-ab)$, and $(-bc)\,\mathcal{R}\,a$.*

Proof. From Theorem 5.20 we know that the equation $-x^2 + (-ba)y^2 + (-ca)z^2 = 0$ has a nontrivial integral solution if and only if the Legendre equation $ax^2 + by^2 + cz^2 = 0$ has one. From Theorem 5.18 we know that $ax^2 + by^2 + cz^2 = 0$ has a nontrivial integral solution if and only if $(-bc)\,\mathcal{R}\,a$, $(-ac)\,\mathcal{R}\,b$, and $(-ab)\,\mathcal{R}\,c$. Observe that $(-ac)\,\mathcal{R}\,(-a)$ and $(-ab)\,\mathcal{R}\,(-a)$ always hold; we see this by taking $x = a$ for both cases in Definition 5.13. Thus, from $(-ac)\,\mathcal{R}\,b$ and $(-ac)\,\mathcal{R}\,(-a)$ and Lemma 5.14 (note that $\gcd(a, b) = 1$ because of (5.5)) we get that $(-ac)\,\mathcal{R}\,(-ab)$. Similarly, from $(-ab)\,\mathcal{R}\,c$, $(-ab)\,\mathcal{R}\,(-a)$ and Lemma 5.14 we get that $(-ab)\,\mathcal{R}\,(-ac)$.
Conversely, we assume that $(-ab)\,\mathcal{R}\,(-ac)$, $(-ac)\,\mathcal{R}\,(-ab)$, and $(-bc)\,\mathcal{R}\,a$. Then from Exercise 5.7 (iv) we deduce that $(-ab)\mathcal{R}c$, $(-ac)\mathcal{R}b$, and $(-bc)\mathcal{R}a$. Theorem 5.18 now implies that $ax^2 + by^2 + cz^2 = 0$ has a nontrivial integral solution. Because of Theorem 5.20 this means that the associated equation $-x^2 + (-ba)y^2 + (-ca)z^2 = 0$ also has a nontrivial integral solution. $\qquad\square$

Remarks. Note that the conditions in Theorems 5.18 and 5.21 are equivalent.

In the previous theorems we have seen how to reduce the study of the Legendre equation to its associated equation. In Theorem 5.22 we prove that if the associated Legendre equation has a nontrivial integral solution, then this solution can be determined algorithmically.

Theorem 5.22. *If the associated equation to the Legendre equation has a nontrivial integral solution, then it can be determined algorithmically.*

Proof. Let us assume that $-x^2 + (-ba)y^2 + (-ca)z^2 = 0$, the associated equation to the Legendre equation, has a nontrivial integral solution. By Theorem 5.21 we deduce that $(-ab)\,\mathcal{R}\,(-ca)$, $(-ca)\,\mathcal{R}\,(-ab)$, and $(-cb)\,\mathcal{R}\,a$. Let us first deal with two special cases.

(1) If $-ca = 1$ (that is $a = 1$ and $c = -1$, see Definition 5.12), then $(1, 0, 1)$ is a nontrivial integral solution, and if $-ba = 1$ (that is $a = 1$ and $b = -1$, see Definition 5.12), then $(1, 1, 0)$ is a nontrivial integral solution.
(2) Now consider the case $-ca = -ba$ (that is $c = b = -1$, see Definition 5.12). $(-cb)\,\mathcal{R}\,a$ means $-1\,\mathcal{R}\,a$, so by the remark to Lemma 5.17 we can determine algorithmically integers r and s, not both zero, such that $a = r^2 + s^2$. Then, $(r^2 + s^2, s, r)$ is a nontrivial solution of $-x^2 + (-ba)y^2 + (-ca)z^2 = 0$.

Now we treat the general case. W.l.o.g. we assume that $-ba < -ca$, i.e., $-b < -c$. Otherwise we only have to interchange the roles of z and y. The strategy will be the following: first, we find a squarefree integer A, with $0 < A < -ca$, and we consider the new equation $Az^2 + (-ba)Y^2 = X^2$, where

$$A\,\mathcal{R}\,(-ba), \quad (-ba)\,\mathcal{R}\,A, \quad \text{and} \quad \frac{-A(-ba)}{\gcd(A, -ba)^2}\,\mathcal{R}\,\gcd(A, -ba) \,.$$

Thus, we reduce the given associated Legendre equation $(-ca)z^2 + (-ba)y^2 = x^2$ to a new equation associated to some Legendre equation (see Remark to Definition 5.19) having a nontrivial solution (see Theorem 5.21). Moreover, we show that a solution of the old equation can be computed from a solution of the new equation. After a finite number of steps, interchanging A and $-ba$ in case A is less than $-ba$ (we are assuming that $-ba < -ca$), we arrive either at the case $A = 1$ or at $A = -ba$, each of which has been treated in (1) or (2). Since $(-ba)\,\mathcal{R}\,(-ca)$, we deduce that there exist $\alpha, k \in \mathbb{Z}$ such that

$$\alpha^2 = -ba + k(-ca) \,.$$

Observe that we can always assume $|\alpha| \leq -ca/2$. We express $k = Am^2$, where $A, m \in \mathbb{Z}$, and A is squarefree; note that A and m can be determined algorithmically from the squarefree factorization of k (see Exercise 5.5). So we have

$$\alpha^2 = -ba + Am^2(-ca) \,.$$

First we show that $0 < A < -ca$. From our assumption $-ba < -ca$ we get

$$0 \leq \alpha^2 = -ba + Am^2(-ca) < -ca + Am^2(-ca) = -ca(1 + Am^2) \,.$$

Neither A nor m can be 0, because otherwise $\alpha^2 = -ba$, which is impossible ($\gcd(a, b) = 1$ and a, b are squarefree). Furthermore, $-ca > 0$ implies that $0 < 1 + Am^2$, so $A > 0$.

The relations $\alpha^2 = -ba + Am^2(-ca)$, $-ba > 0$, and $|\alpha| \leq -ca/2$ imply

$$Am^2(-ca) = \alpha^2 + ba < \alpha^2 \leq \frac{(-ca)^2}{4} .$$

This finishes the proof of $0 < A < -ca$.

Now we consider the new equation

$$AZ^2 + (-ba)Y^2 = X^2 ,$$

and we prove that this equation satisfies the same hypothesis as the original equation $(-ca)z^2 + (-ba)y^2 = x^2$. First, note that $A, -ba \in \mathbb{Z}^+$, and they are squarefree. Now we prove that

$$A\,\mathcal{R}\,(-ba), \quad (-ba)\,\mathcal{R}\,A, \quad \text{and} \quad \frac{-A(-ba)}{\gcd(A, -ba)^2}\,\mathcal{R}\,\gcd(A, -ba) , \qquad (5.7)$$

which implies that $AZ^2 + (-ba)Y^2 = X^2$ is associated to some Legendre equation (see remark to Definition 5.19). Observe that, by Theorem 5.21, we deduce that the new equation has a nontrivial solution.

Let us prove that each of these relations hold.

(i) First we prove that $A\mathcal{R}(-ba)$. For this purpose, we show that $A\mathcal{R}(-a)$ and $A\mathcal{R}b$ which implies, by Lemma 5.14, that $A\,\mathcal{R}\,(-ba)$. From $\alpha^2 = -ba + Am^2(-ca)$ and the squarefreeness of a we deduce that a divides α. So if we set $\alpha_1 = -\alpha/a$, then $\alpha_1 \in \mathbb{Z}$ and

$$-a\alpha_1^2 = b + Am^2c .$$

Hence,

$$Am^2c^2 \equiv_{-a} -cb .$$

Moreover, from $-a\alpha_1^2 = b + Am^2c$ and $\gcd(a, b) = 1$ we get $\gcd(m, a) = 1$. Because of $(-cb)\,\mathcal{R}\,(-a)$ there exists $y_1 \in \mathbb{Z}$ such that $y_1^2 \equiv_{-a} -cb$. Thus,

$$A \equiv_{-a} (m^*)^2(c^*)^2 y_1^2 ,$$

where m^* and c^* are the inverses of m and c modulo $-a$, respectively. m and a are relatively prime, so are c and a. Therefore, $A\,\mathcal{R}\,(-a)$.

Now we show that $A\,\mathcal{R}\,b$. Because of $(-ca)\,\mathcal{R}\,(ab)$ there exists $\beta \in \mathbb{Z}$ such that $\beta^2 \equiv_b (-ca)$. So from $\alpha^2 = -ab + Am^2(-ca)$ we get

$$\alpha^2 \equiv_b Am^2(-ca) \equiv_b Am^2\beta^2 .$$

Observe that $\gcd(\beta, b) = 1$. Indeed, assume $1 \neq d = \gcd(\beta, b)$. Because of $\beta^2 \equiv_b (-ca)$ there exists $\lambda \in \mathbb{Z}$ such that $\beta^2 = (-ca) + \lambda b$. Therefore, d divides ca, which is impossible because of $\gcd(a, b) = \gcd(c, b) = 1$. Furthermore, note that by hypothesis a, b, c are pairwise relatively prime, so also $\gcd(ca, b) = \gcd(m, b) = 1$ (see Exercise 5.8). Putting all this together, we get

$$\alpha^2(m^*)^2(u^*)^2 \equiv_b A ,$$

where u^*, and m^* are the inverses of β, and m modulo b, respectively. Therefore, $A \mathcal{R} b$.

(ii) The condition $(-ba) \mathcal{R} A$ follows from $\alpha^2 = -ba + Am^2(-ca)$.

(iii) Finally, we show that $\frac{-A(-ba)}{\gcd(A,-ba)^2} \mathcal{R} \gcd(A, -ba)$ holds. Let $r = \gcd(A, -ba)$, $A_1 = A/r$, and $b_1 = -ba/r$. Then, we have to show that $(-A_1 b_1) \mathcal{R} r$. From $\alpha^2 = -ba + Am^2(-ca)$ we deduce that

$$\alpha^2 = b_1 r + A_1 r m^2 (-ca) .$$

A is squarefree, so also r is squarefree, and hence that r divides α. So

$$A_1 m^2 (-ca) \equiv_r -b_1 ,$$

which implies that

$$-A_1 b_1 m^2 (-ca) \equiv_r b_1^2 .$$

Note that by hypothesis a, b, c are pairwise relatively prime, so by the same reasoning as above $\gcd(ca, r) = \gcd(m, r) = 1$. From $(-ca) \mathcal{R} (-ba)$ and $b_1 r = -ab$ we obtain $(-ca) \mathcal{R} r$. Thus, there exists $w \in \mathbb{Z}$ such that $w^2 \equiv_r (-ca)$. Observe that $\gcd(w, r) = 1$. Indeed, assume $1 \neq d = \gcd(w, r)$. Because of $w^2 \equiv_r (-ca)$ there exists $\lambda \in \mathbb{Z}$ such that $w^2 = (-ca) + \lambda r$. Therefore, d divides to ca, which is impossible because of $\gcd(ac, r) = 1$. Putting all this together, we get

$$-A_1 b_1 \equiv_r b_1^2 (m^*)^2 v^* \equiv_r b_1^2 (m^*)^2 (w^*)^2 ,$$

where v^*, m^*, and w^* are the inverses of $-ca$ and m and w modulo r, respectively. Therefore, $(-A_1 b_1) \mathcal{R} r$.

So all the relations in (5.7) hold.

Finally, we show that if we have a nontrivial solution $(\overline{X}, \overline{Y}, \overline{Z})$ of $AZ^2 + (-ba)Y^2 = X^2$, we can algorithmically determine a nontrivial solution $(\overline{x}, \overline{y}, \overline{z})$ of $(-ca)z^2 + (-ba)y^2 = x^2$. So assume

$$A\overline{Z}^2 = \overline{X}^2 - (-ba)\overline{Y}^2 .$$

Then, taking into account that $Am^2(-ca) = \alpha^2 - (-ba)$, we get

$$(-ca)(A\overline{Z}m)^2 = (\overline{X}^2 - (-ba)\overline{Y}^2)(\alpha^2 - (-ba))$$
$$= (\overline{X}\alpha + (-ba)\overline{Y})^2 - (-ba)(\alpha\overline{Y} + \overline{X})^2 .$$

Thus,

$$\overline{x} = \overline{X}\alpha + (-ba)\overline{Y}, \quad \overline{y} = \alpha\overline{Y} + \overline{X}, \quad \overline{z} = A\overline{Z}m ,$$

is a solution of the equation $(-ca)z^2 + (-ba)y^2 = x^2$. Clearly, $(\overline{x}, \overline{y}, \overline{z}) \in \mathbb{Z}^3$, but we still have to prove that the solution is nontrivial. For this purpose, we write the above equalities in matrix notation as:

$$
\begin{pmatrix} \overline{x} \\ \overline{y} \\ \overline{z} \end{pmatrix} = \begin{pmatrix} \alpha & -ba & 0 \\ 1 & \alpha & 0 \\ 0 & 0 & Am \end{pmatrix} \cdot \begin{pmatrix} \overline{X} \\ \overline{Y} \\ \overline{Z} \end{pmatrix}.
$$

The determinant of the matrix is $Am(\alpha^2 - (-ba))$. $A > 0$, $m \neq 0$ and $\alpha^2 \neq -ba$ because $\gcd(a, b) = 1$ and a, b are squarefree. So the determinant is nonzero. Since $(\overline{X}, \overline{Y}, \overline{Z})$ is nontrivial, $(\overline{x}, \overline{y}, \overline{z})$ is also nontrivial.   □

In Theorems 5.20 and 5.21, we have seen how to decide whether the Legendre equation has nontrivial solutions (see also the algorithmic comments given after Definition 5.12), and the proof of Theorem 5.22 shows how to compute a nontrivial solution of the Legendre equation, if there exists one. In the following, assuming that the existence of nontrivial solutions has already been checked, we outline an algorithm (derived from the proof of Theorem 5.22) for determining a nontrivial solution of the Legendre equation. To be more precise, we assume that the equation is given in Legendre form (see (5.5) and (5.6)). As above, we assume an algorithm "qr," which for given inputs m, n decides whether $m \mathcal{R} n$, and in the affirmative case outputs $x \in \mathbb{Z}$ such that $x^2 \equiv_n m$. Furthermore, we represent by "oddf" an algorithmic procedure such that if $k \in \mathbb{Z}^*$, then oddf(k) is a squarefree integer A satisfying $k = Am^2$ with $m \in \mathbb{Z}$.

Algorithm ASSOCIATED LEGENDRE SOLVE
Given positive squarefree integers B, C such that the equation $-x^2 + By^2 + Cz^2 = 0$ has a nontrivial solution, the algorithm computes a nontrivial integral solution $(\overline{x}, \overline{y}, \overline{z})$ of the equation $-x^2 + By^2 + Cz^2 = 0$.

1. If $C = 1$, then set $(\overline{x}, \overline{y}, \overline{z}) = (1, 0, 1)$, and go to Step 6.
2. If $B = 1$, then set $(\overline{x}, \overline{y}, \overline{z}) = (1, 1, 0)$, and go to Step 6.
3. If $C = B$, then compute $r, s \in \mathbb{Z}^*$ such that $B = C = r^2 + s^2$ (see Lemma 5.17 and the following remark). Set $(\overline{x}, \overline{y}, \overline{z}) = (r^2 + s^2, s, r)$, and go to Step 6.
4. If $C < B$, then apply Algorithm ASSOCIATED LEGENDRE EQUATION to the inputs C, B. Let (x_1, y_1, z_1) be the solution obtained. Set $(\overline{x}, \overline{y}, \overline{z}) = (x_1, z_1, y_1)$, and go to Step 6.
5. If $B < C$, then compute

$$
\alpha := \mathrm{qr}(B, C), \quad k := (\alpha^2 - B)/C, \quad A := \mathrm{oddf}(k), \quad m := \sqrt{k/A}.
$$

Apply Algorithm ASSOCIATED LEGENDRE SOLVE to the inputs B, A. Let (x_1, y_1, z_1) be the solution obtained. Set $(\overline{x}, \overline{y}, \overline{z}) = (\alpha x_1 + By_1, \alpha y_1 + x_1, Amz_1)$, and go to Step 6.
6. Return the point $(\overline{x}, \overline{y}, \overline{z})$.

Algorithm LEGENDRE SOLVE

Given integers a, b, c defining the Legendre equation $ax^2 + by^2 + cz^2 = 0$, having nontrivial solutions (see Definition 5.12), the algorithm computes a nontrivial integral solution $(\overline{x}, \overline{y}, \overline{z})$ of the Legendre equation.

1. Compute the point (x_1, y_1, z_1) obtained by applying the Algorithm ASSOCIATED LEGENDRE SOLVE to the pair $(-ba, -ca)$.
2. Return the point $(\overline{x}, \overline{y}, \overline{z}) = (x_1, ay_1, az_1)$.

Example 5.23. Consider the Legendre Equation

$$(i) \quad 7x^2 - y^2 - 3z^2 = 0.$$

We show how to solve this equation by the Algorithm LEGENDRE SOLVE. LEGENDRE SOLVE $(7, -1, -3)$:

$$a = 7, \qquad b = -1, \qquad c = -3.$$

(STEP 1) ASSOCIATED LEGENDRE SOLVE $(7, 21)$:
  $B = 7, \quad C = 21$
  (ii) $\quad -x^2 + 7y^2 + 21z^2 = 0.$
  (STEP 5) $B < C$, so
    $\alpha = \mathrm{qr}(7, 21) = 14, \quad k = \frac{196-7}{21} = 9, \quad A = 1, \quad m = 3.$
    ASSOCIATED LEGENDRE SOLVE $(7, 1)$:
      $B = 7, \quad C = 1$
      (iii) $\quad -x^2 + 7y^2 + z^2 = 0.$
      (STEP 1) $C = 1$, so
      Return $(1, 0, 1)$ (Solution of (iii))
    $(\overline{x}, \overline{y}, \overline{z}) := (14 \cdot 1 + 7 \cdot 0, 14 \cdot 0 + 1, 1 \cdot 3 \cdot 1) = (14, 1, 3)$
  Return $(14, 1, 3)$ (Solution of (ii))
(STEP 2) $(\overline{x}, \overline{y}, \overline{z}) := (14, 7, 21)$
Return $(14, 7, 21)$ (Solution of (i))

5.3 Optimal Parametrization of Rational Curves

We have seen how the theorem of Hilbert–Hurwitz (Theorem 5.8) can be used to classify optimal fields of parametrization of a rational curve, and in addition we have outlined an algorithm derived from the constructive proof of its Corollary 5.9. But the algorithm, although theoretically interesting, does not have very satisfactory performance in practice. The reason is that, in general, $\mathcal{O}(d)$ birational transformations are required in order to reach a conic or a line and to invert either the simple points or the parametrization. Furthermore $\mathcal{O}(d)$ adjoint computations have to be carried out. From Theorem 5.8 we know that the birationality check in Step 3.2 is very unlikely to yield a negative result. Avoiding it leads to a probabilistic parametrization algorithm. Instead of directly applying $\mathcal{T}$, or actually its inverse, for computing the implicit equation of $\mathcal{T}(\mathcal{D})$, we might use interpolation methods combined with the technique of families of conjugate simple points (compare Chaps. 3 and 4). Note that if $\mathcal{T}$ is birational, then we must have $\deg(\mathcal{T}(\mathcal{D})) = \deg(\mathcal{D}) - 2$. In any case, these difficulties render the method all but impossible to use in practical applications, unless the curve under consideration has extremely low degree. In this section we show how the algorithm HILBERT–HURWITZ can be improved to avoid most of these difficulties.

Throughout this section, in addition to the notation introduced at the beginning of the chapter, we assume $d > 2$ for the degree of the rational curve $\mathcal{C}$, and we fix $k \in \{d, d-1, d-2\}$. Furthermore, we consider only linear subsystems of $\mathcal{A}_k(\mathcal{C})$ generated as follows.

Definition 5.24. *If* $k \in \{d-2, d-1\}$ *and* $\mathcal{S} \subset \mathcal{C} \setminus \mathrm{Sing}(\mathcal{C})$ *is such that* $\mathrm{card}(\mathcal{S}) \leq kd - (d-1)(d-2) - 1$, *then*

$$\mathcal{A}_k^{\mathcal{S}}(\mathcal{C}) := \mathcal{A}_k(\mathcal{C}) \cap \mathcal{H}(k, \sum_{P \in \mathcal{S}} P)$$

is called the proper linear subsystem of $\mathcal{A}_k(\mathcal{C})$ generated by $\mathcal{S}$.
If $k = d$, *and* $\tilde{\mathcal{S}} \subset \mathcal{C} \setminus \mathrm{Sing}(\mathcal{C})$ *is such that* $\mathrm{card}(\tilde{\mathcal{S}}) \leq 3d - 3$, *and* $\mathcal{S} = \tilde{\mathcal{S}} \cup \{Q\}$, *where* $Q \notin \mathcal{C}$, *then*

$$\mathcal{A}_d^{\mathcal{S}}(\mathcal{C}) := \mathcal{A}_d(\mathcal{C}) \cap \mathcal{H}(d, \sum_{P \in \mathcal{S}} P)$$

is called the proper linear subsystem of $\mathcal{A}_d(\mathcal{C})$ generated by $\mathcal{S}$.
Moreover, if all curves in a proper linear subsystem have $\mathbb{K}$ *as ground field we say that it is* rational.

Before we prove the generalization of the theorem of Hilbert–Hurwitz, we first state some technical lemmas. By the same reasoning as in Sect. 4.8 we can immediately derive the following result.

Lemma 5.25. *If the simple points in* $\mathcal{S}$ *are distributed in families of simple conjugate points over* $\mathbb{K}$ *(see Definitions 3.15 and 3.16) then* $\mathcal{A}_k^{\mathcal{S}}(\mathcal{C})$ *is rational.*

So in the sequel we will make use of families of conjugate simple points. We use the notation introduced just before Lemma 4.59.

Lemma 5.26. *Let* $\mathcal{A}_k^{\mathcal{S}}(\mathcal{C})$ *be a proper linear subsystem of* $\mathcal{A}_k(\mathcal{C})$. *Then the following hold:*

(1) $\dim(\mathcal{A}_k^{\mathcal{S}}(\mathcal{C})) = \dim(\mathcal{A}_k(\mathcal{C})) - \mathrm{card}(\mathcal{S}) = \frac{k(k+3)}{2} - \frac{(d-1)(d-2)}{2} - \mathrm{card}(\mathcal{S})$.
(2) For almost all curves $\mathcal{C}' \in \mathcal{A}_k^{\mathcal{S}}(\mathcal{C})$ *we have*

$$\sum_{P \in \mathrm{Sing}(\mathcal{C}) \cup \mathcal{S}} \mathrm{mult}_P(\mathcal{C}, \mathcal{C}') = dk - \dim(\mathcal{A}_k^{\mathcal{S}}(\mathcal{C})) \, .$$

(3) For all curves $\mathcal{C}' \in \mathcal{A}_k^{\mathcal{S}}(\mathcal{C})$ *satisfying Statement (2), and for almost all* $\lambda \in K$ *we have*

$$\sum_{P \in \mathrm{Sing}(\mathcal{C}) \cup \mathcal{S}} \mathrm{mult}_P(\mathcal{C}, \mathcal{C}' + \lambda \mathcal{C}'') = dk - \dim(\mathcal{A}_k^{\mathcal{S}}(\mathcal{C})) \ ,$$

where $\mathcal{C}'' \in \mathcal{A}_k^{\mathcal{S}}(\mathcal{C})$.

Proof. Statement (1) follows from Theorem 4.58 by the same reasoning as in the proofs of Theorems 4.61 and 4.62 (Condition (1)).

Statement (2) is derived by the same reasoning as in the proofs of Theorems 4.61 and 4.62 (Condition (2)).

Let us prove Statement (3). Let $s = \dim(\mathcal{A}_k^{\mathcal{S}}(\mathcal{C}))$. We identify $\mathcal{A}_k^{\mathcal{S}}(\mathcal{C})$ with $\mathbb{P}^s(K)$ (see Sect. 2.4). Then by Statement (2) there exists a Zariski closed subset V of $\mathbb{P}^s(K)$, where $V \neq \mathbb{P}^s(K)$, such that for all $\mathcal{C}' \in \mathbb{P}^s(K) \setminus V$ we have

$$\sum_{P \in \mathrm{Sing}(\mathcal{C}) \cup \mathcal{S}} \mathrm{mult}_P(\mathcal{C}, \mathcal{C}') = dk - \dim(\mathcal{A}_k^{\mathcal{S}}(\mathcal{C})) \ .$$

Now let $\mathcal{C}' \in \mathbb{P}^s(K) \setminus V$, and consider all the lines in $\mathbb{P}^s(K)$ passing through $\mathcal{C}'$. That is, consider the lines of equation $\mathcal{L} = \mathcal{C}' + t\mathcal{C}''$, where $\mathcal{C}'' \in \mathbb{P}^s(K)$. Since $\mathcal{C}' \notin V$, no line $\mathcal{L}$ is included in V. Thus, each line $\mathcal{L}$ only intersects V in finitely many points. Thus, for every $\mathcal{C}'' \in \mathbb{P}^s(K)$ and all but finitely many values λ of t in K we get that $\mathcal{C}' + \lambda \mathcal{C}'' \notin V$. This finishes the proof. $\square$

Lemma 5.27. *Let* $\mathcal{A}_k^{\mathcal{S}}(\mathcal{C})$ *be a proper linear subsystem of* $\mathcal{A}_k(\mathcal{C})$. *Then there always exist* $\Phi_1, \Phi_2, \Phi_3 \in \mathcal{A}_k^{\mathcal{S}}(\mathcal{C})$ *such that the mapping*

$$\mathcal{T} = \{y_1 : y_2 : y_3 = \Phi_1 : \Phi_2 : \Phi_3\}$$

is birational on $\mathcal{C}$.

Proof. Let $\ell_k = kd - (d-1)(d-2) - 1$ if $k \in \{d-2, d-1\}$, and let $\ell_d = 3d - 2$. Then, if $\mathrm{card}(\mathcal{S}) < \ell_k$, we take a new set $\mathcal{S}'$ of simple points on $\mathcal{C}$ such that $\mathrm{card}(\mathcal{S} \cup \mathcal{S}') = \ell_k$, and we consider the new linear system

$$\mathcal{H}' = \mathcal{A}_k^{\mathcal{S}}(\mathcal{C}) \cap \mathcal{H}\left(k, \sum_{P \in \mathcal{S}'} P\right) .$$

Then, applying Theorems 4.61 and 4.62, we see that $\mathcal{H}'$ parametrizes $\mathcal{C}$ (see Definition 4.51). Since $\dim(\mathcal{H}') = 1$, we may express the defining polynomial of $\mathcal{H}'$ as $H'(x, y, z, t) = \Psi_1 + t\Psi_2$, where $\{\Psi_1, \Psi_2\}$ is a basis of $\mathcal{H}'$. Let $\mathcal{P}(t)$ be the proper parametrization of $\mathcal{C}$ generated by $\mathcal{H}'$ (see Lemma 4.52). Now, we take $a, b \in K$ such that $a \neq b$, and we consider

$$\Phi_1 = \Psi_1 + a\Psi_2, \quad \Phi_2 = \Psi_1 + b\Psi_2, \quad \Phi_3 = \Psi_1 \ .$$

We prove that the mapping $\mathcal{T} = \{y_1 : y_2 : y_3 = \Phi_1 : \Phi_2 : \Phi_3\}$ is birational on $\mathcal{C}$. For this purpose and w.l.o.g. we dehomogenize with respect to z and we reason on the affine plane. Let $\phi_i = \Phi_i(x, y, 1)$, $\psi_i = \Psi_i(x, y, 1)$ and let

$$
\begin{aligned}
\phi : \mathcal{C}_{*,z} &\longrightarrow \phi(\mathcal{C}_{*,z}) \\
(x, y) &\longmapsto \left(\frac{\phi_1(x,y)}{\phi_3(x,y)}, \frac{\phi_2(x,y)}{\phi_3(x,y)} \right) .
\end{aligned}
$$

Note that $\phi_3 \notin I(\mathcal{C})$, because $\phi_3 \in \mathcal{H}'$ and $\mathcal{H}'$ parametrizes $\mathcal{C}$ (see Definition 4.51). Because $\mathcal{H}'$ generates $\mathcal{P}(t)$, we get $H'(\mathcal{P}(t), 1, t) = 0$. On the other hand, observe that $\Psi_2(\mathcal{P}(t), 1)$ cannot be identically zero, since otherwise Bézout's Theorem would imply that $\mathcal{C}$ and the adjoint Ψ_2 have a common component. This, however, is impossible because $\mathcal{H}'$ parametrizes $\mathcal{C}$. Thus, by Lemma 4.36, $-\psi_1/\psi_2$ is the inverse of $\mathcal{P}(t)$, and for almost all points $P \in \mathcal{C}_{*,z}$ we have $P = \mathcal{P}(-\psi_1(P)/\psi_2(P))$.

In order to see that ϕ is birational we now explicitly compute its inverse. We consider the rational map

$$
\begin{aligned}
\rho : \phi(\mathcal{C}_{*,z}) &\longrightarrow \mathcal{C}_{*,z} \\
(y_1, y_2) &\longmapsto \mathcal{P}\left(\frac{b-a}{y_1 - y_2} \right) .
\end{aligned}
$$

First we show that ρ is well defined. Almost all points in $\phi(\mathcal{C}_{*,z})$ are of the form $(\frac{\phi_1(x,y)}{\phi_3(x,y)}, \frac{\phi_2(x,y)}{\phi_3(x,y)})$, where $(x, y) \in \mathcal{C}_{*,z}$. But $\frac{\phi_1(x,y)}{\phi_3(x,y)} - \frac{\phi_2(x,y)}{\phi_3(x,y)}$ cannot be zero, since otherwise we would have $(a - b)\psi_2(x, y) = 0$ for almost all $(x, y) \in \mathcal{C}$. Therefore, since $a \neq b$, from Bézout's Theorem we would get that $\mathcal{C}$ and Ψ_2 have a common component, which is impossible. Reasoning as above, we can also prove that the mapping $\frac{b-a}{y_1 - y_2}$ is not constant over $\phi(\mathcal{C}_{*,z})$, since otherwise Ψ_1 and Ψ_2 would be linearly dependent, which is impossible. Thus ρ is not a constant map and it is defined for almost all points. Finally we observe that for almost all points $P \in \mathcal{C}_{*,z}$ the map ρ inverts the action of ϕ, i.e.,

$$
\rho(\phi(P)) = \rho\left(\frac{\phi_1(P)}{\phi_3(P)}, \frac{\phi_2(P)}{\phi_3(P)} \right) = \mathcal{P}\left(\frac{(b-a)\phi_3(P)}{\phi_1(P) - \phi_2(P)} \right) = \mathcal{P}\left(-\frac{\psi_1(P)}{\psi_2(P)} \right) = P .
$$

Thus, we conclude that $\mathcal{T}$ is birational. $\qquad\square$

Now, we are ready for a generalization of the theorem of Hilbert–Huritwz.

Theorem 5.28. *Let $\mathcal{A}_k^{\mathcal{S}}(\mathcal{C})$ be a proper linear subsystem of $\mathcal{A}_k(\mathcal{C})$, and let $\dim(\mathcal{A}_k^{\mathcal{S}}(\mathcal{C})) = s$. Then for almost all $\Phi_1, \Phi_2, \Phi_3 \in \mathcal{A}_k^{\mathcal{S}}(\mathcal{C})$ the mapping*

$$
\mathcal{T} = \{y_1 : y_2 : y_3 = \Phi_1 : \Phi_2 : \Phi_3\}
$$

transforms $\mathcal{C}$ birationally to an irreducible curve of degree s.

Proof. First we prove that for almost all $\Phi_1, \Phi_2, \Phi_3 \in \mathcal{A}_k^{\mathcal{S}}(\mathcal{C})$ the mapping $\mathcal{T} = \{y_1 : y_2 : y_3 = \Phi_1 : \Phi_2 : \Phi_3\}$ is birational on $\mathcal{C}$. For this purpose, we identify $\mathcal{A}_k^S(\mathcal{C})$ with $\mathbb{P}^s(K)$, and we prove that there exists a nonempty open Zariski subset Ω in $\mathbb{P}^s(K)$ such that for all different $\Phi_1, \Phi_2, \Phi_3 \in \Omega$, $\mathcal{T}$ is birational. We consider three generic elements $\Phi_1, \Phi_2, \Phi_3 \in \mathcal{A}_k^S(\mathcal{C})$. Thus, each Φ_i depends on a different set of undetermined coefficients Λ_i, namely its coordinates as elements in $\mathbb{P}^s(K)$. Then, we consider the formal rational mapping $\mathcal{T}$ defined by Φ_1, Φ_2, Φ_3, and let $\mathcal{D} = \mathcal{T}(\mathcal{C})$. In addition, we take a rational proper parametrization $\mathcal{P}(t)$ of $\mathcal{C}$. So we have the following commutative diagram

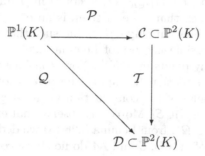

Thus, $\mathcal{T}$ is birational if and only if $\mathcal{Q}$ is birational. Note that, because of Lemma 5.27, there exist parameter values for Λ_i such that $\mathcal{T}$ is birational. Hence, $\mathcal{Q}(t) = \mathcal{T}(\mathcal{P}(t))$ does depend on t, and therefore it is a rational parametrization. So, $\mathcal{T}$ is birational if and only the rational parametrization $\mathcal{Q}(t) = \mathcal{T}(\mathcal{P}(t))$ is proper. In order to prove that $\mathcal{Q}(t)$ is proper, we reason over the affine plane, considering $\mathcal{Q}(t)$ as an affine rational parametrization. We write its components as

$$\mathcal{Q}(t) = \left(\frac{\chi_{11}(t)}{\chi_{12}(t)}, \frac{\chi_{21}(t)}{\chi_{22}(t)} \right),$$

where $\chi_{ij}(t) \in K[\Lambda_1, \Lambda_2, \Lambda_3][t]$ and $\gcd(\chi_{1i}, \chi_{2i}) = 1$. We note that $\mathcal{Q}(t)$ depends on the undetermined coefficients Λ_i. By Theorem 4.30 the parametrization $\mathcal{Q}(t)$ is proper if and only if the gcd of the polynomials

$$G_1^{\mathcal{Q}}(t, h) = \chi_{11}(h)\chi_{12}(t) - \chi_{12}(h)\chi_{11}(t),$$

$$G_2^{\mathcal{Q}}(t, h) = \chi_{21}(h)\chi_{22}(t) - \chi_{22}(h)\chi_{21}(t),$$

with $G_1^{\mathcal{Q}}, G_2^{\mathcal{Q}} \in K[\Lambda_1, \Lambda_2, \Lambda_3][t, h]$, is linear in t. On the other hand, by Lemma 5.27 we know that there exist specific values of $\Lambda_1, \Lambda_2, \Lambda_3$ for which the gcd is in fact linear in t. So the expression of the formal gcd of $G_1^{\mathcal{Q}}, G_2^{\mathcal{Q}}$ cannot have degree higher than 1, and the parameter values under which the gcd has higher degree satisfy certain algebraic conditions. So for a Zariski open subset Ω' of $(\mathbb{P}^s(K))^3$ this gcd has exactly degree 1. Ω' cannot be empty because of Lemma 5.27. Now we can finish the proof of birationality by taking a nonempty open subset Ω included in the projection of Ω' over $\mathbb{P}^s(K)$.

We proceed to prove the second part of the theorem, concerning the irreducibility and the degree of the transformed curve. Let Ω^* be a nonempty open subset of $\mathcal{A}_k^{\mathcal{S}}(\mathcal{C})$ for which Statement (2) in Lemma 5.26 holds. Then, we take $\Sigma = \Omega^* \cap \Omega$. Note that Σ is open and since $\mathbb{P}^s(K)$ is irreducible it is nonempty. Let us see that for all different $\Phi_1, \Phi_2, \Phi_3 \in \Sigma$ the transformed curve $\mathcal{D} = \mathcal{T}(\mathcal{C})$ is irreducible of degree s. Clearly, $\mathcal{D}$ is rational and therefore irreducible. Let us see that $\deg(\mathcal{D}) = s$. Let $n = \deg(\mathcal{D})$. $\mathcal{T}$ determines a 1–1 relation between the points of $\mathcal{C}$ and $\mathcal{D}$, except for finitely many points on these curves. We call these points the exception points. Now take $b \in K$ such that the line $\mathcal{L} = \{(b : t : 1)\}_{t \in K}$ intersects $\mathcal{D}$ in n different simple points $\{(b : \lambda_i : 1)\}_{i=1,\ldots,n}$ such that none of them is an exception point on $\mathcal{D}$ nor the imagine by $\mathcal{T}$ of a point in $\mathrm{Sing}(\mathcal{C}) \cup \mathcal{S}$ and such that $\Phi_1 - b\Phi_3 \in \Omega^*$. Note that this is always possible because of Lemma 5.26 (3) and because we are excluding finitely many points in $\mathbb{P}(K)^2$. Now, consider the curve $\mathcal{M}$ defined by $\Phi_1 - b\Phi_3$. In this situation, we apply the inverse of $\mathcal{T}$ to the n simple points in $\mathcal{L} \cap \mathcal{D}$. Because of the construction we must get n different points $\{P_1, \ldots, P_n\}$ in $\mathcal{C} \setminus [\mathrm{Sing}(\mathcal{C}) \cup \mathcal{S}]$. Moreover, observe that every P_i is also on $\mathcal{M}$. Then, since $\Phi_1 - b\Phi_3 \in \Omega^*$, from Lemma 5.26 (3) we deduce that $n \leq s$; note that by construction of $\mathcal{A}_k^{\mathcal{S}}(\mathcal{C})$, $\mathcal{C}$ and $\mathcal{M}$ do not have common components.

Suppose n were actually less than s. Then, we could take an additional common point P_{n+1} on $\mathcal{C}$ and $\mathcal{M}$, not being in $\{P_1, \ldots, P_n\} \cup \mathrm{Sing}(\mathcal{C}) \cup \mathcal{S}$ and not being an exception point for $\mathcal{T}$. Note that this is always possible, because by Lemma 5.26 (3) there are s free intersection points (i.e., other than those in $\mathrm{Sing}(\mathcal{C}) \cup \mathcal{S}$), and by Lemma 5.26 (1) we have $s \geq 1$. Now, $\{\mathcal{T}(P_1), \ldots, \mathcal{T}(P_{n+1})\}$ contains $n + 1$ different points, and all of them are in $\mathcal{L} \cap \mathcal{D}$, which is impossible. Hence $n = s$. $\qquad \square$

The theorem of Hilbert–Hurwitz is a particular case of Theorem 5.28. Apply Theorem 5.28 taking $k = d - 2$ and $\mathcal{S}$ as the empty set, and use Theorem 4.58.

Corollary 5.29. *If the proper linear system $\mathcal{A}_k^{\mathcal{S}}(\mathcal{C})$ in Theorem 5.28 is rational, then the ground field of the transformed curves $\mathcal{T}(\mathcal{C})$ is $\mathbb{K}$.*

Proof. If $\mathcal{A}_k^{\mathcal{S}}(\mathcal{C})$ is rational then the ground field of all curves in $\mathcal{C}$ is $\mathbb{K}$. But $\mathcal{T}$ does not extend the ground field. $\qquad \square$

Based on this corollary, the strategy for optimal parametrization consists in generating rational proper subsystems of the system of adjoints of low dimension. Because of Corollary 5.9 to Theorem 5.8 we cannot expect to generate, in general, rational subsystems of dimension 1, since it would imply that all rational curves can be parametrized over the ground field. What we may actually try to generate are proper subsystems of dimension at most 2. This means that we generate birational maps over $\mathbb{K}$ mapping the original curve into a line or conic. Lemma 5.26 (1) tells us that we can reduce the dimension of the proper linear subsystem by increasing the cardinality of the set $\mathcal{S}$ of new base points. Of course, we have to do that in such a way that the new linear subsystem is still proper and rational. For this purpose, we apply the technique of families of conjugate points. Obviously, we could reach our goal if we had a method for computing rational points on $\mathcal{C}$. But before we have a parametrization, this is not possible in general. Therefore, the best we can do is to provide a method for computing families of two conjugate simple points. We will see that this is always possible.

Proposition 5.30. *Let $\mathcal{F}$ be a family of ℓ conjugate simple points on $\mathcal{C}$ over $\mathbb{K}$, where $\ell < \dim(\mathcal{A}_k(\mathcal{C}))$, and let*

$$\mathcal{H} = \mathcal{A}_k(\mathcal{C}) \cap \sum_{P \in \mathcal{F}} P \, .$$

Then

(1) $\mathcal{H}$ is a proper rational linear subsystem of $\mathcal{A}_k(\mathcal{C})$, and
(2) $\dim(\mathcal{H}) = \dim(\mathcal{A}_k(\mathcal{C})) - \ell$.

Proof. By the hypothesis and by Theorem 4.58 we have

$$\mathrm{card}(\mathcal{F}) = \ell < \dim(\mathcal{A}_k) = \frac{k(k+3)}{2} - \frac{(d-1)(d-2)}{2}.$$

Therefore $\mathrm{card}(\mathcal{F}) \leq kd - (d-1)(d-2) - 1$ if $k \in \{d-1, d-2\}$, and $\mathrm{card}(\mathcal{F}) \leq kd - (d-1)(d-2)$ if $k = d$. Thus, $\mathcal{H}$ is a proper linear subsystem, and by Lemma 5.25 it is rational. This proves Statement (1). Statement (2) now follows from Lemma 5.26 (1). $\qquad\square$

Theorem 5.31. *Let $\mathcal{A}_k^{\mathcal{S}}(\mathcal{C})$ be a proper rational linear subsystem of dimension s of $\mathcal{A}_k(\mathcal{C})$. Then almost all curves in $\mathcal{A}_k^{\mathcal{S}}(\mathcal{C})$ generate, by intersection with $\mathcal{C}$, families of s conjugate simple points over $\mathbb{K}$.*

Proof. Let $F(x, y, z)$ be the form defining $\mathcal{C}$ and $H(x, y, z, \lambda_1, \dots, \lambda_{s+1})$ the form defining $\mathcal{A}_k^{\mathcal{S}}(\mathcal{C})$. Since H does depend on some of the variables x, y, z, let us assume w.l.o.g. that $\deg_y(H) > 0$. Let us also assume that $(0 : 1 : 0) \notin \mathcal{C}$. Note that undoing the necessary change of coordinates will not change the ground field of the conjugate families of points. Since $\mathcal{A}_k^{\mathcal{S}}(\mathcal{C})$ is a proper linear subsystem, from Lemma 5.26 (2) we get the following relation for almost all curves $\mathcal{C}' \in \mathcal{A}_k^{\mathcal{S}}(\mathcal{C})$:

$$\sum_{P \in \mathrm{Sing}(\mathcal{C}) \cup \mathcal{S}} \mathrm{mult}_P(\mathcal{C}, \mathcal{C}') = dk - \dim(\mathcal{A}_k^{\mathcal{S}}(\mathcal{C})) \, .$$

We identify $\mathcal{A}_k^{\mathcal{S}}(\mathcal{C})$ with $\mathbb{P}^s(K)$, and let $\Omega_1 \subset \mathbb{P}^s(K)$ be the nonempty open subset where the above property is satisfied. Now, we proceed by constructing a chain of nonempty open subset of Ω_1. Let $\mathcal{B}$ be equal to $\mathcal{S} \cup \mathrm{Sing}(\mathcal{C})$ if $k < d$ and to $\mathcal{S} \cup \mathrm{Sing}(\mathcal{C})$ minus the point of $\mathcal{S}$ not in $\mathcal{C}$ if $k = d$.

(i) We prove that there exists a nonempty open subset $\Omega_2 \subset \Omega_1$, such that, for $\mathcal{C}' \in \Omega_2$, no point in $[\mathcal{C} \cap \mathcal{C}'] \setminus \mathcal{B}$ is on a line passing through $(0 : 1 : 0)$ and a point in $\mathcal{B}$; note that $(0 : 1 : 0) \notin \mathcal{C}$ and therefore all these lines are well defined. Indeed, Ω_2 is the intersection of Ω_1 with a finite union of open subsets, each for them generated by a point in $\mathcal{B}$.

In order to show that $\Omega_2 \neq \emptyset$, take s points on $\mathcal{C}$ not lying on a line passing through $(0 : 1 : 0)$ and a point in $\mathcal{B}$. Consider a curve $\mathcal{C}'$ in Ω_1 passing through them. This is possible because $\dim(\mathcal{A}_k^{\mathcal{S}}(\mathcal{C})) = s$, and $\sum_{P \in \mathcal{B}} \mathrm{mult}_P(\mathcal{C}, \mathcal{C}') = dk - s$. Clearly $\mathcal{C}' \in \Omega_2$, and hence it is not empty.

(ii) Now, we prove that there exists a nonempty open subset $\Omega_3 \subset \Omega_2$, such that, for $\mathcal{C}' \in \Omega_3$, the x-coordinates of all points in $[\mathcal{C} \cap \mathcal{C}'] \setminus \mathcal{B}$ are different. Indeed, compute the resultant of F and H w.r.t. y, and cross out the factors corresponding to points in $\mathcal{B}$. Let us require that this new polynomial, seen as a polynomial in $K[\lambda_1, \ldots, \lambda_{s+1}][x, z]$, is squarefree. This leads to inequalities in λ_i defining an open set $\tilde{\Omega}_2$. Now set $\Omega_3 = \Omega_2 \cap \tilde{\Omega}_2$. Note that, because of Ω_2, intersection points out of $\mathcal{B}$ and points in $\mathcal{B}$ generate different factors in the resultant.

In order to show that $\Omega_3 \neq \emptyset$ reason as in (i), by taking s points on $\mathcal{C}$ not lying on a line passing through $(0 : 1 : 0)$ and a point in $\mathcal{B}$ and having all x-coordinates different.

(iii) Next we prove that there exists a nonempty open subset $\Omega_4 \subset \Omega_3$, such that, for $\mathcal{C}' \in \Omega_4$, no point in $[\mathcal{C} \cap \mathcal{C}'] \setminus \mathcal{B}$ is of the type $(0 : 1 : c)$ or $(a : 1 : 0)$. Indeed, take the resultant of F and H w.r.t. y and cross out the factors corresponding to points in $\mathcal{B}$. Let us call the result $\tilde{R}(x, z, \lambda_1, \ldots, \lambda_{s+1})$. Now consider the open subset $\tilde{\Omega}_3$ defined by the inequalities $\tilde{R}(0, z, \lambda_1, \ldots, \lambda_{s+1}) \neq 0$ and $\tilde{R}(x, 0, \lambda_1, \ldots, \lambda_{s+1}) \neq 0$, and define $\Omega_4 = \Omega_3 \cap \tilde{\Omega}_3$.

In order to show that $\Omega_4 \neq \emptyset$ reason as in (i), by taking s points on $\mathcal{C}$ not on the lines $x = 0$ and $z = 0$, not on a line passing through $(0 : 1 : 0)$ and a point in $\mathcal{B}$, and having all x-coordinates different.

(iv) Finally we prove that there exists a nonempty open subset $\Omega_5 \subset \Omega_4$, such that, for $\mathcal{C}' \in \Omega_5$ its defining polynomial has positive degree w.r.t. y. This is always possible because by assumption $\deg_y(H) > 0$.

Let us demonstrate that all curves in Ω_5 have the desired property. Let G be the defining polynomial of $\mathcal{C}' \in \Omega_5$, and $R(x, z) = \mathrm{res}_y(F, G)$. Let $\tilde{R}(x, z)$ be the result of crossing out in R all factors generated by points in $\mathcal{B}$. By Lemma 19 in [SSeS05], and because of Ω_1 and Ω_2, we have $\deg(\tilde{R}) = s$ (note that $\deg_y(F) > 0$ because $F(0, 1, 0) \neq 0$, and $\deg_y(G) > 0$ by Ω_5). In addition, because of the properness of $\mathcal{A}_k^{\mathcal{S}}(\mathcal{C})$, $\tilde{R}$ is a polynomial over $\mathbb{K}$. $\tilde{R}(t, 1)$ has degree s because of Ω_4 and it is squarefree because of Ω_3. We take any irreducible factor $A(t)$ over $\mathbb{K}$ of $\tilde{R}(t, 1)$, and let α be one of its roots. Because of Ω_3, the x-coordinate of the intersection points out of $\mathcal{B}$ are all different. Thus $\gcd_{\mathbb{K}(\alpha)[y]}(H(\alpha, y, 1), F(\alpha, y, 1))$ must be linear. Therefore, the corresponding y-coordinate can be expressed rationally in α. So we have generated a family of s conjugate simple points on $\mathcal{C}$.   □

The previous proof provides a probabilistic process to generate families of conjugate points over $\mathbb{K}$. We outline this process:

Let $F(x, y, z)$ be the form defining $\mathcal{C}$ and $H(x, y, z, \lambda_1, \ldots, \lambda_{s+1})$ the form defining $\mathcal{A}_k^{\mathcal{S}}(\mathcal{C})$, where $\mathcal{A}_k^{\mathcal{S}}(\mathcal{C})$ is as in Theorem 5.31, and $\dim(\mathcal{A}_k^{\mathcal{S}}(\mathcal{C})) = s$.

(1) Consider a variable v in $\{x, y, z\}$ such that $\deg_v(H) > 0$. Say $v = y$.
(2) If $F(0, 1, 0) = 0$, perform a linear change of coordinates $\mathcal{T}$ over $\mathbb{K}$ such that $(0 : 1 : 0) \notin \mathcal{T}(\mathcal{C})$.
(3) Give random values to λ_i in H to generate a polynomial $H'(x, y, z)$.
(4) Compute $R := \operatorname{res}_y(F, H')$ and cross out the factors corresponding to points in $\operatorname{Sing}(\mathcal{C}) \cup \mathcal{S}$. Let $\tilde{R}(x, z)$ be the resulting polynomial.
(5) If either $\deg(\tilde{R}(t, 1)) \neq s$ or $\tilde{R}(t, 1)$ is not squarefree go back to (4).
(6) For each irreducible factor $A(t)$ of $\tilde{R}(t, 1)$ compute

$$\gcd_{\mathbb{K}(\alpha)[y]} (F(\alpha, y, 1), H'(\alpha, y, 1)),$$

where α is a root of $A(t)$. Let $B(\alpha)$ be the root of this gcd, and take $\mathcal{F}_A = \{(t : B(t) : 1)\}_{A(t)}$. Take $\mathcal{F}$ as the union of all $\mathcal{F}_A$.
(7) Return $\mathcal{T}^{-1}(\mathcal{F})$.

We illustrate this process by an example, where a family of three conjugate points over a quintic curve is computed.

Example 5.32. We consider the rational curve $\mathcal{C}$ of degree 5, given by the polynomial:

$$F(x, y, z) = y^3 z^2 - 2y^4 z + y^5 + 8x^2 y^2 z + x^3 z^2 - 2x^4 z + x^5 .$$

The ground field is $\mathbb{K} = \mathbb{Q}$. $\mathcal{C}$ has a triple point at $(0 : 0 : 1)$ and three double points at $(1 : 0 : 1), (0 : 1 : 1), (-1 : -1 : 1)$. So $\mathcal{C}$ is rational. In addition, note that $F(0, 1, 0) \neq 0$. Now, we consider $k = d - 2 := 3$ and we take $\mathcal{S}$ as the empty set. That is $\mathcal{A}_3^{\mathcal{S}}(\mathcal{C}) = \mathcal{A}_3(\mathcal{C})$, and its dimension is $s = 3$. The defining polynomial of $\mathcal{A}_3(\mathcal{C})$ is

$$H(x, y, z, \lambda_1, \ldots, \lambda_4) = \lambda_1 y^2 z - \lambda_1 y^3 + \lambda_3 xyz + \lambda_2 xy^2 + \lambda_4 x^2 z +$$
$$(2\lambda_1 + \lambda_3 - \lambda_2 + 2\lambda_4)x^2 y - \lambda_4 x^3 .$$

Note that $\deg_y(H) > 0$. We take a curve $\mathcal{C}' \in \mathcal{A}_3(\mathcal{C})$ with defining polynomial

$$H'(x, y, z) := H(x, y, z, 0, -1, 2, 0) = 2xzy - xy^2 + 3x^2 y .$$

Then

$$R(x, z) := \operatorname{res}_y(F, H') = -2x^8(x - z)^2(122x^3 + 115x^2 z + 34z^2 x + 4z^3)(z + x)^2 .$$

$\tilde{R}(x, z) = 122x^3 + 115x^2z + 34z^2x + 4z^3$ is one of the irreducible factors. Note that $\deg(\tilde{R}) = 3$ and it is squarefree. We set

$$A(t) := \tilde{R}(t, 1) = 122t^3 + 115t^2 + 34t + 4 \ ,$$

and we let α be a root of $A(t)$. We compute $\underset{\mathbb{K}(\alpha)[y]}{\gcd} \ (F(\alpha, y, 1), H'(\alpha, y, 1)) =$

$$122(105229 \, \alpha^2 + 41478 \, \alpha + 6028)y - 4552819\alpha^2 - 1593522\alpha - 208084.$$

The root of this univariate polynomial is

$$B(\alpha) = \frac{208084 + 4552819 \, \alpha^2 + 1593522 \, \alpha}{122(105229 \, \alpha^2 + 41478 \, \alpha + 6028)} \ .$$

Therefore the family of three conjugate points generated by $\mathcal{A}_3(\mathcal{C})$ is

$$\left\{ \left(t : \frac{208084 + 4552819 \, t^2 + 1593522 \, t}{122(105229 \, t^2 + 41478 \, t + 6028)} : 1 \right) \right\}_{A(t)}$$

Corollary 5.33 states that there always exist families of conjugate simple points of certain cardinality. In fact, the above theorem may be applied to find families of two conjugate simple points on $\mathcal{C}$ (see Statement (ii) of corollary).

Corollary 5.33. *We can algorithmically produce families of conjugate simple points on $\mathcal{C}$ over $\mathbb{K}$ with the following cardinality:*

(i) *families of $(d - 2)$, $(2d - 2)$, and $(3d - 2)$ conjugate simple points,*
(ii) *families of two conjugate simple points (i.e., points over an algebraic extension of degree 2 over $\mathbb{K}$),*
(iii) *individual simple points, if d is odd.*

Proof. (i) This is a consequence of Theorem 5.31 with $\mathcal{S} = \emptyset$, i.e., $\mathcal{A}_k(\mathcal{C})$, and Theorem 4.58.

(ii) We first apply Statement (i) to obtain two different families, $\mathcal{F}_1$ and $\mathcal{F}_2$, of $(d-2)$ simple points on $\mathcal{C}$. By Proposition 5.30, the proper rational linear subsystem $\mathcal{A}_{d-1}^{\mathcal{F}_1 \cup \mathcal{F}_2}(\mathcal{C})$ has dimension 2. Thus, applying Theorem 5.31 to $\mathcal{A}_{d-1}^{\mathcal{F}_1 \cup \mathcal{F}_2}(\mathcal{C})$ we obtain families of two simple points.

(iii) Let $\ell = \frac{d-3}{2}$. Applying Statement (ii) we can determine ℓ different families, $\mathcal{F}_1, \ldots, \mathcal{F}_\ell$, of two simple points each on $\mathcal{C}$. By Proposition 5.30, the proper rational linear subsystem $\mathcal{A}_{d-2}^{\mathcal{F}_1 \cup \cdots \cup \mathcal{F}_\ell}(\mathcal{C})$ has dimension 1. Now the statement follows from Theorem 5.31. $\square$

Obviously, the proof of Corollary 5.33 provides algorithmic processes. We briefly outline the corresponding methods for Statements (ii) and (iii).

Let $\mathcal{C}$ be rational of degree d. The following process provides families of two conjugate points of $\mathcal{C}$ over the ground field:

(1) Apply to $\mathcal{A}_{d-2}^{\emptyset}$ the algorithmic method derived from the proof of Theorem 5.31 to generate two different families $\mathcal{F}_1$, $\mathcal{F}_2$ of $(d-2)$ conjugate simple points on $\mathcal{C}$ over $\mathbb{K}$.

(2) Apply to $\mathcal{A}_{d-2}^{\mathcal{F}_1 \cup \mathcal{F}_2}$ the algorithmic method derived from the proof of Theorem 5.31 to generate a family $\mathcal{F}$ of two conjugate simple points on $\mathcal{C}$ over $\mathbb{K}$, and return $\mathcal{F}$.

Let $\mathcal{C}$ be rational of odd degree d. The following process provides rational simple points on $\mathcal{C}$ over the ground field.

(1) If $d = 1$ it is trivial. If $d = 3$, simply intersect $\mathcal{C}$ with a line through the double point on $\mathcal{C}$ and return the resulting rational intersection point.

(2) Otherwise, apply the previous method to compute $\ell = \frac{d-3}{2}$ different families, $\mathcal{F}_1, \ldots, \mathcal{F}_\ell$, of two simple points each on $\mathcal{C}$ over $\mathbb{K}$.

(3) Apply to $\mathcal{A}_{d-2}^{\mathcal{F}_1 \cup \cdots \cup \mathcal{F}_\ell}$ the algorithmic method derived from the proof of Theorem 5.31 to generate a family $\mathcal{F}$ of one simple point on $\mathcal{C}$ over $\mathbb{K}$. Return this simple rational point.

In the sequel we put together all the ideas that have been developed in this chapter to finally derive an algebraically optimal parametrization algorithm for rational curves.

In Sect. 5.1 (see Corollary (5.9) to Theorem 5.8) we have seen that every rational plane curve is parametrizable over an algebraic extension of the ground field of degree at most 2. Furthermore, if the curve has odd degree, then parametrizations over the ground field exist. However, when the curve is of even degree a decision problem appears, and the existence of parametrizations over the ground field depends directly on the existence of simple points on the curve over the ground field. Furthermore, the algorithm HILBERT–HURITWZ shows how to reduce this problem to conics. In Sect. 5.2 we have analyzed the problem of finding rational points on conics over the ground field $\mathbb{Q}$, and we have given an algorithm for solving this problem. However, as we have already mentioned before, the direct application of the algorithm HILBERT–HURITWZ requires, in general, $\mathcal{O}(d)$ birational transformations, where $d = \deg(\mathcal{C})$. In the following we show how Theorem 5.28 and Corollary 5.33 can be applied to solve the problem using only one birational transformation in case d is even, and no one in case d is odd. Moreover, when the birational transformation is required, then the image curve is computed by simply solving a linear system of five equations over $\mathbb{K}$.

To be more precise, let $F \in \mathbb{K}[x, y, z]$ be the homogeneous form defining $\mathcal{C}$. By Theorem 5.28, the problem of computing optimal parametrizations of $\mathcal{C}$ can be reduced to the problem of computing a rational proper linear subsystem of $\mathcal{A}_{d-2}(\mathcal{C})$ of dimension 1 or 2. If d is odd, applying Corollary 5.33, we can compute $\frac{d-3}{2}$ families of two conjugate points over $\mathbb{K}$ that can be used to construct a rational proper linear subsystem of $\mathcal{A}_{d-2}(\mathcal{C})$ of dimension 1 (see also

Proposition 5.30). Therefore, a parametrization over the ground field can be determined. If d is even, applying Corollary 5.33, we can compute $\frac{d-4}{2}$ families of two conjugate points over $\mathbb{K}$ that can be used to construct a rational linear subsystem of $\mathcal{A}_{d-2}(\mathcal{C})$ of dimension 2 (see also Proposition 5.30). Applying Theorem 5.28 to this subsystem, we can always find a birational transformation defined by elements of the linear subsystem mapping $\mathcal{C}$ onto a conic. Hence, the optimality question is reduced to the existence and computation of optimal parametrizations of the corresponding conic. Indeed, since we have a subsystem of dimension 2, we only need to lift a single point on the conic with coordinates over an optimal field extension to obtain a new subsystem of dimension 1, and therefore to parametrize $\mathcal{C}$ over an optimal extension. Thus, the question now is how to compute the birationally equivalent conic, and how to invert a rational point, when it exists.

Let us consider how we can determine the birationally equivalent conic to a rational curve $\mathcal{C}$ of even degree d. Let $\mathcal{F}$ be the union of $\frac{d-4}{2}$ different families of two conjugate simple points on $\mathcal{C}$ over $\mathbb{K}$. We consider the proper linear subsystem $\mathcal{A}_{d-2}^{\mathcal{F}}(\mathcal{C})$ of $\mathcal{A}_{d-2}(\mathcal{C})$. Because of Theorem 5.28 and its Corollary 5.29, we may take $\Phi_1, \Phi_2, \Phi_3 \in \mathcal{A}_{d-2}^{\mathcal{F}}(\mathcal{C})$ such that $\{y_1 : y_2 : y_3 = \Phi_1 : \Phi_2 : \Phi_3\}$ $\mathbb{K}$-birationally transforms $\mathcal{C}$ onto a conic. Let $G(y_1, y_2, y_3)$ be the equation of this conic. The goal is to compute this conic by interpolation. For this purpose, we take a line $\mathcal{L}$ not passing through any point in $\mathrm{Sing}(\mathcal{C}) \cup \mathcal{F}$. Then, we consider the conjugate family defined by $\mathcal{L} \cap \mathcal{C}$; if $d = 4$, we determine two families of four points on $\mathcal{C}$. Now, applying the birational transformation to the constructed family, we get a family $\mathcal{G}$ of d simple conjugate points over $\mathbb{K}$ on the conic. Therefore, we have detected more than five points on the conic and we can interpolate it.

Now, let us consider the problem of inverting points on the conic, i.e., mapping them back to the original curve. First of all, we observe that we are only interested in inverting rational points on the conic, because, if no rational point on the conic exists, then we take a point on the original curve over an algebraic extension of degree 2 as described in Corollary 5.33. Let us assume that $Q = (q_1 : q_2 : 1)$ is a rational point on the conic. Then, we want to compute the inverse rational point P on $\mathcal{C}$. Thus, we have to solve the system

$$F(x, y, 1) = 0$$
$$\Phi_3(x, y, 1)q_1 - \Phi_1(x, y, 1) = 0$$
$$\Phi_3(x, y, 1)q_2 - \Phi_2(x, y, 1) = 0.$$

We know that the system has a unique solution. Therefore, we can solve the system by computing resultants and rational roots of univariate polynomials over $\mathbb{K}$.

We summarize all these ideas in algorithm OPTIMAL-PARAMETRIZATION. In this algorithm we do not consider the trivial case of rational curves that can be parametrized by lines, since the direct application of Algorithm PARAMETRIZATION-BY-LINES (see Sect. 4.6) always provides an algebraically optimal output (see Theorem 5.7). Furthermore, we use the results presented in Sect. 5.2 to compute rational points on conics. Therefore the algorithm only works over $\mathbb{Q}$. If one is working over other fields, the algorithm can be adapted to output a "nearly" optimal parametrization in the sense that the output is over the ground field or over a field extension of degree at most 2, but without guaranteeing that the field of parametrization is optimal.

Algorithm OPTIMAL-PARAMETRIZATION.

Given $F(x, y, z) \in \mathbb{Q}[x, y, z]$, an irreducible homogeneous polynomial defining a rational plane curve $\mathcal{C}$, the algorithm computes an optimal rational parametrization of $\mathcal{C}$.

1. The case of a linear F is trivial. So, in the subsequent steps we assume that $\deg(F) > 1$.
2. Let d be the degree of the polynomial F.
   2.1 If d is odd, we determine $\mathcal{A}_{d-2}(\mathcal{C})$. Then, we apply Statement (ii) of Corollary 5.33 to produce $\frac{d-3}{2}$ different families of 2 simple conjugate points each over $\mathbb{Q}$. Let $\mathcal{F}$ be the union of all these families. Determine the defining polynomial H of the proper linear subsystem $\mathcal{A}_{d-2}^{\mathcal{F}}(\mathcal{C})$.
   2.2 If $d = 2$, analyze the existence of rational points on $\mathcal{C}$ (see Sect. 5.2). Apply Steps 4 and 5 of the algorithm CONIC-PARAMETRIZATION (see Sect. 4.6) with an optimal point.
   2.3 If d is even and $d \neq 2$, determine $\mathcal{A}_{d-2}(\mathcal{C})$.
       2.3.1 Apply Statement (ii) of Corollary 5.33 to produce $\frac{d-4}{2}$ different families of 2 simple conjugate points over $\mathbb{Q}$. Let $\mathcal{F}$ be the union of all these families. Determine $\mathcal{A}_{d-2}^{\mathcal{F}}(\mathcal{C})$.
       2.3.2 Using adjoints in $\mathcal{A}_{d-2}^{\mathcal{F}}(\mathcal{C})$ find the birationally equivalent conic $\mathcal{D}$ to $\mathcal{C}$.
       2.3.3 Analyze the existence of rational points on $\mathcal{D}$ (see Sect. 5.2).
       2.3.4 If there exist rational points on $\mathcal{D}$, then compute one (see Sect. 5.2), map it back to obtain a rational point P on $\mathcal{C}$, and determine $\mathcal{A}_{d-2}^{\mathcal{F}}(\mathcal{C}) \cap \mathcal{H}(d-2, P)$; let H be its defining polynomial.
       2.3.5 If there are no rational points on $\mathcal{D}$, then apply Statement (ii) of Corollary 5.33 to produce a new different family $\mathcal{F}'$ of 2 simple conjugate points on $\mathcal{C}$ over $\mathbb{Q}$. Take one point Q in $\mathcal{F}'$. This implies a field extension of degree 2. Compute $\mathcal{A}_{d-2}^{\mathcal{F} \cup \{Q\}}(\mathcal{C})$; let H be its defining polynomial.
3. Set one of the parameters in H to 1 and let t be the remaining parameter of H. Return the solution in $\mathbb{P}^2(\mathbb{C})$ of $\{\mathrm{pp}_t(\mathrm{res}_y(F, H))) = 0, \mathrm{pp}_t(\mathrm{res}_x(F, H))) = 0\}$.

In general the coefficients of the birationally equivalent conic can be extremely large, and therefore the computation of rational points can be time consuming. To avoid this problem, and for practical implementations, one may also consider an algorithm that provides "nearly" optimal parametrizations (see above), i.e., in which Steps 2.2, 2.3, and 2.4 are omitted. In addition, the algorithm can be simplified for some special cases, e.g., if the curve $\mathcal{C}$ has singularities of certain multiplicities. For instance, in Example 5.32, taking lines through one of the double points one generates directly families of two conjugate points. In the following example we illustrate the algorithm as well as the previous remark.

Example 5.34. Let $\mathcal{C}$ be the curve of degree 5 defined by the polynomial

$$F(x, y, z) = y^5 - xy^2z^2 + xy^3z + xy^4 + 4x^2y^2z + x^2y^3 + x^3z^2 + x^3yz - x^3y^2$$

(see Fig. 5.1). $\mathcal{C}$ has a triple point at $(0 : 0 : 1)$, a double point at $(1 : 0 : 0)$, and two double points in the family $\{(1 : t : 1)\}_{t^2+1}$. So $\mathcal{C}$ is rational.
We apply the Algorithm OPTIMAL-PARAMETRIZATION to $\mathcal{C}$. We have to construct two families $\mathcal{F}_1, \mathcal{F}_2$ of $d - 2 = 3$ conjugate simple points each. For this purpose, we use $\mathcal{A}_3(\mathcal{C})$. Next, we compute $\mathcal{A}_4^{\mathcal{F}_1 \cup \mathcal{F}_2}(\mathcal{C})$ and use it for generating a family $\mathcal{F}_3$ of two conjugate simple points. Finally, constructing $\mathcal{A}_3^{\mathcal{F}_3}(\mathcal{C})$ we get a proper rational linear subsystem of dimension 1 parametrizing $\mathcal{C}$.

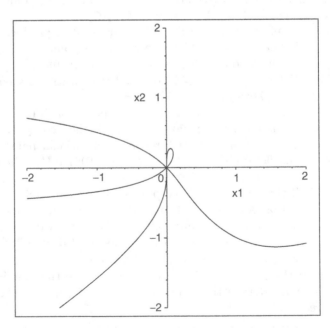

Fig. 5.1. Real part of $\mathcal{C}$

(i) Construction of $\mathcal{F}_1, \mathcal{F}_2$: The defining polynomial of $\mathcal{A}_3(\mathcal{C})$ is

$$H_3(x, y, z, \lambda_1, \ldots, \lambda_4) = \lambda_1 y^2 z - \lambda_2 y^2 z + \lambda_3 y^3 + \lambda_4 y^3 + \lambda_4 xyz +$$

$$\lambda_2 xy^2 + \lambda_1 x^2 z + \lambda_3 x^2 y .$$

Now, using the adjoint

$$H_3(x, y, z, 1, -1, 1, 1) = 2y^2 z + 2y^3 + yzx - xy^2 + x^2 z + x^2 y ,$$

we generate the family $\mathcal{F}_1 =$

$$\left\{ \left(t, \frac{-4(428117873\, t^2 + 370615860\, t - 237870392)}{11(205830037\, t^2 + 178336836\, t - 114428472)}, 1 \right) \right\}_{11\, t^3 - 48\, t^2 - 56\, t + 32} .$$

The family $\mathcal{F}_1$ consists of three real affine points (see rectangles in Fig. 5.2) Similarly, using the adjoint

$$H_3(x, y, z, 1, -1, 1, -1) = 2y^2 z - yzx - xy^2 + x^2 z + x^2 y ,$$

we generate the family $\mathcal{F}_2 =$

$$\left\{ \left(t, \frac{512 + 880t^4 + 76t^6 - 3t^8 - 343t^5 + 1856t^2 - 1544t^3 - 1408t}{-1165t^4 + 430t^5 - 512 + 3t^8 - 89t^6 + 2096t^3 - 2440t^2 + 1664t}, 1 \right) \right\}_{A(t)}$$

where $A(t) = t^3 + 4\, t^2 - 24\, t + 32$. The family $\mathcal{F}_2$ consists of one real affine point and two complex affine points (see rhombus in Fig. 5.2).

(ii) Construction of $\mathcal{F}_3$: The implicit equation of $\mathcal{A}_4^{\mathcal{F}_1 \cup \mathcal{F}_2}(\mathcal{C})$ is

$$H_4(x, y, z, \lambda_1, \lambda_2, \lambda_3) = -8\, x^2 z^2 \lambda_2 + 4\, x^2 z^2 \lambda_3 + 2\, x^2 y^2 \lambda_1 + 5\, x^2 y^2 \lambda_2 +$$
$$2\, x^2 y^2 \lambda_3 + 12\, y^2 z^2 \lambda_3 + 4\, y^3 z \lambda_1 - 3\, xy^3 \lambda_1 + 16\, y^3 z \lambda_3 - 7\, xy^3 \lambda_3 + 4\, y^2 z^2 \lambda_1 -$$
$$4\, xy^3 \lambda_2 - 8\, y^3 z \lambda_2 + \lambda_1 x^2 yz + \lambda_2 x^3 z + \lambda_3 x^3 y - 20\, xy^2 z \lambda_2 - 6\, xy^2 z \lambda_3 -$$
$$6\, xy^2 z \lambda_1 - 8\, y^4 \lambda_2 + 4\, y^4 \lambda_3 + 8\, xyz^2 \lambda_3 - 12\, xyz^2 \lambda_2.$$

We consider the adjoint

$$H_4(x, y, z, 0, 1, 1) = -4\, x^2 z^2 + 7\, x^2 y^2 + 12\, y^2 z^2 + 8\, y^3 z - 11\, xy^3 + x^3 z +$$
$$x^3 y - 26\, xy^2 z - 4\, y^4 - 4\, xyz^2,$$

and we generate the family $\mathcal{F}_3 = \{(tq(t), p(t), q(t))\}_{81 + 330\, t + 53\, t^2}$, where

$$p(t) = 3(2494124925978455383603257572871324049939t$$

$$+ 638441695230869281812342822978067045125)$$

$$q(t) = 53(1325488021962561501279224821352326432 01t$$

$$+ 339296083898977021696022930426 58593727) .$$

The family $\mathcal{F}_3$ consists of two real affine points (see circles in Fig. 5.2).

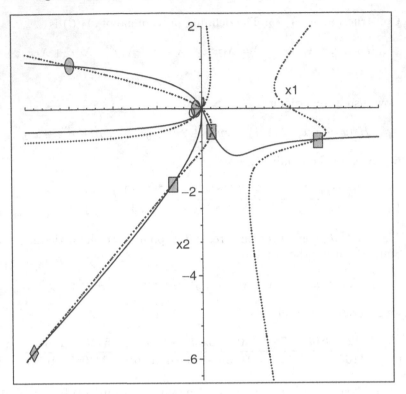

Fig. 5.2. $\mathcal{C}$ in continuous trace; $H_4(x, y, z, 0, 1, 1)$ in *dot*; points in $\mathcal{F}_3$ are *circles*; points in $\mathcal{F}_2$ are *rhombi*; points in $\mathcal{F}_1$ are *rectangles*

(iii) Parametrization of $\mathcal{C}$: The implicit equation of $\mathcal{A}_3^{\mathcal{F}_3}(\mathcal{C})$ is
$$H(x, y, z, t) = -3\,y^2 z - 3\,t y^2 z + 4\,t y^3 + y^3 + xyz + 3\,yzxt + 4\,xy^2 + 3\,xy^2 t + x^2 z + tx^2 y.$$
In Step (3) we finally determine the parametrization

$$\mathcal{P}(t) = \left(-\frac{7t^3 + 5t + 9t^2 + 1 + 2t^4}{-3t + 2t^5 - 2t^3 - 3t^2 - 1}, -\frac{1 + 3t + 2t^2}{2t + 2t^2 + 2t^3 + 1} \right).$$

In Fig. 5.3, one may observe how different elements in $H(x, y, z, t)$ generate the points on $\mathcal{C}$. In the first picture (top left box) $\mathcal{C}$ and $\mathcal{F}_3$ are plotted; in the second picture (top right box) $\mathcal{C}$, $\mathcal{F}_3$, and one element of $H(x, y, z, t)$ are plotted; in the third picture (bottom left box) $\mathcal{C}$, $\mathcal{F}_3$, and two elements of $H(x, y, z, t)$ are plotted; in the forth picture (bottom right box) $\mathcal{C}$, $\mathcal{F}_3$, and three elements of $H(x, y, z, t)$ are plotted. Note how, in all cases, points of $\mathcal{F}_3$ are fixed, as well as the singularities (the origin is a real 3-fold point, the other singularities are either complex or at infinity) and always a new intersection point appears, namely the point generated by the parametrization.

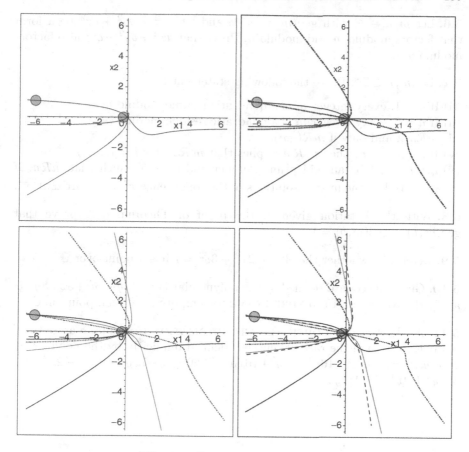

Fig. 5.3. Parametrization process

Exercises

5.1. Compute the implicit equation of the curve $\mathcal{D}$ in Example 5.10 using interpolation.

5.2. Prove that if d is even and $\mathcal{C}$ has a singularity over $\mathbb{K}$ of odd multiplicity, then $\mathcal{C}$ has simple points over $\mathbb{K}$.

5.3. Apply Exercise 5.2 to compute a rational point on the curve given in Example 5.32.

5.4. Decide whether $7\mathcal{R}18$ by using the algorithm suggested in the paragraph before Lemma 5.15.

5.5. Let $a \in \mathbb{Z}$. Prove that a can be expressed as $a = br^2$, where b is squarefree.

5.6. Let $m, n \in \mathbb{N}$ with $\gcd(m, n) = 1$, and let $ax^2 + by^2 + cz^2$ be a form that factors modulo m and modulo n. Prove that $ax^2 + by^2 + cz^2$ also factors modulo mn.

5.7. Let $m, n \in \mathbb{Z}^\star$. Prove the following statements:

(i) If $n = 1$, every integer m is a quadratic residue modulo n.
(ii) If $n = 2$, every integer m is a quadratic residue modulo n.
(iii) $m \mathcal{R} n$ if and only if $m \mathcal{R}(-n)$.
(iv) If $n = n_1 \cdots n_r$ then $m \mathcal{R} n$ implies that $m \mathcal{R} n_i$ for $i = 1, \ldots, r$.
(v) Let $n \in \mathbb{Z}^\star$ be an odd prime number and $m \in \mathbb{Z}^\star$ such that $m \mathcal{R} n$. If $n \equiv_4 3$, then the integer solutions of the congruence $x^2 \equiv_n m$ are $\pm m^{\frac{n+1}{4}}$.

5.8. With the notation given in the proof of Theorem 5.22 prove that $\gcd(ca, b) = \gcd(m, b) = 1$.

5.9. Determine whether the ellipse $2x^2 + 3y^2 = 1$ has a point over $\mathbb{Q}$.

5.10. Given the conic $\mathcal{C}$ defined by the polynomial $4x^2 - 8xy - 3y^2 + 8x + 8y - 5$, apply the Algorithm LEGENDRE SOLVE to compute a rational point on $\mathcal{C}$.

5.11. Apply Algorithm OPTIMAL-PARAMETRIZATION to compute an optimal parametrization of the rational curve $\mathcal{C}$ defined implicitly by the polynomial $F(x, y, z) = -9y^3z^2 + 16y^4z + 3y^5 + 16xy^2z^2 + 20xy^3z - 7xy^4 - x^2yz^2 + 2x^2y^2z - 5x^2y^3 + 9x^3z^2 + 4x^3yz$.

6

Rational Reparametrization

Summary. In Chaps. 4 and 5 we have studied different problems related to rational curves, assuming that the original curve was given implicitly. In this chapter we consider similar problems but now from another point of view, namely, we assume the curve is given in parametric form. But this parametric representation might not be optimal with respect to various criteria. So we want to transform such a rational parametrization into a better one. This new statement of the problem is specially interesting in some practical applications in CAGD, where objects are often given and manipulated parametrically. More precisely, we will focus on three different types of criteria: properness of parametrizations in Section 6.1, polynomiality of parametrizations in Section 6.2, and normality of parametrizations in Section 6.3. There are other criteria that might be considered. For instance, as we did in Chap. 5, one may ask for the algebraic optimality of the parametrization. For this case we refer to [ARS97], [ARS99], [ARS04], [RSV04], [SeV01], [SeV02].

As in previous chapters, we restrict the discussion to curves in the affine or projective plane over an algebraically closed field K of characteristic 0. Most of the results presented in this chapter can easily be extended to space curves. The results in Sects. 6.1 and 6.2 are also valid without the assumption of an algebraically closed field.

Given a rational parametrization of the curve, a solution to the problem of optimizing the parametrization might consist in first implicitizing the curve and afterward applying the algorithms developed in previous chapters. This solution might be too time consuming, and we would like to approach the problem by means of rational *"reparametrizations."* This means we want to avoid implicitizing the curve. Instead, our aim is to find a rational change of parameter, which transforms the given parametrization to a new and optimal parametrization of the same curve w.r.t. these criteria. Note that any reparametrization of a rational parametrization by means of a nonconstant rational change of parameter is again a parametrization of the same curve (see Exercise 6.1).

6.1 Making a Parametrization Proper

In Lemma 4.13 we have seen that any rational curve can be properly parametrized. Furthermore, in Theorem 4.14 the properness was characterized by means of the function field, and in Theorem 4.21 we also have characterized the properness by means of the degree. Moreover, in Theorem 4.30 an algorithmic criterion, based on the tracing index, was given for deciding the properness of a parametrization. In this section, we deal with the problem of determining a proper parametrization of a parametrically given rational curve. The *proper reparametrization problem* can be stated as follows: given a rational improper parametrization $\mathcal{P}(t) \in K(t)^2$ of an algebraic plane curve $\mathcal{C}$, find a rational proper parametrization $\mathcal{Q}(t) \in K(t)^2$ of $\mathcal{C}$.

Note that the existence of such a proper parametrization $\mathcal{Q}(t)$ is guaranteed by Lemma 4.13. Furthermore, by Lemma 4.17, we know that there exists a rational function $R(t) \in K(t)$ such that $\mathcal{P}(t) = \mathcal{Q}(R(t))$, namely $R(t) = \mathcal{Q}^{-1}(\mathcal{P}(t))$. We will develop an algorithm which, in addition to the proper parametrization, will also provide the transforming rational function $R(t)$. In the following, we first show how constructive proofs of Lüroth's Theorem (see Theorem 4.8) provide an algorithmic solution, and afterward we develop Sederberg's approach to the reparametrization problem (see [Sed86]).

In the sequel we use the same notation as in Chap. 4; i.e., if $\mathcal{P}(t)$ is a rational affine parametrization of $\mathcal{C}$ over K in reduced form, we write its components either as

$$\mathcal{P}(t) = \left(\frac{\chi_{11}(t)}{\chi_{12}(t)}, \frac{\chi_{21}(t)}{\chi_{22}(t)} \right),$$

where $\chi_{ij}(t) \in K[t]$, or as

$$\mathcal{P}(t) = (\chi_1(t), \chi_2(t)),$$

where $\chi_i(t) \in K(t)$. Furthermore, associated to $\mathcal{P}(t)$ we consider the polynomials

$$G_1^{\mathcal{P}}(s,t) = \chi_{11}(s)\chi_{12}(t) - \chi_{12}(s)\chi_{11}(t),$$
$$G_2^{\mathcal{P}}(s,t) = \chi_{21}(s)\chi_{22}(t) - \chi_{22}(s)\chi_{21}(t).$$

as well as

$$G^{\mathcal{P}}(s,t) = \gcd(G_1^{\mathcal{P}}(s,t), G_2^{\mathcal{P}}(s,t)).$$

6.1.1 Lüroth's Theorem and Proper Reparametrizations

Let us see now that a constructive proof of Lüroth's Theorem provides an algorithmic solution to the problem. We consider $\mathcal{P}(t)$, a rational parametrization of a curve $\mathcal{C}$ over K. Then

$$K \subset K(\mathcal{P}(t)) \subset K(t)$$

and clearly $K(\mathcal{P}(t))$ is different from K. Therefore, by Lüroth's Theorem, $K(\mathcal{P}(t))$ is isomorphic to $K(t)$. Let

$$\Phi : K(t) \longrightarrow K(\mathcal{P}(t)),$$

be an isomorphism which fixes the field K. Then

$$\Phi(K(t)) = K(\Phi(t)) = K(\mathcal{P}(t)).$$

Thus, there exists a rational function $R(t) \in K(t)$, namely $R(t) = \Phi(t)$, such that $K(R(t)) = K(\mathcal{P}(t))$. In particular, there exist rational functions $\xi_1, \xi_2 \in K(t)$ such that

$$\frac{\chi_{i\,1}(t)}{\chi_{i\,2}(t)} = \xi_i(R(t)), \quad i = 1, 2.$$

Hence for $\mathcal{Q}(t) = (\xi_1(t), \xi_2(t))$ we have

$$\mathcal{P}(t) = \mathcal{Q}(R(t)).$$

Let us prove that $\mathcal{Q}(t)$ is, in fact, a proper parametrization.

Lemma 6.1. $\mathcal{Q}(t) = (\xi_1(t), \xi_2(t))$ *is a proper parametrization of* C.

Proof. First, since $\mathcal{P}(t)$ is a parametrization, at least one of its components is not constant. Thus, the equality $\mathcal{P}(t) = \mathcal{Q}(R(t))$ implies that at least one component of $\mathcal{Q}(t)$ is not constant, from which we deduce that $\mathcal{Q}(t)$ is a parametrization, and that $R(t)$ is not constant either. So, applying Exercise 6.1 we see that $\mathcal{Q}(t)$ and $\mathcal{P}(t)$ parametrize the same curve, namely C.

Now, let us see that $\mathcal{Q}(t)$ is proper. By Theorem 4.33 we get that $\text{index}(\mathcal{P}(t)) = \deg(R(t)) \cdot \text{index}(\mathcal{Q}(t))$. Thus, if we can prove that $\text{index}(\mathcal{P}(t)) = \deg(R(t))$, then the properness of $\mathcal{Q}(t)$ follows by Theorem 4.30. For this purpose, we consider the rational maps

$$R := K \to K; t \mapsto R(t), \quad \text{and} \quad \mathcal{P} : K \to C; t \mapsto \mathcal{P}(t).$$

Note that, since $R(t)$ is not constant, by Exercise 4.5 the rational map R is dominant. Also, it is clear that $\mathcal{P}$ is dominant. Moreover, note that by Lemma 4.32, $\text{degree}(R) = \deg(R(t))$. Then, by Definitions 2.40 and 4.24

$$\deg(R(t)) = \text{degree}(R) = [K(t) : \tilde{\phi}_R(K(t))] = [K(t) : K(R(t))]$$

and

$$\text{index}(\mathcal{P}(t)) = \text{degree}(\mathcal{P}) = [K(t) : \tilde{\phi}_{\mathcal{P}}(K(C))] = [K(t) : K(\mathcal{P}(t))],$$

where $\tilde{\phi}_R$ and $\tilde{\phi}_P$ are the induced homomorphisms between the functions fields (see Sect. 2.2). But $K(R(t)) = K(P(t))$, so index$(P(t)) = \deg(R(t))$.   □

Therefore, any isomorphism between $K(t)$ and $K(P(t))$ provides a solution of the reparametrization problem. Conversely, any solution of the reparametrization problem provides an isomorphism between $K(t)$ and $K(P(t))$. More precisely, we have the following result.

Lemma 6.2. *Let $P(t)$ be a rational parametrization of an algebraic curve C, let $Q(t)$ be a proper parametrization of C, and let $R(t) \in K(t)$ be such that $P(t) = Q(R(t))$. Then, the homomorphism defined as*

$$\Phi: \begin{array}{ccc} K(t) & \longrightarrow & K(P(t)) \\ t & \longmapsto & Q^{-1}(P(t)) = R(t) \end{array}$$

is an isomorphism.

Proof. First we observe that Φ is injective, since Φ is not the zero homomorphism. Now, let us see that $\mathrm{Im}(\Phi) = K(P(t))$. Since $Q(t)$ is proper, Theorem 4.14 implies that $K(t) = K(Q(t))$.

Now, let us prove that $K(R(t)) = K(Q(R(t)))$. Indeed, since $Q(t)$ is proper, there exists Q^{-1}, say that a representant of it is the rational function $M(x,y)$. Then $t = M(Q(t))$. So, $R(t) = M(Q(R(t)))$. Thus, $R(t) \in K(Q(R(t)))$. The other inclusion is trivial. From these equalities of fields we get that

$$\mathrm{Im}(\Phi) = K(R(t)) = K(Q(R(t))) = K(P(t)).$$ □

6.1.2 Proper Reparametrization Algorithm

There are many approaches to the problem, most of them based on constructive proofs of Lüroth's Theorem, see for instance [GuR92], [Sed86], [Zip91], etc. For contributions to the problem in the surface case see for instance [CGS06] and [PeD06]. In [Sed86] a simple method based on evaluations of the polynomial $G^P(s,t)$ and subsequent gcd computations is proposed. But Sederberg is not very precise about which evaluations actually work. In the following we present a detailed analysis of this approach.

We start showing how the results on the tracing index stated in Sect. 4.3 give information on the reparametrization problem.

Theorem 6.3. *Let $P(t)$ be a rational parametrization of an algebraic curve C, let $Q(t)$ be a proper parametrization of C, and let $R(t) \in K(t)$ be such that $P(t) = Q(R(t))$. Then the following hold:*

(1) $\deg(R(t)) = \mathrm{index}(P(t))$;

(2) if $f(x, y)$ is the implicit equation of C then $\deg(R(t)) = \frac{\deg(\mathcal{P}(t))}{\max\{\deg_x(f), \deg_y(f)\}}$;

(3) $\deg(\mathcal{Q}(t)) = \frac{\deg(\mathcal{P}(t))}{\text{index}(\mathcal{P}(t))} = \frac{\deg(\mathcal{P}(t))}{\deg(R(t))}$.

Proof. These statements follow from Theorems 4.30, 4.33, and 4.35.   □

In order to deal with the proper reparametrization problem, we first observe that Lüroth's theorem guarantees the existence of solutions to the problem. But the solution is not unique; note that because of Lemma 4.17 any linear reparametrization of a proper parametrization is again proper. However, for a given proper parametrization $\mathcal{Q}(t)$ there exists a unique rational function $R(t)$ such that $\mathcal{P}(t) = \mathcal{Q}(R(t))$; namely $R(t) = \mathcal{Q}^{-1}(\mathcal{P}(t))$. Moreover, by Theorem 6.3, we know how to relate the degrees of $\mathcal{P}(t)$, $\mathcal{Q}(t)$, and $R(t)$. Therefore, if we were able to describe the set of all possible "proper reparametrizing rational functions" $R(t)$, taking an element there we would deduce $\deg(\mathcal{Q}(t))$. Then, introducing undetermined coefficients in the expression of $\mathcal{Q}(t)$, and applying the equality $\mathcal{P}(t) = \mathcal{Q}(R(t))$, we would generate a linear system of equations proving $\mathcal{Q}(t)$.

In Theorem 6.4, for a given parametrization $\mathcal{P}(t)$ we describe the set containing all the associated reparametrizing rational functions by means of the polynomials $G^{\mathcal{P}}(s, t)$ (see introduction to Sect. 6.1).

Theorem 6.4. *Let $\mathcal{P}(t)$ be a parametrization of C. Then, the set of all possible proper reparametrizing rational functions $R(t)$ associated to $\mathcal{P}(t)$ is*

$$\left\{ \frac{a\,G^{\mathcal{P}}(\alpha, t) + b\,G^{\mathcal{P}}(\beta, t)}{c\,G^{\mathcal{P}}(\alpha, t) + d\,G^{\mathcal{P}}(\beta, t)} \;\middle|\; \text{where } G^{\mathcal{P}}(\alpha, \beta) \neq 0, \text{ and } ad - bc \neq 0 \right\}$$

Proof. Let us denote the proposed set in the statement of the theorem by $\mathcal{R}$. Let $R(t) = M(t)/N(t)$, where $\gcd(M, N) = 1$, be a proper reparametrizing rational function associated to $\mathcal{P}(t)$; i.e., there exists a proper rational parametrization $\mathcal{Q}(t)$ such that $\mathcal{Q}(R(t)) = \mathcal{P}(t)$. By Lemma 3 in [PeD06], we have

$$G^{\mathcal{P}}(s, t) = M(t)N(s) - M(s)N(t)$$

up to multiplication by nonzero constants. Now, let $(\alpha, \beta) \in K^2$ be such that $G^{\mathcal{P}}(\alpha, \beta) \neq 0$. Then

$$\begin{pmatrix} N(\alpha) & -M(\alpha) \\ N(\beta) & -M(\beta) \end{pmatrix} \cdot \begin{pmatrix} M(t) \\ N(t) \end{pmatrix} = \begin{pmatrix} G^{\mathcal{P}}(\alpha, t) \\ G^{\mathcal{P}}(\beta, t) \end{pmatrix}.$$

Since $G^{\mathcal{P}}(\alpha, \beta) \neq 0$, the above 2×2 matrix is invertible, and hence

$$M(t) = aG^{\mathcal{P}}(\alpha, t) + bG^{\mathcal{P}}(\beta, t), \quad N(t) = cG^{\mathcal{P}}(\alpha, t) + dG^{\mathcal{P}}(\beta, t),$$

where

$$\begin{pmatrix} a & b \\ c & d \end{pmatrix} = \begin{pmatrix} N(\alpha) & -M(\alpha) \\ N(\beta) & -M(\beta) \end{pmatrix}^{-1}.$$

Thus, $ad \neq bc$. Therefore, $R(t) \in \mathcal{R}$.

Conversely, let us see that if $R(t) \in \mathcal{R}$, then $R(t)$ is a proper reparametrizing rational function. So there exist $\alpha, \beta \in K$ such that $G^{\mathcal{P}}(\alpha, \beta) \neq 0$, and there exist $a, b, c, d \in K$ such that $ab - cd \neq 0$, satisfying that $R(t) = \frac{a\,G^{\mathcal{P}}(\alpha,t)+b\,G^{\mathcal{P}}(\beta,t)}{c\,G^{\mathcal{P}}(\alpha,t)+d\,G^{\mathcal{P}}(\beta,t)}$. Let $\phi(t) = (at + b)/(ct + d)$ (note that this is an invertible rational function; see Exercise 4.1), and let $H = G^{\mathcal{P}}(\alpha,t)/G^{\mathcal{P}}(\beta,t)$. Then, $R(t) = \phi(H(t))$. Now, let $\mathcal{Q}(t)$ be a proper parametrization of $\mathcal{C}$. Then, there exists a proper reparametrizing function $S(t) \in K(t)$ such that $\mathcal{Q}(S(t)) = \mathcal{P}(t)$. So, $S(t) \in \mathcal{R}$. Furthermore, note that in the first part of the proof we have seen that for every $(\alpha', \beta') \in K^2$ such that $G^{\mathcal{P}}(\alpha', \beta') \neq 0$, there exist $a', b', c', d' \in K$ such that $a'b' - c'd' \neq 0$ satisfying that $S(t) = \frac{a'\,G^{\mathcal{P}}(\alpha',t)+b'\,G^{\mathcal{P}}(\beta',t)}{c'\,G^{\mathcal{P}}(\alpha',t)+d'\,G^{\mathcal{P}}(\beta',t)}$. Applying the above remark to the α, β generating $R(t)$, i.e., taking $\alpha' = \alpha, \beta' = \beta$, we get some $a', b', c', d' \in K$, with $a'd' \neq b'c'$ such that if $\tilde{\phi}(t) = (a't + b')/(c't + d')$, then $S(t) = \tilde{\phi}(H(t))$. Thus, taking into account that $\phi(t)$ is invertible, one has that

$$\mathcal{Q}(\tilde{\phi}(\phi^{-1}(R)(t))) = \mathcal{P}(t).$$

Finally, observing that $\tilde{\phi}(\phi^{-1}(t))$ is an invertible rational function, by Lemma 4.17, we get that $\tilde{\mathcal{Q}}(t) := \mathcal{Q}(\tilde{\phi}(\phi^{-1}(t)))$ is proper and $\tilde{\mathcal{Q}}(R(t)) = \mathcal{P}(t)$. $\square$

Remarks. Once the set of all proper reparametrizing rational functions have been described, one may (as mentioned before) compute the corresponding proper parametrization by introducing undetermined coefficients and solving a linear system of equations. Alternatively, one may proceed as follows: let $\mathcal{P}(t) = (\frac{\chi_{1,1}(t)}{\chi_{1,2}(t)}, \frac{\chi_{2,1}(t)}{\chi_{2,2}(t)})$ be a parametrization in reduced form, let $R(t)$ be a proper reparametrizing rational function for $\mathcal{P}(t)$, and let $\mathcal{Q}(t) = (\frac{\xi_{1,1}(t)}{\xi_{1,2}(t)}, \frac{\xi_{2,1}(t)}{\xi_{2,2}(t)})$ be a proper parametrization in reduced form, such that $\mathcal{Q}(R(t)) = \mathcal{P}(t)$. We consider the polynomials $g_i(x,y) := \xi_{i,2}(y) - x\xi_{i,1}(y)$. If $\xi_{i,1}(t) \neq 0$, since $\xi_{i,2}(t) \neq 0$ and $\gcd(\xi_{i,1}, \xi_{i,2}) \neq 1$, $g_i(x,y)$ defines an irreducible curve. Moreover $g_i(\frac{\chi_{i,1}(t)}{\chi_{i,2}(t)}, R(t)) = 0$. Thus, the curve defined by g_i is indeed rational. In this situation, we may implicitize the parametrizations $(\frac{\chi_{i,1}(t)}{\chi_{i,2}(t)}, R(t))$, and from there determine $\mathcal{Q}(t)$. If for some i we have $\chi_{i,1}(t) = 0$, then the curve is either the line $x = 0$ or $y = 0$, and we can obviously determine $\mathcal{Q}(t)$.

From these results we can derive the following algorithm.

Algorithm PROPER-REPARAMETRIZATION
Given a rational parametrization

$$\mathcal{P}(t) = \left(\frac{\chi_{11}(t)}{\chi_{12}(t)}, \frac{\chi_{21}(t)}{\chi_{22}(t)} \right)$$

in reduced form, of a plane algebraic curve $\mathcal{C}$, the algorithm computes a proper rational parametrization $\mathcal{Q}(t)$ of $\mathcal{C}$, and a rational function $R(t)$ such that $\mathcal{Q}(R(t)) = \mathcal{P}(t)$.

1. Compute the polynomials

$$G_1^{\mathcal{P}}(s,t) = \chi_{11}(s)\chi_{12}(t) - \chi_{12}(s)\chi_{11}(t),$$

$$G_2^{\mathcal{P}}(s,t) = \chi_{21}(s)\chi_{22}(t) - \chi_{22}(s)\chi_{21}(t),$$

and

$$G^{\mathcal{P}}(s,t) = \gcd(G_1^{\mathcal{P}}(s,t), G_2^{\mathcal{P}}(s,t)).$$

2. If $\deg_t(G^{\mathcal{P}}(s,t)) = 1$, then $\mathcal{P}$ is proper. Thus, return $\mathcal{Q}(t) = \mathcal{P}(t)$ and $R(t) = t$. Otherwise, go to Step 3.
3. Choose $\alpha, \beta \in K$ such that $G^{\mathcal{P}}(\alpha, \beta) \neq 0$, and choose $a, b, c, d \in K$ such that $ad - bc \neq 0$. Consider the rational function

$$R(t) = \frac{aG^{\mathcal{P}}(\alpha, t) + bG^{\mathcal{P}}(\beta, t)}{cG^{\mathcal{P}}(\alpha, t) + dG^{\mathcal{P}}(\beta, t)}.$$

4. (Option 1). Let $r = \deg(\mathcal{P}(t))/\deg(R(t))$. Introduce a generic rational parametrization $\mathcal{Q}(t)$ of degree r with undetermined coefficients. From the equality $\mathcal{P}(t) = \mathcal{Q}(R(t))$ derive a linear system of equations in the undetermined coefficients, and solving it determine $\mathcal{Q}(t)$.
4. (Option 2). For $i = 1, 2$, compute the implicit equations $g_i(x, y) = 0$ of the curves defined by the parametrizations $(\frac{\chi_{i1}(t)}{\chi_{i2}(t)}, R(t))$. From $g_i(x, y)$ determine $\mathcal{Q}(t)$ (see remark to Theorem 6.4).
5. Return the proper parametrization $\mathcal{Q}(t)$, and the rational function $R(t)$.

Example 6.5. Let $\mathcal{C}$ be the rational curve defined by the parametrization

$$\mathcal{P}(t) = \left(\frac{8t^6 - 12t^5 + 32t^3 + 24t^2 + 12t}{t^6 - 3t^5 + 3t^4 + 3t^2 + 3t + 1}, \frac{24t^5 + 54t^4 - 54t^3 - 54t^2 + 30t}{t^6 - 3t^5 + 3t^4 + 3t^2 + 3t + 1} \right).$$

First, we compute the polynomial

$$G^{\mathcal{P}}(s,t) = \gcd(G_1^{\mathcal{P}}(s,t), G_2^{\mathcal{P}}(s,t)) = (-t + st - 1 - s)(-s + t).$$

Since $\deg_t(G^{\mathcal{P}}(s,t)) \neq 1$, we apply Step 3 of the algorithm, and we arbitrarily select $\alpha = 2$, and $\beta = 4$ (note that $G^{\mathcal{P}}(\alpha, \beta) \neq 0$), and we consider $a = 1$,

$b = 2, c = 3, d = 4$ (observe that $ad - bc \neq 0$). So we get the rational function

$$R(t) = \frac{aG^{\mathcal{P}}(\alpha, t) + bG^{\mathcal{P}}(\beta, t)}{cG^{\mathcal{P}}(\alpha, t) + dG^{\mathcal{P}}(\beta, t)} = \frac{-39t + 7t^2 + 46}{-83t + 15t^2 + 98}.$$

In Step 4 (Option 1) of the algorithm, we have that $r = 3$, and we consider

$$\mathcal{Q}(t) = \left(\frac{a_3t^3 + a_2t^2 + a_1t + a_0}{b_3t^3 + b_2t^2 + b_1t + b_0}, \frac{c_3t^3 + c_2t^2 + c_1t + c_0}{d_3t^3 + d_2t^2 + d_1t + d_0} \right).$$

Solving the linear system derived from $\mathcal{Q}(R(t)) = \mathcal{P}(t)$, we get that

$$\mathcal{Q} = \left(\frac{q_1(t)}{q(t)}, \frac{q_2(t)}{q(t)} \right)$$

where

$$q_1(t) = 8(325997t^3 - 460641t^2 + 216963t - 34063)$$

$$q_2(t) = 12(343245t^3 - 482261t^2 + 225859t - 35259),$$

$$q(t) = 205757t^3 - 291399t^2 + 137559t - 21645.$$

If we apply Option 2 in Step 4 of the algorithm, we compute the implicit equations of the parametrizations

$$\left(\frac{8t^6 - 12t^5 + 32t^3 + 24t^2 + 12t}{t^6 - 3t^5 + 3t^4 + 3t^2 + 3t + 1}, \frac{-39t + 7t^2 + 46}{-83t + 15t^2 + 98} \right),$$

and

$$\left(\frac{24t^5 + 54t^4 - 54t^3 - 54t^2 + 30t}{t^6 - 3t^5 + 3t^4 + 3t^2 + 3t + 1}, \frac{-39t + 7t^2 + 46}{-83t + 15t^2 + 98} \right).$$

This leads to

$$g_1(x, y) = 2607976y^3 - 3685128y^2 + 1735704y - 272504 - 205757xy^3$$
$$+ 291399xy^2 - 137559xy + 21645x ,$$
$$g_2(x, y) = 4118940y^3 - 5787132y^2 + 2710308y - 423108 - 205757xy^3$$
$$= + 291399xy^2 - 137559xy + 21645x .$$

Solving $g_1 = 0$ and $g_2 = 0$ for the variable x, we get the parametrization $\mathcal{Q}(t)$.

6.2 Making a Parametrization Polynomial

Till now, we have been dealing with rational parametrizations. However, one may ask whether the curve admits a polynomial parametrization, i.e., a rational affine parametrization where all components are polynomial. This type of representation is specially interesting when dealing with applications. For instance, if one is giving numerical values to the parameter, and the parametrization is not polynomial, one may have a numerically unstable behavior when getting close to the roots of a denominator.

Clear examples of curves having polynomial parametrizations are, for instance, curves defined by polynomials of the type $x - G(y)$, like the parabolas. This class of curves can be extended to a more general family, namely those affine rational curves having only one place at infinity (see [Abh66]).

In this section, we show how to decide whether a given affine rational parametrization can be reparametrized into a polynomial parametrization. Moreover, if this is the case, we explicitly compute the change of parameter to be performed. Although we will present this reparametrization method for plane algebraic curves, one may observe that it can be easily extended to space curves. A similar analysis can be done for quasipolynomial curves, i.e., rational affine curves that can be parametrized with a parametrization having at least one polynomial component, as for instance the hyperbola $yx - 1$. Here we do not consider this extension, for further details we refer to [SeV02].

In the sequel we follow the reasoning scheme in [CaM91]. Similar results can be found in [AGR95]. For further reading on the topic we refer to [GRS02].

Definition 6.6. *A rational affine parametrization $\mathcal{P}(t)$ of a rational affine curve $\mathcal{C}$ is called a* polynomial parametrization *if all its components are polynomial. Furthermore, the affine curve $\mathcal{C}$ is called a* polynomial curve *if it is rational and can be parametrized by means of a polynomial parametrization.*

We start our analysis with some lemmas on rational functions.

Lemma 6.7. *Let $R(t) = \frac{p(t)}{(bt-a)^s} \in K(t)$, with $a, b \in K$, $b \neq 0$, $s \in \mathbb{N}$, and $\deg(p) \leq s$. Then $R\left(\frac{b+at}{bt}\right)$ is a polynomial.*

Proof. Let $\deg(p) = r$ and $p(t) = a_r t^r + \cdots + a_0$ then

$$R\left(\frac{b+at}{bt}\right) = \frac{1}{b^{s+r}}\left(a_r(b+at)^r + a_{r-1}(b+at)^{r-1}bt + \cdots + a_0(bt)^r\right) \cdot t^{s-r}$$

is a polynomial since $s \geq r$. $\qquad\square$

Lemma 6.8. *Let $p(t) \in K[t]$ be a nonconstant polynomial, and let $\varphi(t) \in K(t)$ be a nonconstant rational function such that $p(\varphi(t))$ is a polynomial. Then $\varphi(t)$ is a polynomial.*

Proof. Let $\varphi = \frac{\varphi_1}{\varphi_2}$ with $\gcd(\varphi_1, \varphi_2) = 1$, let $p(t) = a_r t^r + \cdots + a_0$ with $r = \deg(p) > 0$, and let $p(\varphi(t)) = A(t) \in K[t]$. Then $a_r \varphi_1^r + a_{r-1}\varphi_1^{r-1}\varphi_2 + \cdots + a_0\varphi_2^r = \varphi_2^r A(t)$. Therefore, φ_2 divides the left-hand side of the equality, and hence it divides φ_1. Thus, since $\gcd(\varphi_1, \varphi_2) = 1$, φ_2 must be constant. $\quad\square$

Lemma 6.9. *Let $R(t) \in K(t)$. If there exists a nonconstant polynomial $\varphi(t) \in K[t]$ such that $R(\varphi(t))$ is a polynomial then $R(t)$ is also a polynomial.*

Proof. Let $R = \frac{p}{q}$, with $\gcd(p, q) = 1$, and let $R(\varphi(t)) = A(t) \in K[t]$. Then, one has the polynomial equality

$$p(\varphi(t)) = q(\varphi(t))A(t).$$

Now assume that $\deg(q) > 0$, and take a root α of q. Since φ is not constant, there exists $t_0 \in K$ such that $\varphi(t_0) = \alpha$. But this implies that $p(\varphi(t_0)) = p(\alpha) = 0$, which is impossible since $\gcd(p, q) = 1$. □

Lemma 6.10. *Let* $R(t) = \frac{p(t)}{q(t)} \in K(t)$ *be a nonpolynomial rational function in reduced form. The following statements are equivalent:*

(1) There exists a nonconstant rational function $\varphi(t) \in K(t)$ *such that* $R(\varphi(t))$ *is a polynomial.*
(2) $q(t) = (bt - a)^s$ *with* $a, b \in K$, $b \neq 0$, $s \in \mathbb{N}$, *and* $\deg(p(t)) \leq s$.

Proof. (2) implies (1) because of Lemma 6.7.
Let us prove that (1) implies (2). For this purpose, let $\varphi = \frac{\varphi_1}{\varphi_2}$ with $\gcd(\varphi_1, \varphi_2) = 1$ and let $\deg(p) = r, \deg(q) = s$, and $p(t) = a_r t^r + \cdots + a_0$, $q(t) = b_s t^s + \cdots + b_0$. Moreover, let $R(\varphi(t)) = A(t) \in K[t]$. First we see that $r \leq s$. Indeed, if $r > s$ then

$$a_r \varphi_1^r + a_{r-1} \varphi_1^{r-1} \varphi_2 + \cdots + a_0 \varphi_2^r = \varphi_2^{r-s}(b_s \varphi_1^s + a_{s-1} \varphi_1^{s-1} \varphi_2 + \cdots + b_0 \varphi_2^s)A(t)$$

where φ_2^{r-s} is a nonzero polynomial. Therefore, φ_2 divides the left-hand side of the equality, and hence it divides φ_1. So, because of $\gcd(\varphi_1, \varphi_2) = 1$, φ_2 must be constant. Thus, we have the polynomial equality

$$p(\varphi(t)) = q(\varphi(t))A(t).$$

Then, since φ is not constant, we can always take $t_0 \in K$ such that $\varphi(t_0)$ is a root of $q(t)$. But this implies that $\varphi(t_0)$ is also a root of $p(t)$, which is impossible since $\gcd(p, q) = 1$ (note that q is not constant because R is not polynomial). Therefore, $r \leq s$.
Now, we prove that $q(t)$ has only one different root. For this, let us write p and q as

$$p(t) = \prod_{i=1}^{r}(\beta_i t - \alpha_i), \quad q(t) = \prod_{j=1}^{s}(\rho_j t - \gamma_j).$$

This leads to the polynomial equality

$$\varphi_2^{s-r} \prod_{i=1}^{r}(\beta_i \varphi_1 - \alpha_i \varphi_2) = \prod_{j=1}^{s}(\rho_j \varphi_1 - \gamma_j \varphi_2)A(t) .$$

Now we prove that for every j, $\rho_j \varphi_1 - \gamma_j \varphi_2$ is a nonzero constant polynomial. First observe that $\rho_j \varphi_1 - \gamma_j \varphi_2$ cannot be zero, since otherwise, taking into account that φ_2 is not zero and that $\rho_j \neq 0$, then $\varphi_1/\varphi_2 = \gamma_j/\rho_j \in K$, but

φ is not constant. Let us assume that there exists j such that $\rho_j\varphi_1 - \gamma_j\varphi_2$ is not constant, and let t_0 be one of its roots. Note that $\varphi_2(t_0) \neq 0$, since $\rho_j \neq 0$ and $\gcd(\varphi_1, \varphi_2) = 1$. Therefore, there exists i such that t_0 is a root of $\beta_i\varphi_1 - \alpha_i\varphi_2$, but this implies that p, q have a common root, namely $\frac{\gamma_j}{\rho_j} = \frac{\alpha_i}{\beta_i}$, which is impossible.

In this situation, let us assume that $q(t)$ has two different roots, say $\frac{\gamma_1}{\rho_1}, \frac{\gamma_2}{\rho_2}$. So

$$\begin{cases} \rho_1\varphi_1 - \gamma_1\varphi_2 \doteq \lambda_1 \in K \\ \rho_2\varphi_1 - \gamma_2\varphi_2 = \lambda_2 \in K \end{cases}$$

Since the two roots are different, then $\Delta := -\rho_1\gamma_2 + \gamma_1\rho_2 \neq 0$, and hence both φ_1 and φ_2 are constant. But this implies that φ is constant too, which is impossible. Therefore, $q(t) = (bt - a)^s$ for some $a, b \in K$, and $b \neq 0$ because R is not polynomial. □

Remarks. Note that if Statement (1) in Lemma 6.10 holds, we have also proved that for every root α of $q(t)$ one gets that $\varphi_1 - \alpha\varphi_2$ is constant, where φ_1 and φ_2 are the numerator and denominator of φ in reduced form.

In Definition 6.6, we have introduced the notion of a polynomial affine curve by requiring the existence of a polynomial affine parametrization. However, we have not imposed on the polynomial parametrization the condition of being proper. In Theorem 6.11, we see that properness can always be achieved simultaneously with polynomiality.

Theorem 6.11. *For every polynomial curve there exist proper polynomial parametrizations.*

Proof. Let $\mathcal{C}$ be polynomial. So there exists a polynomial parametrization $\mathcal{P}(t)$ of $\mathcal{C}$. By Lemma 4.17 there exists a nonconstant rational function $\varphi(t) \in K(t)$ and a proper rational parametrization $\mathcal{Q}(t)$ such that $\mathcal{Q}(\varphi(t)) = \mathcal{P}(t)$. If $\mathcal{Q}(t)$ is polynomial, then the theorem holds. Let $\mathcal{Q}(t)$ be nonpolynomial. We distinguish two cases depending on whether one component of $\mathcal{Q}(t)$ is polynomial or not. If only one component of $\mathcal{Q}(t)$ is polynomial, then it must be constant. Indeed, if it is not constant, applying Lemma 6.8, one gets that φ is polynomial, and applying Lemma 6.9 to the other component one gets that the other component is also polynomial, which is impossible because $\mathcal{Q}(t)$ is not polynomial. Therefore if only one component of $\mathcal{Q}(t)$ is polynomial, say for instance the first one, then (λ, t), with $\lambda \in K$ parametrizes polynomially and properly the curve $\mathcal{C}$. Now, let us assume that no component of $\mathcal{Q}(t)$ is polynomial. Applying Lemma 6.10 to each component, we can express $\mathcal{Q}(t)$ as

$$\mathcal{Q}(t) = \left(\frac{p_1(t)}{(b_1t - a_1)^{s_1}}, \frac{p_2(t)}{(b_2t - a_2)^{s_2}} \right) ,$$

where p_i is a polynomial of degree smaller or equal to s_i, and $b_i \neq 0$. Furthermore, by the remark to Lemma 6.10, one also has

$$\begin{cases} b_1\varphi_1 - a_1\varphi_2 = \lambda_1 \\ b_2\varphi_1 - a_2\varphi_2 = \lambda_2 \end{cases}$$

where λ_i are constants and φ_1, φ_2 are the numerator and denominator of φ expressed in reduced form. If the roots of the denominators in $\mathcal{Q}(t)$ are different, then $\Delta := a_1b_2 - a_2b_1 \neq 0$. Therefore φ_1 and φ_2 are both constant, that φ must be constant, which is impossible. Thus, $a_1b_2 = a_2b_1$. In this situation, the reparametrization suggested in Lemma 6.7 is the same for both components of $\mathcal{Q}(t)$, namely

$$\Phi(t) = \frac{b_1 + a_1t}{b_1t} = \frac{1}{t} + \frac{a_1}{b_1} = \frac{1}{t} + \frac{a_2}{b_2} = \frac{b_2 + a_2t}{b_2t} \, .$$

Hence, since $\Phi(t)$ is linear, and $\mathcal{Q}(\Phi(t))$ is a proper polynomial rational parametrization of C. □

Theorem 6.12. *If C is a polynomial curve, then every proper nonpolynomial rational parametrization in reduced form of C is of the type*

$$\mathcal{P}(t) = \left(\frac{\chi_{11}(t)}{(bt - a)^r}, \frac{\chi_{21}(t)}{(bt - a)^s} \right),$$

where $\deg(\chi_{11}) \leq r$ and $\deg(\chi_{21}) \leq s$, and $b \neq 0$.

Proof. Since C is polynomial, by Theorem 6.11, one knows that there exists a proper polynomial parametrization $\mathcal{Q}(t)$ of C. Moreover, since $\mathcal{P}(t)$ is proper, by Lemma 4.17, there exists a linear rational function $\varphi(t) \in K(t)$ such that $\mathcal{P}(\varphi(t)) = \mathcal{Q}(t)$. Therefore, reasoning as in the proof of Theorem 6.11, we get the result. □

Theorem 6.13. *If*

$$\mathcal{P}(t) = \left(\frac{\chi_{11}(t)}{(bt - a)^r}, \frac{\chi_{21}(t)}{(bt - a)^s} \right),$$

where $\deg(\chi_{11}) \leq r$ and $\deg(\chi_{21}) \leq s$ is a rational parametrization of an affine curve C, then C is polynomial and can be polynomially parametrized as

$$\mathcal{P}\left(\frac{b + at}{bt} \right).$$

Proof. This follows from Lemma 6.7. □

Corollary 6.14. *Let C be a rational curve, and let*

$$\mathcal{P}(t) = \left(\frac{\chi_{1,1}(t)}{\chi_{1,2}(t)}, \frac{\chi_{2,1}(t)}{\chi_{2,2}(t)} \right)$$

be a proper nonpolynomial parametrization of C, in reduced form. Then the following statements are equivalent

(1) C is polynomial;
(2) $\deg(\chi_{i,1}) \le \deg(\chi_{i,2})$ for $i = 1, 2$ and both $\chi_{1,2}$ and $\chi_{2,2}$ are powers of the same linear polynomial.

Proof. That (1) implies (2) follows from Theorem 6.12, and that (2) implies (1) follows from Theorem 6.13. □

These results lead to the following reparametrization algorithm for polynomiality.

Algorithm POLYNOMIAL-REPARAMETRIZATION

Given a rational (nonpolynomial) parametrization $\mathcal{P}(t) = \left(\frac{\chi_{11}(t)}{\chi_{12}(t)}, \frac{\chi_{21}(t)}{\chi_{22}(t)} \right)$ in reduced form, of a plane affine algebraic curve C, the algorithm decides whether C is polynomial, and in the affirmative case it computes a polynomial parametrization of C.

1. Apply algorithm PROPER-REPARAMETRIZATION and if $\mathcal{P}(t)$ is not proper replace it by a proper parametrization.
2. Compute the square-free part $\chi_{i2}^*(t)$ of $\chi_{i2}(t)$ for $i = 1, 2$.
3. If χ_{12}^* and χ_{22}^* are both linear and have the same root β, then return "A polynomial proper parametrization of C is $\mathcal{P}\left(\frac{1+\beta t}{t} \right)$" else return "$C$ is not polynomial".

Example 6.15. Let C be the rational affine curve parametrized as

$$\mathcal{P}(t) = \left(\frac{35\,t^3 + 66\,t^2 + 42\,t + 9}{27\,t^3 + 54\,t^2 + 36\,t + 8}, \frac{97\,t^4 + 248\,t^3 + 240\,t^2 + 104\,t + 17}{81\,t^4 + 216\,t^3 + 216\,t^2 + 96\,t + 16} \right).$$

Algorithm PROPER-REPARAMETRIZATION tells us that $\mathcal{P}(t)$ is proper. Moreover, we have

$$\chi_{12}^*(t) = \chi_{22}^*(t) = 3t + 2.$$

Therefore, C is polynomial and a proper polynomial parametrization of C is

$$\mathcal{P}\left(\frac{3 - 2t}{3t} \right)$$

$$= \left(\frac{35}{27} - \frac{4}{27}t + \frac{2}{81}t^2 - \frac{1}{729}t^3, \frac{97}{81} - \frac{32}{243}t + \frac{8}{243}t^2 - \frac{8}{2187}t^3 + \frac{1}{6561}t^4 \right).$$

We also observe that the implicit equation of $\mathcal{C}$ is

$$f(x, y) = -2 + 4x + 3y - 6x^2 - 3y^2 + 4x^3 + y^3 - x^4.$$

Note that $\mathcal{C}$ has only one point at infinity, namely $(0 : 1 : 0)$. This is a simple point and consequently $\mathcal{C}$ only has only one place at infinity.

Example 6.16. Let $\mathcal{C}$ be the rational affine curve parametrized as

$$\mathcal{P}(t) = \left(\frac{35\,t^3 - 99\,t^2 + 279\,t - 98}{27\,t^3 - 135\,t^2 + 225\,t - 125}, \frac{35\,t^3 - 99\,t^2 + 279\,t - 98}{18\,t^3 - 33\,t^2 - 40\,t + 75} \right).$$

Algorithm PROPER-REPARAMETRIZATION tells us that $\mathcal{P}(t)$ is proper. Moreover, we have

$$\chi^*_{1\,2}(t) = 3t - 5, \quad \chi^*_{2\,2}(t) = 6t^2 - t - 15.$$

Therefore, $\mathcal{C}$ is not polynomial. We also observe that the implicit equation of $\mathcal{C}$ is

$$f(x, y) = -x^3 + xy^3 - y^3.$$

Note that $\mathcal{C}$ has two points at infinity, namely $(0 : 1 : 0)$ and $(1 : 0 : 0)$, and consequently $\mathcal{C}$ has more than one place at infinity. In addition, one may observe that

$$\chi^*_{1\,2}(t) = 3t - 5, \quad \chi^*_{2\,2}(t) = (2t + 3)(3t - 5).$$

If we consider the reparametrization corresponding to the common root $\frac{5}{3}$ of $\chi^*_{i\,2}(t)$, we get

$$\mathcal{P}\left(\frac{3 + 5t}{3t} \right) = \left(\frac{35}{27} + \frac{76}{27}t + \frac{722}{81}t^2 + \frac{6859}{729}t^3, \right.$$
$$\left. \frac{1}{81} \frac{945 + 2052\,t + 6498\,t^2 + 6859\,t^3}{6 + 19\,t} \right),$$

which is not polynomial but quasipolynomial. For an explanation of this phenomenon see [SeV02].

6.3 Making a Parametrization Normal

In the previous sections we have reparametrized a given parametrization w.r.t. the criteria of properness and polynomiality. In this section we analyze another property of the parametrization, namely its normality. In Chap. 4, we have seen that any rational parametrization $\mathcal{P}(t)$ induces a natural dominant rational mapping $\mathcal{P}$ from the affine line onto the curve. In fact, in studying the properness of a parametrization $\mathcal{P}(t)$ we have analyzed the injectivity of $\mathcal{P}$ over almost all values in $\mathbb{A}^1(K)$. Now, we focus on the surjectivity. The mapping $\mathcal{P}$ is dominant. Thus, in general, it might not be surjective (i.e., normal), and hence some points of the algebraic set might be missed. This phenomenon may generate unexpected complications in applications; for instance in the problem of plotting of geometric objects on the screen of a computer. Therefore, the question of deciding whether a rational parametrization is normal and if not computing a normal parametrization, if possible, arises.

In this section we show how to decide whether a given rational parametrization of a curve over K is normal or not. Furthermore, if it is not normal we show how to reparametrize the given parametrization into a normal one. Here we consider the problem over algebraically closed fields. In Sect. 7.3, we will treat the case of the field of the real numbers. Most of the results in this section can be found in [Sen02]. For further reading on this problem we refer to [AnR06], [BaR95], and [ChG91].

Throughout this section we will use the same notation as in Sect. 6.1. $\mathcal{P}(t)$ is a rational affine parametrization in reduced form, not necessarily proper, of a curve $\mathcal{C}$ over K. We write the components of $\mathcal{P}(t)$ as

$$\mathcal{P}(t) = \left(\frac{\chi_{11}(t)}{\chi_{12}(t)}, \frac{\chi_{21}(t)}{\chi_{22}(t)} \right).$$

In addition, we denote by $\mathcal{P}$ the rational map

$$\mathcal{P} : \mathbb{A}^1(K) \longrightarrow \mathcal{C}$$
$$t \longmapsto \mathcal{P}(t).$$

Furthermore, associated to $\mathcal{P}(t)$ we also consider the polynomials (see Chap. 4)

$$H_1^{\mathcal{P}}(x,t) = x \cdot \chi_{12}(t) - \chi_{11}(t), \quad H_2^{\mathcal{P}}(y,t) = y \cdot \chi_{22}(t) - \chi_{21}(t).$$

In this situation, we introduce the notion of normality as follows.

Definition 6.17. *A rational affine parametrization $\mathcal{P}(t)$ is* normal *iff the rational mapping $\mathcal{P}$ is surjective, or equivalently iff for all $P \in \mathcal{C}$ there exists $t_0 \in K$ such that $\mathcal{P}(t_0) = P$. Furthermore, if there exists a normal parametrization of $\mathcal{C}$ we say that $\mathcal{C}$ can be* normally parametrized.

An obvious example of normality is the parabola parametrization (t, t^2). It is also easy to check that the circle parametrization

$$\mathcal{P}(t) = \left(\frac{2t}{t^2 + 1}, \frac{t^2 - 1}{t^2 + 1} \right)$$

is not normal, since the point $(0, 1)$ is on the circle $x^2 + y^2 = 1$ but no value of t generates $(0, 1)$. In the following we show a less trivial example of a normal parametrization.

Example 6.18. We consider the parametrization

$$\mathcal{P}(t) = \left(\frac{t^2 - 1}{t^3}, \frac{t - 1}{t^2} \right)$$

of the plane cubic defined by the polynomial

$$f(x, y) = y^3 + 2y^2 - 3xy + x^2.$$

Applying methods from Sect. 4.3, one may check that $\mathcal{P}(t)$ is a proper parametrization. We compute the inverse of $\mathcal{P}(t)$ (see Sect. 4.4), which can be expressed as

$$\mathcal{P}^{-1}(t) = \frac{2y - x}{y^2}.$$

Thus, we deduce that for every $(a, b) \in \mathcal{C} \setminus \{(0, 0)\}$

$$(a, b) = \mathcal{P}\left(\frac{2b - a}{b^2}\right).$$

Furthermore, $\mathcal{P}(1) = (0, 0)$. Therefore, $\mathcal{P}(t)$ is normal.

We will prove that any affine rational parametrization generates, when the parameter takes values in an algebraically closed field, all affine points on the curve with the exception of at most one point, and we will show that any affine parametrization can always be reparametrized into a normal one. The basic idea is that the image of a projective variety under a regular map is closed (see, e.g., [Sha94] Vol. 1., Chap. 1, Sect. 5.2, Theorem 2). Thus, if the parametrization is seen projectively, the induced rational map is defined from the projective line over K onto the projective closure of $\mathcal{C}$. Hence, in this case, the image is the whole curve. This means that when considering again the affine parametrization, only the corresponding point on the curve corresponding to the point at infinity of K may not be generated. Moreover, taking a projective parametrization of the curve that sends the infinity of K into a point of $\mathcal{C}$ at infinity one may generate a normal parametrization of the curve.

In the following we describe these ideas in details giving constructive proofs that will provide algorithmic procedures. We start our analysis with Lemma 6.19. This result can also be found in [GRY02], Remark 1.6. The approach in [GRY02] is derived using rational function field theory.

For technical reasons, in the following we consider the leading coefficient of the zero polynomial to be 0.

Lemma 6.19. *Let $\ell_1(x), \ell_2(y)$ be the leading coefficients w.r.t. t of the polynomials $H_1^{\mathcal{P}}(x, t), H_2^{\mathcal{P}}(y, t)$, respectively. Then*

$$\mathcal{P}(K) = \{(a, b) \in \mathcal{C} \mid \gcd(H_1^{\mathcal{P}}(a, t), H_2^{\mathcal{P}}(b, t)) \neq 1\}.$$

Furthermore,

$$\mathcal{C} \setminus \mathcal{P}(K) \subset \{(a, b) \in \mathcal{C} \mid \ell_1(a) = \ell_2(b) = 0\}.$$

Proof. Note that if $\mathcal{P}(t)$ has a constant component the result is trivial. So, in the following, we assume that no component of $\mathcal{P}(t)$ is constant. First, we prove that

$$\mathcal{P}(K) = \{(a,b) \in \mathcal{C} \mid \gcd(H_1^{\mathcal{P}}(a,t), H_2^{\mathcal{P}}(b,t)) \neq 1\} \, .$$

Clearly, if $(a,b) \in \mathcal{P}(K)$ then $(a,b) \in \mathcal{C}$, and there exists $t_0 \in K$ such that $\mathcal{P}(t_0) = (a,b)$. Thus, t_0 is a common root of $H_1^{\mathcal{P}}(a,t)$ and $H_2^{\mathcal{P}}(b,t)$. Conversely, if $(a,b) \in \mathcal{C}$ and $t_0 \in K$ is a common root of $H_1^{\mathcal{P}}(a,t)$ and $H_2^{\mathcal{P}}(b,t)$, then $\chi_{12}(t_0)\chi_{22}(t_0) \neq 0$, since otherwise either $H_1^{\mathcal{P}}(a,t_0) = \chi_{11}(t_0) = 0$ or $H_2^{\mathcal{P}}(b,t_0) = \chi_{21}(t_0) = 0$, which is impossible because of $\gcd(\chi_{11}, \chi_{12}) = \gcd(\chi_{21}, \chi_{22}) = 1$. Therefore, $\mathcal{P}(t_0) = (a,b)$, and hence $(a,b) \in \mathcal{P}(K)$.

Now, for every $(a,b) \in \mathcal{C}$ such that $\ell_1(a) \neq 0$ or $\ell_2(b) \neq 0$ we consider the evaluation homomorphism $\psi_{(a,b)} : K[x,y][t] \rightarrow K[t]$ defined as $\psi_{(a,b)}(M(x,y,t)) = M(a,b,t)$. W.l.o.g we assume that $\ell_1(a) \neq 0$. Taking into account the behavior of the resultant under a homomorphism (see, e.g., Lemma 4.3.1 in [Win96]), one has that

$$\psi_{(a,b)}(\mathrm{res}_t(H_1^{\mathcal{P}}(x,t), H_2^{\mathcal{P}}(y,t)))$$

$$= \ell_1(a)^{\deg_t(H_2^{\mathcal{P}}(y,t)) - \deg_t(H_2^{\mathcal{P}}(b,t))} \mathrm{res}_t(H_1^{\mathcal{P}}(a,t), H_2^{\mathcal{P}}(b,t)) \, ,$$

since $\ell_1(a) \neq 0$. Moreover, up to multiplication by a nonzero constant,

$$\mathrm{res}_t(H_1^{\mathcal{P}}(x,t), H_2^{\mathcal{P}}(y,t)) = f(x,y)^{\mathrm{index}(\mathcal{P})}$$

(see Theorem 4.41), where $f(x,y)$ is the defining polynomial of $\mathcal{C}$. Therefore, since $(a,b) \in \mathcal{C}$ and, $\ell_1(a) \neq 0$, one gets that $\mathrm{res}_t(H_1^{\mathcal{P}}(a,t), H_2^{\mathcal{P}}(b,t)) = 0$. Thus, $\gcd(H_1^{\mathcal{P}}(a,t), H_2^{\mathcal{P}}(b,t)) \neq 1$. So, one deduces that

$$\{(a,b) \in \mathcal{C} \mid \ell_1(a) \neq 0 \text{ or } \ell_2(b) \neq 0\} \subset$$

$$\{(a,b) \in \mathcal{C} \mid \gcd(H_1^{\mathcal{P}}(a,t), H_2^{\mathcal{P}}(b,t)) \neq 1\} = \mathcal{P}(K),$$

and therefore $\mathcal{C} \setminus \mathcal{P}(K) \subset \{(a,b) \in \mathcal{C} \mid \ell_1(a) = \ell_2(b) = 0\}$. □

From Lemma 6.19, one immediately gets the following corollaries. Note that for the case of parametrizations with constant components the corollaries hold trivially.

Corollary 6.20. *If one of the denominators in $\mathcal{P}(t)$ has degree less than the degree of its numerator, then $\mathcal{P}(t)$ is normal.*

Proof. Note that the corresponding leading coefficient ℓ_i of the parametrization component with higher degree in the numerator, is a nonzero constant. Thus, $\mathcal{C} = \mathcal{P}(K)$. □

Corollary 6.21. *Any polynomial parametrization is normal.*

Example 6.18 shows that the conditions in Corollary 6.20 to Lemma 6.19 do not characterize the normality of a parametrization. In Theorem 6.22 we give a complete characterization. For this purpose, we use the following notation: if $p \in K[t]$ and $k \in \mathbb{N}$, $\mathrm{coeff}(p, k)$ denotes the coefficient of the term t^k in $p(t)$.

Theorem 6.22. *Let* $n = \deg(\chi_{11}), m = \deg(\chi_{12}), r = \deg(\chi_{21}), s = \deg(\chi_{22})$, *and let* $a = \mathrm{coeff}(\chi_{11}, m)$, $b = \mathrm{coeff}(\chi_{12}, m)$, $c = \mathrm{coeff}(\chi_{21}, s)$, $d = \mathrm{coeff}(\chi_{22}, s)$. *Then*

(1) if $n > m$ or $r > s$ then $\mathcal{P}(t)$ is normal;
(2) if $n \leq m$ and $r \leq s$ then $\mathcal{P}(t)$ is normal if and only if

$$\deg(\gcd(a\chi_{12}(t) - b\chi_{11}(t), c\chi_{22}(t) - d\chi_{21}(t))) \geq 1 .$$

Furthermore, if $\mathcal{P}(t)$ is not normal, all points in C are generated by $\mathcal{P}(t)$ with the exception of $\left(\frac{a}{b}, \frac{c}{d}\right)$, which is a point on C.

Proof. Note that if $\mathcal{P}(t)$ has a constant component, the result is trivial. So, in the following, we assume that none component of $\mathcal{P}(t)$ is constant. Statement (1) is Corollary 6.20 to Lemma 6.19. In order to prove Statement (2), let

$$\mathcal{P}(t) = \left(\frac{a_n t^n + \cdots + a_0}{b_m t^m + \cdots + b_0}, \frac{c_r t^r + \cdots + c_0}{d_s t^s + \cdots + d_0} \right) .$$

Also, let $\ell_1(x), \ell_2(y)$ be the leading coefficients w.r.t. t of $H_1^{\mathcal{P}}(x, t)$ and $H_2^{\mathcal{P}}(y, t)$, respectively, and let $\mathcal{Q}(t) = \mathcal{P}\left(\frac{1}{t}\right)$. Clearly

$$\mathcal{Q}(t) = \left(\frac{a_n + \cdots + a_0 t^n}{b_m + \cdots + b_0 t^m} t^{m-n}, \frac{c_r + \cdots + c_0 t^r}{d_s + \cdots + d_0 t^s} t^{s-r} \right) .$$

In this situation, we distinguish the following cases:

(i) Let $n < m$ and $r < s$. First observe that the denominators of $\mathcal{Q}(t)$ do not vanish at $t = 0$, since $b_m \neq 0$, $d_s \neq 0$, and $m - n > 0$, $s - r > 0$. Thus, the parametrization $\mathcal{Q}(t)$ is defined for $t = 0$, and $\mathcal{Q}(0) = (0, 0) \in C$. Moreover, $\ell_1(x) = b_m x, \ell_2(y) = d_s y$. Therefore, by Lemma 6.19, the only point that might not be generated by $\mathcal{P}(t)$ is the origin. Furthermore, by Lemma 6.19, $(0, 0)$ is generated by $\mathcal{P}(t)$ if and only if $\gcd(H_1^{\mathcal{P}}(0, t), H_2^{\mathcal{P}}(0, t)) \neq 1$. Finally, note that $\gcd(H_1^{\mathcal{P}}(0, t), H_2^{\mathcal{P}}(0, t)) = \gcd(a\chi_{12}(t) - b\chi_{11}(t), c\chi_{22}(t) - d\chi_{21}(t))$ and $(0, 0) = \left(\frac{a}{b}, \frac{c}{d}\right)$. Therefore the statement holds.

(ii) Let $m = n$, and $r = s$. Since $b_m \neq 0, d_s \neq 0$, we have $\mathcal{Q}(0) = \left(\frac{a_n}{b_m}, \frac{c_r}{d_s}\right) \in C$. Moreover, $\ell_1(x) = b_m x - a_n, \ell_2(y) = d_s y - c_r$. Therefore, applying Lemma 6.19 as in case (i), we get the result.

(iii) Let $m > n$, and $r = s$. Since $b_m \neq 0, d_s \neq 0$, we have $\mathcal{Q}(0) = \left(0, \frac{c_r}{d_s}\right) \in C$. Moreover, $\ell_1(x) = b_m x, \ell_2(y) = d_s y - c_r$. Therefore, applying Lemma 6.19 as in case (i), we get the result.

(iv) Let $m = n$, and $r < s$. Since $b_m \neq 0, d_s \neq 0$, we have $\mathcal{Q}(0) = \left(\frac{a_n}{b_m}, 0\right) \in \mathcal{C}$. Moreover, $\ell_1(x) = b_m x - a_n, \ell_2(y) = d_s y$. Therefore, applying Lemma 6.19 as in case (i), we get the result. $\qquad \square$

Corollary 6.23. *A rational parametrization of an affine plane curve reaches all points on the curve with the exception of at most one point. Moreover, the only possible unreachable point is the one described in Statement (2) of Theorem 6.22.*

Definition 6.24. *Let $\mathcal{P}(t)$ be a rational parametrization in reduced form such that the degree of each numerator is not higher than the degree of the corresponding denominator, i.e., $\deg(\chi_{i2}(t)) \geq \deg(\chi_{i1}(t))$ for $i = 1, 2$. Then, the only possible missing point of the parametrization is called the* critical point *of $\mathcal{P}(t)$.*

Remarks. Observe that the notion of critical point is not defined for parametrizations such that at least one of the numerators has higher degree than its denominator.

Theorem 6.22 leads to the following algorithm for deciding whether a given parametrization is normal.

Algorithm NORMALITY-TEST

Given a rational parametrization $\mathcal{P}(t) = \left(\frac{\chi_{11}(t)}{\chi_{12}(t)}, \frac{\chi_{21}(t)}{\chi_{22}(t)}\right)$ in reduced form, of a plane affine algebraic curve $\mathcal{C}$, the algorithm decides whether $\mathcal{P}(t)$ is normal.

1. $n := \deg(\chi_{11}(t)), m := \deg(\chi_{12}(t)), r := \deg(\chi_{21}(t)), s := \deg(\chi_{22}(t))$.
2. If $n > m$ or $r > s$ then return "$\mathcal{P}(t)$ is normal".
3. $a := \text{coeff}(\chi_{11}, m), b := \text{coeff}(\chi_{12}, m), c := \text{coeff}(\chi_{21}, s), d := \text{coeff}(\chi_{22}, s)$.
4. Compute $M(t) := \gcd(a\chi_{12} - b\chi_{11}, c\chi_{22} - d\chi_{21})$.
5. If $\deg(M) \geq 1$ then return "$\mathcal{P}(t)$ is normal" else return "$\mathcal{P}(t)$ is not normal and the only point on $\mathcal{C}$ that is not reachable is $\left(\frac{a}{b}, \frac{c}{d}\right)$".

Example 6.25. Let us consider the affine conic defined by

$$\frac{x^2}{u^2} + \frac{y^2}{v^2} = 1, \quad u \neq 0, \ v \neq 0$$

and the standard proper parametrization

$$\mathcal{P}(t) = \left(\frac{u\left(t^2 - 1\right)}{1 + t^2}, \ 2\frac{vt}{1 + t^2}\right).$$

Applying Algorithm NORMALITY-TEST, in Step 4 we get that

$$M(t) := \gcd(2u, -2vt) = 1 .$$

Therefore, $\mathcal{P}(t)$ is not normal and the only point on the affine conic that is not reachable is $(u, 0)$. That is the critical point of the parametrization.

We have already described a method for deciding whether a given parametrization is normal. In the last part of this section we deal with the problem of computing normal parametrizations of a rational curve over K. We start with the following theorem due to T. Recio, in a personal communication (1994). Geometrically, the idea is to take a projective parametrization sending the infinity of K into a point of $\mathcal{C}$ at infinity.

Theorem 6.26. *Every rational affine curve over K can be properly and normally parametrized.*

Proof. Let $\mathcal{C}$ be an affine rational curve over K, and let $\mathcal{P}(t)$ be a proper nonnormal parametrization, in reduced form, of $\mathcal{C}$ (see Lemma 4.13). Then, by Corollary 6.21 to Lemma 6.19, $\mathcal{P}(t)$ is not polynomial, and hence either χ_{12} or χ_{22} is nonconstant. W.l.o.g. we assume that χ_{12} is nonconstant. Then, we take $\alpha \in K$ such that $\chi_{12}(\alpha) = 0$. Note that $\gcd(\chi_{11}, \chi_{12}) = 1$ and hence $\chi_{11}(\alpha) \neq 0$. By the linear change of parameter $\phi(t) = \frac{\alpha t + 1}{t}$, the parametrization $\mathcal{P}(t)$ is transformed to a normal parametrization. Furthermore, since $\phi(t)$ is invertible, the reparametrization preserves the properness (see Lemma 4.17). $\qquad\square$

Remarks. In Theorem 6.26 we have emphasized the fact that the normal parametrization is proper. If the parametrization $\mathcal{P}(t)$ in the proof has tracing index k then, since the reparametrization function $\phi(t)$ is linear, the tracing index of the normal parametrization is also k (see Theorem 4.33).

Combining algorithm NORMALITY-TEST and the constructive proof of Theorem 6.26, one can derive the following algorithm.

Algorithm NORMAL-PARAMETRIZATION

Given a rational parametrization $\mathcal{P}(t) = \left(\frac{\chi_{11}(t)}{\chi_{12}(t)}, \frac{\chi_{21}(t)}{\chi_{22}(t)} \right)$ in reduced form, of a plane affine algebraic curve $\mathcal{C}$, the algorithm computes a normal parametrization of $\mathcal{C}$.

1. Apply algorithm NORMALITY-TEST to check whether $\mathcal{P}(t)$ is normal. If so, return $\mathcal{P}(t)$.
2. If $\chi_{12}(t)$ is not constant take $\alpha \in K$ such that $\chi_{12}(\alpha) = 0$, else take $\alpha \in K$ such that $\chi_{22}(\alpha) = 0$.
3. Return $\mathcal{P}\left(\frac{\alpha t + 1}{t} \right)$.

Example 6.27. From the algorithm NORMAL-PARAMETRIZATION applied to the proper parametrizations given in Example 6.25, we get

$$\left(-\frac{u\left(2\,t^2 - 2\,it - 1\right)}{2\,it + 1}, 2\,\frac{v\left(it + 1\right)t}{2\,it + 1} \right).$$

This is a normal and proper parametrization of the affine conics.

Remarks. The situation is different when the curve is seen over a subfield of K. The difficulty arises when a point on the curve, although reachable by the parametrization, might only be generated by parameter values in $K \setminus \mathbb{L}$ ($\mathbb{L}$ being a subfield of K). For instance, the parametrization $\mathcal{P}(t) = (t^4 + 1, t^3 + 2t + 1)$, which is proper and normal over $\mathbb{C}$, is not normal over $\mathbb{R}$ because the point $(5, 1)$, which is on the curve, is only reachable via $\mathcal{P}(t)$ taking $t = \pm i\sqrt{2}$. Thus, in certain cases normality and optimal field parametrization may not be reachable at the same time. This type of phenomena will be studied in Chap. 7.

Exercises

6.1. Let $\mathcal{P}(t)$ be a rational parametrization of a curve $\mathcal{C}$ over K, and let $\phi(t) \in K(t) \setminus K$. Prove that $\mathcal{Q}(t) = \mathcal{P}(\phi(t))$ also parametrizes $\mathcal{C}$.

6.2. Apply algorithm PROPER-REPARAMETRIZATION to compute a rational proper parametrization of the curve parametrized by

$$\mathcal{P}(t) = \left(\frac{3t^4 + 4t^3 + 32t^2 + 28t + 99}{(t^2 + t + 7)(t^2 + 1)}, \frac{(t^2 + t + 7)^3}{(t + 6)(t^2 + 1)^2} \right).$$

6.3. Apply algorithm POLYNOMIAL-REPARAMETRIZATION to compute a rational polynomial parametrization of the curve parametrized by

$$\mathcal{P}(t) = \left(\frac{7t^2 + 17t + 13}{4t^2 + 4t + 1}, \frac{2t^2 + 11t + 15}{8t^3 + 12t^2 + 6t + 1} \right).$$

Check that $\mathcal{C}$ has only one place at infinity.

6.4. Let $\mathcal{C}$ be an irreducible affine curve of degree d having an affine $(d-1)$-fold point. Prove that $\mathcal{C}$ is polynomial if and only if the homogeneous form of maximum degree of the defining polynomial of $\mathcal{C}$ is the power of a linear homogeneous polynomial.

6.5. Construct a nonpolynomial rational affine curve such that the homogeneous form of maximum degree of its defining polynomial is the power of a linear homogeneous polynomial.

6.6. Give an example of each of the cases analyzed in the proof of Theorem 6.22.

6.7. Apply algorithm NORMAL-PARAMETRIZATION to compute a rational normal parametrization of the curve parametrized by

$$\mathcal{P}(t) = \left(\frac{7t^2 + 17t + 13}{t^3 - 3t + 2}, \frac{7t^2 + 42t + 35}{8t^3 + 12t^2 + 6t + 1} \right).$$

7

Real Curves

Up to now we have dealt with curves in affine or projective planes over algebraically closed fields. For practical applications this may be unsatisfactory. So in this chapter we consider curves in the real plane. It turns out that a real curve which can be parametrized over an algebraically closed field can actually be parametrized with real coefficients. In Sect. 7.1 we adapt the previously introduced parametrization algorithm to find such a real parametrization. The material in this part of the chapter follows the ideas in [SeW97] and [SeW99]. In Sect. 7.2 we consider the following problem: given a complex parametrization of a curve $\mathcal{C}$ we decide whether $\mathcal{C}$ is a real curve and in the affirmative case we determine a real parametrization of $\mathcal{C}$. The ideas and results developed in this section follow essentially the material presented in [ReS97a]. In Sect. 7.3 we consider the problem of determining real normal parametrizations; i.e., real parametrizations $\mathcal{P}(t)$ covering the whole curve and not just a dense subset of it, with real values of the parameter t. The material in this part of the chapter follows the ideas in [Sen02]. For further reading on algorithms in real algebraic geometry we refer to [BPR03].

7.1 Parametrization

In this section, we introduce the notion of real plane curve and we analyze some basic properties related to defining polynomials, irreducibility, and rationality. We finish the section showing how the algorithms in Chaps. 4 and 5 can be adapted to the real case.

Definition 7.1. A real affine plane curve *is an affine plane curve over* $\mathbb{C}$ *with infinitely many points in the affine plane over* $\mathbb{R}$*. Similarly, we define a real projective plane curve* *as a projective plane curve over* $\mathbb{C}$ *with infinitely many points in the projective plane over* $\mathbb{R}$*.*

Observe that the projective closure of any real affine plane curve is a real projective curve, and that, in general, the dehomogenization of a real projective plane curve is a real affine plane curve. In fact, if the curve is irreducible, the only exception is the line at infinity. Therefore, in the sequel, we can choose freely between projective and affine situations, whatever we find more convenient.

Real curves are not necessarily defined by real polynomials. For instance, the polynomial $f(x, y) = x^2 + ixy \in \mathbb{C}[x, y]$ defines a real curve according to Definition 7.1, since all the points $(0, b)$, with $b \in \mathbb{R}$, are on the curve. Clearly, in this example, what is happening is that the curve decomposes over $\mathbb{C}$ into two curves, namely the real line defined by x, and the complex line defined by $x + iy$. On the other hand, real polynomials may not define real curves as for instance $f(x, y) = x^2 + y^2 + 1 \in \mathbb{R}[x, y]$. However, in Lemma 7.2 we see that if the curve is real and irreducible over $\mathbb{C}$, then it can be defined by a real polynomial. Here, when we say that the real curve is irreducible, we mean that it is irreducible over $\mathbb{C}$; i.e., its defining polynomial is irreducible as a polynomial in $\mathbb{C}[x, y]$.

Lemma 7.2. *Every irreducible real curve has a real defining polynomial.*

Proof. Obviously the line at infinity has a real defining polynomial. So now, let C be different from the line at infinity, and let $f(x, y) = f_1(x, y) + if_2(x, y) \in \mathbb{C}[x, y]$, $f_i(x, y) \in \mathbb{R}[x, y]$, $i = 1, 2$ be the defining polynomial for the affine version of C. By Definition 7.1 there exist infinitely many points $(a, b) \in \mathbb{R}^2$ such that $f_1(a, b) + if_2(a, b) = 0$; i.e., $f_1(a, b) = f_2(a, b) = 0$. Thus, the curves defined over $\mathbb{C}$ by f_1 and f_2, have infinitely many points in common. Hence, by Bézout's Theorem, they must have a common component. Let $g = \gcd(f_1, f_2) \in \mathbb{R}[x, y]$ be the polynomial defining the common component. Then, $f(x, y) = g(x, y) \cdot \tilde{f}(x, y)$. Since f is irreducible over $\mathbb{C}$, $\tilde{f}(x, y)$ must be constant, and therefore $g(x, y)$ is a real defining polynomial of C.　□

Observe that a plane curve over $\mathbb{C}$ is real if and only if at least one component of the curve is real. Since we can always decompose any curve into irreducible components over $\mathbb{C}$, we get the following result.

Lemma 7.3. *The defining polynomial of a real plane curve has a real factor.*

Proof. Let C be a real plane curve defined by $f(x, y) \in \mathbb{C}[x, y]$. One of the irreducible factors of f, say $\tilde{f}$, has infinitely many real roots. So $\tilde{f}$ defines a real curve $\tilde{C}$. Because of Lemma 7.2, $\tilde{C}$ has a real defining polynomial $g(x, y) \in \mathbb{R}[x, y]$. Then, $\tilde{f}$ and g can only differ by a constant complex factor. So g is a factor of f.　□

Of course, since an algebraic curve can have only finitely many singularities, a real algebraic curve must have real simple points. But also the converse is true, so the existence of a real simple point on C is a criterion for the realness of C.

Theorem 7.4. *Let C be a plane algebraic curve defined by a real squarefree polynomial. Then, C is real if and only if C has at least one real simple point.*

Proof. The left to right implication is trivial. Conversely, let $f(x, y) \in \mathbb{R}[x, y]$ be a squarefree polynomial defining $\mathcal{C}$, and let $P \in \mathbb{A}^2(\mathbb{R})$ be a simple point on $\mathcal{C}$. Then, not both partial derivatives of f vanish at P. Therefore, since $f(x, y) \in \mathbb{R}[x, y]$, we apply the Implicit Function Theorem (see Appendix B), and we deduce that $\mathcal{C}$ has infinitely many points in $\mathbb{A}^2(\mathbb{R})$. Thus, $\mathcal{C}$ is a real curve. □

Note that Theorem 7.4 is not true without the assumption of squarefreeness; for instance the curve of equation x^2 has infinitely many real points, but all of them are singular. Also, note that the condition on the simplicity of the real point is necessary. For instance, one may consider the curve $\mathcal{C}$ defined by the irreducible real polynomial

$$f(x, y) = x^2 + 2y^2 + 2x^2y^2,$$

which is not real, but it has a real point, namely the origin. However, in this example the origin is a double point of the curve.

Now that real curves have been analyzed, we focus on irreducible real curves. In Lemma 7.5, we prove that the irreducibility of real curves follows from the irreducibility over $\mathbb{R}$ of its defining polynomial.

Lemma 7.5. *A real curve $\mathcal{C}$ is irreducible if and only if it is defined by a real polynomial irreducible over $\mathbb{R}$.*

Proof. This follows from Lemma 7.2, and taking into account that a real polynomial irreducible over $\mathbb{R}$ with infinitely many real roots is irreducible in $\mathbb{C}[x, y]$ (see Exercise 7.1). □

So from now on we will assume that the real curves we work with are irreducible and therefore defined by real irreducible polynomials. Concerning rationality, since a real curve can be seen as a curve over $\mathbb{C}$, by Theorem 4.63, we know that the real curve can be parametrized if and only if its genus is zero. However, strictly speaking this characterization ensures that real curves with genus zero can be parametrized over $\mathbb{C}$; i.e., by means of rational functions with complex coefficients. But for practical applications, we are actually interested in the possibility of parametrizing the real rational curve over $\mathbb{R}$. That is, we want to know whether $\mathbb{R}$ is a field of parametrization of a given rational real curve (see Definition 5.1). This is always possible, and it can be seen as a consequence of the Real Lüroth Theorem (see [ReS97b]). We can also deduce this result directly from Theorems 5.4 and 7.4, from which we get Theorem 7.6.

Theorem 7.6. *Every rational real curve can be properly parametrized over the reals.*

For actually carrying out the parametrization process we have several options. We might use a slightly generalized version of the Algorithm

OPTIMAL-PARAMETRIZATION (see Sect. 5.3). We input a defining real polynomial $f(x, y)$. All the steps in the algorithm can be executed over computable subfields of $\mathbb{R}$. We still can determine whether the conic in Step 2.3.5 has a real point, and if so determine one. Note that there exist elementary methods in linear algebra to check whether a real quadratic polynomial defines a real conic, for instance analyzing the signature and rank of the associated quadratic form.

Alternatively, we can try to determine a finite number of regular real points on $\mathcal{C}$ directly from the defining polynomial. This can be achieved by a standard application of cylindrical algebraic decomposition as shown in Theorem 7.7 (see [Joh98]). Once we have these regular real points on $\mathcal{C}$, we can apply any one of the variants of Algorithm PARAMETRIZATION-BY-ADJOINTS (see Sects. 4.7 and 4.8).

Theorem 7.7. *Let $f(x, y) \in \mathbb{R}[x, y]$ be a squarefree polynomial, not having a linear factor independent of y. Let $\mathcal{C}$ be the affine plane curve defined over $\mathbb{C}$ by the polynomial $f(x, y)$, and let $D(x)$ be the discriminant of f w.r.t the variable y. Then the following hold:*

1. *If $D(x)$ does not have a real root, then $\mathcal{C}$ is a real plane curve if and only if the univariate polynomial $f(0, y)$ has real roots. Furthermore, if $\beta \in \mathbb{R}$ is a root of $f(0, y)$, then $(0, \beta) \in \mathbb{R}^2$ is a real simple point of $\mathcal{C}$.*
2. *Let $a_1, \ldots, a_n$ be the real roots of $D(x)$, and let $\lambda_0, \ldots, \lambda_n \in \mathbb{R}$ be such that*

$$-\infty = a_0 < \lambda_0 < a_1 < \lambda_1 < a_2 < \cdots < \lambda_{n-1} < a_n < \lambda_n < a_{n+1} = +\infty.$$

Then, $\mathcal{C}$ is a real plane curve if and only if there exists $i \in \{0, \ldots, n\}$ such that the univariate polynomial $f(\lambda_i, y)$ has real roots. Furthermore, if $\beta \in \mathbb{R}$ is a root of $f(\lambda_i, y)$, then $(\lambda_i, \beta) \in \mathbb{R}^2$ is a real simple point of $\mathcal{C}$.

In the following examples, we illustrate various methods in the full parametrization process.

Example 7.8. Let $\mathcal{C}$ be the plane curve defined over $\mathbb{C}$ by the polynomial

$$F(x, y, z) = 10y^3z^2 - 15xy^2z^2 + 6xy^3z - 12x^2y^2z + 8x^3z^2 + 3x^3y^2.$$

From applying Theorem 7.7 to $F(x, y, 1)$, we get that $\mathcal{C}$ is a real curve. In addition, $P_1 = (-1 : \sqrt[3]{2} : 1)$ and $P_2 = (5 : -\sqrt[3]{25} : 1)$, are two real simple points on $\mathcal{C}$. In order to check whether $\mathcal{C}$ is rational, we compute the genus of $\mathcal{C}$ (see Chap. 3). For this purpose, first we determine the singularities of $\mathcal{C}$. We get

$$Q_1 = (0 : 0 : 1), \quad Q_2 = (0 : 1 : 0), \quad Q_3 = (1 : 0 : 0), \quad Q_4 = (1 : 1 : 1),$$

where Q_1 is a triple point, and Q_i, $i = 2, 3, 4$, are double points. Thus, genus$(\mathcal{C}) = 0$, and therefore $\mathcal{C}$ is rational. Now, we proceed to compute a real rational parametrization. For this purpose, we apply Algorithm PARAMETRIZATION-BY-ADJOINTS, taking $k = d - 2$ (see Sect. 4.7). We consider a form in x, y, z of degree $d - 2 = 3$ defining the linear system $\mathcal{H}$ of adjoint curves to $\mathcal{C}$:

$$H(x, y, z, \lambda_1, \lambda_2, \lambda_3, \lambda_4) = \lambda_2 y^2 z + (-\lambda_2 - \lambda_1 - \lambda_3 - \lambda_4)xyz + \lambda_1 xy^2$$
$$+\lambda_3 x^2 z + \lambda_4 x^2 y.$$

Now, we determine the equation of the subsystem of $\mathcal{H}$ having P_1 and P_2 as simple base points. We get this subsystem as the solution of the linear equations $H(P_i) = 0$, $i = 1, 2$, namely

$$H(x, y, z, t) = 64tx^2 - 3x^2 y \beta^2 \alpha t + 9\beta^2 \alpha^2 x^2 yt - 6x^2 y \alpha^2 t \beta - 7\beta^2 \alpha^2 xyt +$$
$29xy\beta \alpha t + 34xy\beta \alpha^2 t - 5xy\beta^2 \alpha t + 24xy\beta \alpha + 16\beta \alpha^2 xy + 36\beta x^2 yt + 6\alpha x^2 yt - 18x^2 y\alpha^2 t + 12x^2 y\beta^2 t + 14xy\alpha^2 t + 20xy\beta^2 t - 116\beta xyt - 66\beta y^2 \alpha^2 t + 10\alpha xyt - 2\beta^2 \alpha^2 tx^2 + 38\beta \alpha^2 tx^2 + 8\beta^2 \alpha tx^2 - 20tx^2 \beta \alpha + 38\beta \alpha^2 xy + 8\beta^2 \alpha xy^2 - 20xy^2 \beta \alpha - 24\beta^2 \alpha^2 xy + 8\beta^2 xy\alpha - 9x^2 y\beta \alpha t - 2\beta^2 \alpha^2 xy^2 - 24\beta^2 y^2 + 120y^2 \alpha - 140\alpha^2 y^2 + 104\beta y^2 + 136\alpha^2 xy - 16tx^2 \alpha + 4\alpha^2 tx^2 - 32\beta^2 tx^2 + 64xy^2 - 24x^2 yt - 184\beta xy - 54\beta \alpha^2 y^2 - 156xy + 132x^2 y - 16\beta^2 \alpha y^2 - 32\beta^2 xy^2 + 80xy^2 \beta + 4\alpha^2 xy^2 - 40xyt - 16xy^2 \alpha + 26\beta^2 \alpha^2 y^2 - 40y^2 - 4\beta y^2 \alpha - 104xy\alpha + 56\beta^2 xy + 80tx^2 \beta,$

where $\alpha := \sqrt[3]{2}$, and $\beta := \sqrt[3]{5}$. Finally, from the primitive parts of the resultants of F and H, w.r.t x and w.r.t y, respectively, we get the real rational parametrization $\mathcal{P}(t)$ over $\mathbb{Q}(\sqrt[3]{2}, \sqrt[3]{5})$:

$$\mathcal{P}(t) = \left(\frac{\chi_{11}(t)}{\chi_{12}(t)}, \frac{\chi_{21}(t)}{\chi_{22}(t)} \right),$$

where

$\chi_{11}(t) = 3(9024\alpha + 2046\beta^2 t^2 - 1122t^2 \beta - 84t^3 \beta + 7248\alpha t + 408t^3 + 1716t^2 - 912\beta \alpha^2 t + 4800\beta^2 \alpha^2 t - 2688\beta \alpha t - 3888\beta^2 \alpha t - 672\alpha^2 + 3264\beta^2 + 6576\beta^2 t - 14976 + 3600\alpha^2 t + 4912\beta^2 \alpha^2 - 11040t + 1020t^3 \alpha + 2550t^3 \alpha^2 - 3680\beta \alpha + 375t^3 \beta^2 \alpha^2 + 150t^3 \beta^2 \alpha^2 - 525t^3 \beta \alpha^2 - 210t^3 \beta \alpha - 3456\beta^2 \alpha + 14016\beta + 7056\beta t + 7458t^2 \alpha^2 + 2112t^2 \alpha + 60t^3 \beta^2 - 5328\beta \alpha^2 - 2112t^2 \alpha^2 \beta + 462t^2 \alpha \beta - 1782t^2 \alpha \beta^2 + 2442t^2 \alpha^2 \beta^2),$

$\chi_{12}(t) = 2(2816 - 704\alpha + 507\beta^2 t^2 - 129t^2 \beta + 3096\alpha t - 618t^2 - 2832\beta \alpha^2 t + 1608\beta^2 \alpha^2 t - 552\beta \alpha t - 1680\beta^2 \alpha t + 176\alpha^2 - 1408\beta^2 + 2232\beta^2 t + 3384\alpha^2 t - 88\beta^2 \alpha^2 - 3408t - 880\beta \alpha + 352\beta^2 \alpha + 3520\beta + 360\beta t - 51t^2 \alpha^2 + 996t^2 \alpha + 1672\beta \alpha^2 - 534t^2 \alpha^2 \beta - 141t^2 \alpha \beta - 3t^2 \alpha \beta^2 + 1263t^2 \alpha^2 \beta^2),$

$\chi_{21}(t) = -24560\alpha - 462\beta^2 t^2 - 2112t^2 \beta - 1020t^3 \beta - 24000\alpha t - 750t^3 + 8910t^2 + 5520\beta \alpha^2 t - 3528\beta^2 \alpha^2 t - 3600\beta \alpha t + 912\beta^2 \alpha t - 8160\alpha^2 + 3680\beta^2 + 2688\beta^2 t - 16440\alpha^2 t - 7008\beta^2 \alpha^2 + 19440t - 1875t^3 \alpha - 150t^3 \alpha^2 + 672\beta \alpha + 42t^3 \beta^2 \alpha^2 + 525t^3 \beta^2 \alpha^2 - 204t^3 \beta \alpha^2 - 2550t^3 \beta \alpha + 5328\beta^2 \alpha - 9024\beta - 7248\beta t - 5115t^2 \alpha^2 - 12210t^2 \alpha + 210t^3 \beta^2 + 17280 + 7488\beta \alpha^2 - 858t^2 \alpha^2 \beta - 7458t^2 \alpha \beta + 2112t^2 \alpha \beta^2 + 561t^2 \alpha^2 \beta^2,$

$\chi_{22}(t) = 4(1108\beta^2\alpha^2 - 2920\alpha^2 + 2220\alpha + 472\beta - 1236\beta^2 + 1024\beta^2\alpha - 1724\beta\alpha + 288\beta\alpha^2 + 256\beta^2\alpha^2 t - 310\beta\alpha^2 t + 1190\alpha t + 296\beta^2\alpha t + 2340 - 608\beta\alpha t + 355t - 293\beta^2 t - 50\alpha^2 t + 386\beta t).$

So, we have determined a real parametrization, but unfortunately the coefficients involve several algebraic numbers.

Example 7.9. In this example, we consider again the curve $\mathcal{C}$ of Example 7.8. We apply Algorithm SYMBOLIC-PARAMETRIZATION-BY-DEGREE-d-ADJOINTS (see Sect. 4.8) and we use $P = P_1 = (-1 : \sqrt[3]{2} : 1)$ to obtain the point $(b_1 : b_2 : \beta)$ in Steps 3 and 4. For this purpose, we first determine the polynomial H defining the linear system $\mathcal{H}$ of adjoints curves to $\mathcal{C}$ of degree $d = 5$. Afterward, we compute a real point $Q \notin \mathcal{C}$, for instance $Q = (1 : 1 : 0)$, and three families of four conjugate points over $\mathbb{Q}(\sqrt[3]{2})$ on lines through P. Note that P satisfies the conditions required in Steps 3 and 4 of Algorithm SYMBOLIC-PARAMETRIZATION-BY-DEGREE-d-ADJOINTS. More precisely, we get the following families over $\mathbb{Q}(\alpha)$:

$$\mathcal{F}_1 = \{(\gamma - 1 : \gamma + \alpha : 1) \mid q_1(\gamma) = 0\}$$

$$\mathcal{F}_2 = \{(-1 : \alpha + \gamma : \gamma + 1) \mid q_2(\gamma) = 0\},$$

and

$$\mathcal{F}_3 = \{(-1 + \gamma : \alpha + \gamma : 2\gamma + 1) \mid q_3(\gamma) = 0\},$$

where $\alpha := \sqrt[3]{2}$, and

$q_1(t) = 36 + 30t^2 + 48\alpha t + 30\alpha^2 - 24t - 24t^2\alpha - 3\alpha^2 t - 15t^3 + 6t^3\alpha + 3t^2\alpha^2 + 3t^4,$

$q_2(t) = 12 + 22t^2 + 48\alpha t + 30\alpha^2 + 12t + 72t^2\alpha + 57\alpha^2 t + 29t^3 + 30t^3\alpha + 30t^2\alpha^2 + 10t^4,$

$q_3(t) = 60 + 66t^2 + 120\alpha t + 66\alpha^2 + 144t + 156t^2\alpha + 129\alpha^2 t - 3t^3 - 6t^3\alpha + 75t^2\alpha^2 + 3t^4.$

Now, we determine the equation H of the subsystem of $\mathcal{H}$ having Q and the points of the families $\mathcal{F}_i$, $i = 1, 2, 3$, as simple base points. This implies to solve the linear system of equations, in the undetermined coefficients of H, given by the conditions

$$H(Q) = 0, \qquad H(t - 1, t + \alpha, 1) = 0 \mod q_1(t),$$

$$H(-1, \alpha + t, t + 1) = 0 \mod q_2(t), \qquad H(-1 + t, \alpha + t, 2t + 1) = 0 \mod q_3(t).$$

Finally, from the primitive parts of the resultants of F and H, w.r.t. x and y, we get the rational parametrization $\mathcal{P}(t)$ over the real extension $\mathbb{Q}(\alpha)$ of $\mathbb{Q}$:

$$\mathcal{P}(t) = \left(\frac{\chi_{11}(t)}{\chi_{12}(t)}, \frac{\chi_{21}(t)}{\chi_{22}(t)} \right),$$

where

$\chi_{11}(t) = -50208t + 12084t^3 + 229068t^2 - 27808\alpha - 230736t\alpha - 134244t^2\alpha - 38448t^2\alpha^2 + 219288t\alpha^2 - 36849t^3\alpha^2 - 27776\alpha^2 + 18678t^3\alpha - 335264,$

$\chi_{12}(t) = 16368t\alpha + 6138t^3\alpha - 11220t^2\alpha - 29920\alpha - 17600\alpha^2 - 6600t^2\alpha^2 - 12320 - 4620t^2 + 20064t + 7524t^3 + 5016t\alpha^2 + 1881t^3\alpha^2,$

$\chi_{21}(t) = -515328t + 1246344t^3 + 179388t^2 - 87192t\alpha + 749241t^3\alpha^2 + 13945824\alpha^2 + 17648672\alpha + 53532t^2\alpha^2 + 975648t^3\alpha - 394176t\alpha - 69804t^2\alpha + 22011616,$

$\chi_{22}(t) = 9164760t - 3083364t^2\alpha - 3911292t^2 - 4234800\alpha^2 + 300699t^3\alpha + 408672t^3 + 7489920t\alpha - 2369988t^2\alpha^2 + 5849640t\alpha^2 + 252558t^3\alpha^2 - 6591200 - 5226400\alpha.$

This second parametrization involves only one algebraic number of degree 3, but the integer coefficients of the parametrization have grown in comparison to the result of Example 7.8.

Example 7.10. In this last example, we consider again the curve $\mathcal{C}$ of Example 7.8. We apply Algorithm OPTIMAL PARAMETRIZATION (see Sect. 5.3) for obtaining an algebraically optimal real parametrization of $\mathcal{C}$. Note that the ground field is $\mathbb{Q}$. For this purpose, since the degree of $\mathcal{C}$ is 5, we apply Step 2.1 and we determine $\mathcal{A}_3(\mathcal{C})$. Thus, we construct two families $\mathcal{F}_1, \mathcal{F}_2$ of $d - 2 = 3$ conjugate simple points over $\mathbb{Q}$ using $\mathcal{A}_3(\mathcal{C})$. Next, we compute $\mathcal{A}_4^{\mathcal{F}_1 \cup \mathcal{F}_2}(\mathcal{C})$, from which we generate a family $\mathcal{F}_3$ of 2 conjugate simple points over $\mathbb{Q}$. Finally, we determine $\mathcal{A}_3^{\mathcal{F}_3}(\mathcal{C})$ and we get a proper rational linear subsystem of dimension 1, which we may use to parametrize $\mathcal{C}$. We start with the construction of $\mathcal{F}_1, \mathcal{F}_2$. The defining polynomial of $\mathcal{A}_3(\mathcal{C})$ is

$$H_3(x, y, z, \lambda_1, \lambda_2, \lambda_3, \lambda_4)$$
$$= -y^2 z \lambda_1 - y^2 z \lambda_2 - y^2 z \lambda_3 - y^2 z \lambda_4 + \lambda_1 xyz + \lambda_2 xy^2 + \lambda_3 x^2 z + \lambda_4 x^2 y \, .$$

Now, using the adjoint

$$H_3(x, y, z, 1, 1, -1, 1) = -2y^2 z + xyz + xy^2 - x^2 z + yx^2$$

we generate the family

$$\mathcal{F}_1 = \left\{ \left(t, \frac{t(3t^2 - 13t - 52)}{3t^3 - 18t^2 - 55t - 40}, 1 \right) \right\}_{33t^3 - 138t^2 - 967t - 1232} .$$

Similarly, using the adjoint

$$H_3(x, y, z, -1, -1, -1, -1) = 4y^2 z - xyz - xy^2 - x^2 z - yx^2$$

we generate the family

$$\mathcal{F}_2 = \left\{ \left(t, \frac{-t(-78 + 81t + 3t^3 - 38t^2)}{53t - 7t^2 - 27t^3 + 10 + 3t^4}, 1 \right) \right\}_{33t^3 - 300t^2 + 197t - 1466}.$$

Now, we construct $\mathcal{F}_3$. The implicit equation of $\mathcal{A}_4^{\mathcal{F}_1 \cup \mathcal{F}_2}(\mathcal{C})$ is

$H_4(x, y, z, \lambda_1, \lambda_2, \lambda_3) = \lambda_1 x^3 z + \lambda_3 y^3 z - 175/162 xyz^2 \lambda_3 + \lambda_2 x^3 y + 8165/891 xyz^2 \lambda_1 - 23/3 xyz^2 \lambda_2 + 7/9 xy^2 z \lambda_3 + 38/11 x^2 yz \lambda_1 + 4/11 x^2 yz \lambda_2 + 17/27 xy^3 \lambda_1 - 5/54 xy^3 \lambda_3 - 79/162 x^2 z^2 \lambda_3 - 8/11 x^2 y^2 \lambda_2 + 302/99 x^2 y^2 \lambda_1 - 1/18 x^2 y^2 \lambda_3 - 280/33 x^2 z^2 \lambda_2 + 10520/891 x^2 z^2 \lambda_1 - 1303/99 xy^2 z \lambda_1 + 118/11 xy^2 z \lambda_2 - 14206/891 y^2 z^2 \lambda_1 + 158/33 y^2 z^2 \lambda_2 - 5/81 y^2 z^2 \lambda_3.$

Then, we consider the adjoint

$H_4(x, y, z, 1, 0, 1) = x^3 z + y^3 z + 14405/1782 xyz^2 - 1226/99 xy^2 z + 38/11 x^2 yz + 29/54 xy^3 + 20171/1782 x^2 z^2 + 593/198 x^2 y^2 - 14261/891 y^2 z^2$

to generate the family

$$\mathcal{F}_3 = \left\{ \left(t, \frac{p(t)}{q(t)}, 1 \right) \right\}_{-14827985 + 5414364t + 8371836t^2},$$

where

$$p(t) = 118059733738080860580474711172411775729 6834t$$
$$+ 19987596870995582233845840971567338905065,$$

and

$$q(t) = -296735076(7649054326953001304599730551 1523t$$
$$+ 12949429253623112273968591801755 5).$$

Then, the implicit equation of $\mathcal{A}_3^{\mathcal{F}_3}(\mathcal{C})$ is

$$H(x, y, z, t) = -\frac{545}{432} y^2 z + \frac{5}{162} y^2 zt + xyz - \frac{167}{162} txyz - \frac{319}{432} xy^2 + x^2 z + tx^2 y.$$

Executing Step 3 of the algorithm OPTIMAL-PARAMETRIZATION (see Sect. 5.3), we finally get the following rational parametrization over $\mathbb{Q}$:

$$\mathcal{P}(t) = \left(\frac{-9 - 72t^2 + 64t^3 + 72t}{-3(8t^2 + 3)}, \frac{-9 - 72t^2 + 64t^3 + 72t}{-18(t - 1)} \right).$$

This parametrization is defined over the ground field $\mathbb{Q}$, which is clearly optimal.

The previous examples show the advantages and disadvantages of the possible variants of the algorithms. If one is not interested in the symbolic manipulation of the problem the best option is the parametrization by $d - 2$ degree adjoints since the degrees of the polynomials involved are smaller. However, from the symbolic point of view the best option to manipulate the output is OPTIMAL PARAMETRIZATION.

7.2 Reparametrization

In Sect. 7.1 we have seen how to decide whether an algebraic plane curve, given by its implicit equation, is real and in the affirmative case, how to parametrize it over the reals. In this section, we deal with a related but slightly different problem. More precisely, we consider a complex rational parametrization of a plane curve C, defined over $\mathbb{C}$, and we want to deduce whether C is a real curve, and in the affirmative case, we would like to compute a real parametrization of C. An obvious approach to this problem consists in implicitizing $\mathcal{P}(t)$ and afterward applying the algorithms in Sect. 7.1. However, we want to solve the problem without using the implicit equation of C. That is, once the reality of C is guaranteed, we look for a change of parameter that transforms $\mathcal{P}(t)$ into a real parametrization. We will see that this problem can be solved by computing a gcd of two univariate polynomials and by parametrizing over $\mathbb{R}$ a line or a circle.

The ideas and results presented in this section are contained in the paper [ReS97a]. We will omit the proofs of some technical results on analytic polynomials and rational functions, and we will present proofs of the main results, following the description in that paper.

Throughout this section, if $f(z) \in \mathbb{C}[z]$ we denote by $\overline{f}(z)$ the conjugation of the polynomial $f(z)$.

7.2.1 Analytic Polynomial and Analytic Rational Functions

In the following, given a bivariate complex polynomial $p(x, y)$ we refer to its real and imaginary parts as the *components* of $p(x, y)$.

Definition 7.11. *A polynomial $p(x, y) \in \mathbb{C}[x, y]$ is called* analytic *if there exists a polynomial $f(z) \in \mathbb{C}[z]$ such that*

$$f(x + iy) = p(x, y).$$

The polynomial $f(z)$ is called the (polynomial) generator *of $p(x, y)$.*

Equivalent definitions can be found in complex analysis textbooks in the context of harmonic and holomorphic functions (see, e.g., [BaN82]). Analytic polynomials can be characterized as follows (we leave the proof as an exercise; Exercise 7.6).

Proposition 7.12. *Let $u(x, y)$ and $v(x, y)$ be the components of $p(x, y) \in \mathbb{C}[x, y]$. Then, the following statements are equivalent:*

1. *$p(x, y)$ is an analytic polynomial,*
2. *$\dfrac{\partial u}{\partial x} = \dfrac{\partial v}{\partial y}$, and $\dfrac{\partial u}{\partial y} = -\dfrac{\partial v}{\partial x}$ (Cauchy–Riemann conditions),*
3. *$p(x, y) = p(x + iy, 0)$,*
4. *$p(x, y) = p(0, -ix + y)$.*

From this proposition, we see that if $p(x, y)$ is an analytic polynomial, then $p(z, 0) = p(0, -iz)$, and $f(z) = p(z, 0)$ generates $p(x, y)$. Consequently, the generator of an analytic polynomial is unique. In addition, one also deduces that constant analytic polynomials are only generated by the constant polynomials in $\mathbb{C}[z]$. See also Exercise 7.7 for additional properties of analytic polynomials. For our purposes, we are specially interested in analyzing properties of the components of an analytic polynomial.

Lemma 7.13. *Let $f(z)$ be the generator of a nonconstant analytic polynomial of components $u(x, y)$, and $v(x, y)$, and let $\mathcal{C}_u$, $\mathcal{C}_v$ be the affine plane curves defined over $\mathbb{C}$ by $u(x, y)$ and $v(x, y)$, respectively. Then $\mathcal{C}_u$ and $\mathcal{C}_v$ do not have common components; i.e., $\gcd(u, v) = 1$.*

Proof. Let $p(x, y)$ be the analytic polynomial generated by $f(z)$, and let $w(x, y) = \gcd(u, v)$. Since $u, v \in \mathbb{R}[x, y]$, one has that $A \in \mathbb{R}[x, y]$, and w is a factor of $p(x, y)$. Because of Exercise 7.7 (4) this polynomial w is analytic, and because of Exercise 7.7 (2) it must be constant. Therefore, $\gcd(u, v) = 1$. □

In Exercise 7.8 one may check that this coprimality does not hold in general for nonanalytic polynomials. Additional properties of the curves $\mathcal{C}_u$, $\mathcal{C}_v$ can be found in Exercise 7.9.

Now that basic properties of analytic polynomials have been studied, we focus on the case of analytic rational functions. Similarly as for bivariate complex polynomials, we call the real and imaginary parts of an element in $r(x, y) \in \mathbb{C}(x, y)$ the *components* of $r(x, y)$.

Definition 7.14. *A rational function $r(x, y) \in \mathbb{C}(x, y)$ is called* analytic *if there exists a rational function $h(z) \in \mathbb{C}(z)$ such that*

$$h(x + iy) = r(x, y).$$

The rational function $h(z)$ is called the (rational function) generator *of $r(x, y)$.*

As in the case of analytic polynomials, equivalent definitions and basic properties of analytic rational functions can be found in complex analysis textbooks in the context of harmonic and holomorphic functions (see [BaN82], [GuR65], [Rud66], or [Kiy93]). In addition, one may characterize analytic rational functions as follows (we leave the proof as an exercise; Exercise 7.6).

Proposition 7.15. *Let $u(x, y)$, and $v(x, y)$ be the components of $r(x, y) \in \mathbb{C}(x, y)$. Then, the following statements are equivalent:*

1. *$r(x, y)$ is an analytic rational function,*
2. *$\dfrac{\partial u}{\partial x} = \dfrac{\partial v}{\partial y}$, and $\dfrac{\partial u}{\partial y} = -\dfrac{\partial v}{\partial x}$ (Cauchy–Riemann conditions),*
3. *$r(x + iy, 0)$ is well defined, and $r(x, y) = r(x + iy, 0)$,*
4. *$r(0, -ix + y)$ is well defined, and $r(x, y) = r(0, -ix + y)$.*

Now, let us see that the generator of an analytic rational function is unique. For this purpose, from Proposition 7.15 one gets that $r(z, 0) = r(0, -iz)$, and that $h(z) = r(z, 0)$ is a generator of $r(x, y)$. Also, one deduces that the generator of a rational function is unique.

The following lemmas analyze different properties of the components of analytic rational functions, and they will be used in Theorems 7.19 and 7.20.

Lemma 7.16. *Let* $\chi(z) = \frac{f(z)}{h(z)} \in \mathbb{C}(z) \setminus \mathbb{C}$, *and* $d(z) = \gcd(f, h)$. *Let*

(i) u_h, v_h *be the components of the analytic polynomial generated by* $h(z)$,
(ii) u_d, v_d *be the components of the analytic polynomial generated by* $d(z)$,
(iii) u, v *be the real and imaginary parts of* $f(x + iy) \cdot \overline{h}(x - iy)$.

Then, the following hold:

1. $u/(u_h^2 + v_h^2), v/(u_h^2 + v_h^2)$ *are the components of the analytic rational function generated by* $\chi(z)$,
2. u *and* v *are not identically zero,*
3. $\gcd(u, v) = \gcd(u, v, u_h^2 + v_h^2) = \gcd(u, u_h^2 + v_h^2) = \gcd(v, u_h^2 + v_h^2) = \beta(u_d^2 + v_d^2)$, *for some* $\beta \in \mathbb{R}^\star$.

Proof. For (1) observe that

$$\chi(x+iy) = \frac{f(x + iy) \cdot \overline{h(x + iy)}}{(u_h^2 + v_h^2)} = \frac{f(x + iy) \cdot \overline{h}(x - iy)}{(u_h^2 + v_h^2)} = \frac{u(x, y) + iv(x, y)}{(u_h^2 + v_h^2)}.$$

(2) follows from (1). For (3) we refer to Lemma 2.2. in [ReS97a].   □

Lemma 7.17. *Let* $\chi_1(z) = \frac{f}{h}, \chi_2(z) = \frac{g}{h} \in \mathbb{C}(z) \setminus \mathbb{C}$ *be such that* $\gcd(f, g, h) = 1$. *Let*

(i) u_h, v_h *be the components of the analytic polynomial generated by* $h(z)$,
(ii) $u_1/(u_h^2 + v_h^2), v_1/(u_h^2 + v_h^2)$ *be the components of the analytic rational function generated by* $\chi_1(z)$,
(iii) $u_2/(u_h^2 + v_h^2), v_2/(u_h^2 + v_h^2)$ *be the components of the analytic rational function generated by* $\chi_2(z)$.

Then $\gcd(v_1, v_2, u_h^2 + v_h^2) = 1$.

Proof. See Corollary 2.1 in [ReS97a].   □

Lemma 7.18. *Let* $\chi(z) = \frac{az+b}{cz+d} \in \mathbb{C}(z)$ *be invertible, and let*

(i) u_1, v_1 *be the components of the analytic polynomial generated by* $cz + d$,
(ii) $u/(u_1^2 + v_1^2), v/(u_1^2 + v_1^2)$ *be the components of the analytic rational function generated by* $\chi(z)$.

Then the following hold:

(1) $u_1^2 + u_2^2$ *is irreducible over* $\mathbb{R}$,
(2) $u(x, y), v(x, y)$ *are nonconstant and irreducible over* $\mathbb{C}$,

(3) $u(x,y)$ defines either a real line or a real circle; similarly for $v(x,y)$.

Proof. For (1) see Lemma 2.3 in [ReS97a]. Now, we prove (2) and (3), simultaneously. Let $\Re(z)$ and $\Im(z)$ denote the real and imaginary part of $z \in \mathbb{C}$, respectively. Then, we consider the expressions

$$A_1 = \Re(a\bar{c}), \; A_2 = \Re(a\bar{d} + b\bar{c}), \; A_3 = \Im(b\bar{c} - a\bar{d}), \; A_4 = \Re(b\bar{d}),$$

$$B_1 = \Im(a\bar{c}), \; B_2 = \Im(a\bar{d} + b\bar{c}), \; B_3 = \Re(a\bar{d} - b\bar{c}), \; B_4 = \Im(b\bar{d}),$$

With these notations

$$u(x,y) = A_1(x^2+y^2)+A_2x+A_3y+A_4, \;\; v(x,y) = B_1(x^2+y^2)+B_2x+B_3y+B_4.$$

Let us see that the statements hold for u; similarly for v. We observe that $|ad - bc|^2 = A_2^2 + A_3^2 - 4A_4A_1$. Now, let us assume that $u(x,y)$ is constant, then $A_1 = A_2 = A_3 = 0$. However, this implies that $|ad - bc| = 0$, which is impossible because $\chi(z)$ is invertible.

Let us see that u is irreducible over $\mathbb{C}$, and that it defines either a real line or a real circle. We distinguish two cases. If $A_1 = 0$, then the result follows using that u is nonconstant. Now let $A_1 \neq 0$. Then,

$$u(x,y) = A_1 \left(x + \frac{A_2}{2A_1} \right)^2 + A_1 \left(y + \frac{A_3}{2A_1} \right)^2 - \frac{|ad - bc|^2}{4A_1}.$$

Therefore, since $|ad - bc| \neq 0$ and $\chi(z)$ is invertible, the polynomial $u(x,y)$ must be irreducible over $\mathbb{C}$. In addition, in this case, u is the real circle of radius $\frac{|ad-bc|}{2}$, centered at $\left(-\frac{A_2}{2A_1}, -\frac{A_3}{2A_1} \right)$. $\qquad\square$

7.2.2 Real Reparametrization

Given a rational parametrization with complex coefficients, we consider the problem of deciding whether it defines a real curve. Moreover, if the curve is real, we want to find a reparametrization transforming the original parametrization into a real parametrization.

For this purpose, throughout this section, $\mathcal{C}$ is an affine plane curve defined over $\mathbb{C}$, and

$$\mathcal{P}(t) = \left(\frac{\chi_{1,1}(t)}{\xi(t)}, \frac{\chi_{2,1}(t)}{\xi(t)} \right) \in \mathbb{C}(t)^2,$$

where $\gcd(\chi_{1,1}, \chi_{2,1}, \xi) = 1$, a proper parametrization of $\mathcal{C}$. Note that if any component of $\mathcal{P}(t)$ is constant the problem is trivial. Thus, we assume that $\mathcal{P}(t)$ does not define a vertical or an horizontal line.

In Theorem 7.19, which reproduces Theorem 3.1 in [ReS97a], we give necessary and sufficient conditions for $\mathcal{C}$ to be real.

Theorem 7.19. *Let $\mathcal{C}$ and $\mathcal{P}(t)$ be as above, and let*

(i) $u_1(x,y), v_1(x,y)$ be the real and imaginary parts of $\chi_{1,1}(x+iy) \cdot \overline{\xi}(x-iy)$,
(ii) $u_2(x,y), v_2(x,y)$ be the real and imaginary parts of $\chi_{2,1}(x+iy) \cdot \overline{\xi}(x-iy)$.

Then, $\mathcal{C}$ is real if and only if $\gcd(v_1, v_2)$ defines either a real line or a real circle.

Proof. Let $u_{\chi_{1,1}}, v_{\chi_{1,1}}$ be the components of the analytic polynomial generated by $\chi_{1,1}(z)$. Similarly for $u_{\chi_{2,1}}, v_{\chi_{2,1}}$, and $\chi_{2,1}(z)$, and for u_ξ, v_ξ, and $\xi(z)$.

We first assume that $\gcd(v_1, v_2)$ is either a real line or a real circle. By Lemmas 7.16 and 7.17, $\gcd(v_1, v_2, u_\xi^2 + v_\xi^2) = 1$. Therefore, if V_1, V_2, H denote the varieties defined over $\mathbb{C}$ by $v_1, v_2, u_\xi^2 + v_\xi^2$, respectively, one has that $\Omega = [\mathbb{R}^2 \setminus H] \cap [V_1 \cap V_2 \cap \mathbb{R}^2]$ has dimension 1. Now, we observe that $\mathcal{N} = \{\mathcal{P}(x + iy) \mid (x,y) \in \Omega\} \subset \mathcal{C} \cap \mathbb{R}^2$, and $\mathrm{card}(\mathcal{N}) = \infty$; note that $\mathcal{P}(t)$ is proper. Therefore, $\mathcal{C}$ has infinitely many real points and so $\mathcal{C}$ is real.

Conversely, let $\mathcal{C}$ be real, and let $f(x,y) \in \mathbb{R}[x,y]$ be the irreducible polynomial defining $\mathcal{C}$ (see Lemma 7.2). Taking into account Theorem 7.6, $\mathcal{C}$ can be properly parametrized over the reals. Let $\mathcal{Q}(t) \in \mathbb{R}(t)^2$ be a real proper rational parametrization of $\mathcal{C}$. Then, by Lemma 4.17 there exists an invertible rational function $\varphi(t) = \frac{at+b}{ct+d} \in \mathbb{C}(t)$, such that $\mathcal{P}(t) = \mathcal{Q}(\varphi(t))$. Also, let $M(x,y) = M_1/M_2 \in \mathbb{R}(x,y)$ be the inverse of $\mathcal{Q}$ with $\gcd(M_1, M_2) = 1$; note that we can take $M(x,y) \in \mathbb{R}(x,y)$. Therefore, $M(\mathcal{Q}(t)) = t$, and hence $M(\mathcal{P}(t)) = \varphi(t)$. We consider the homogenization $M_j^h(x,y,w) \in \mathbb{R}[x,y,w]$ of the polynomials $M_j(x,y) \in \mathbb{R}[x,y]$, for $j = 1,2$. We have

$$M_j(\mathcal{P}(x+iy)) = \frac{M_j^h(\chi_{1,1}(x+iy), \chi_{2,1}(x+iy), \xi(x+iy))}{\xi(x+iy)^{\alpha_j}},$$

where $\alpha_j = \deg(M_j^h)$, for $j = 1, 2$. In these conditions, we get that

$$M_j(\mathcal{P}(x+iy)) = \frac{M_j^h(\chi_{1,1}(x+iy), \chi_{2,1}(x+iy), \xi(x+iy))\overline{\xi}(x-iy)^{\alpha_j}}{(u_\xi^2 + v_\xi^2)^{\alpha_j}}$$

$$= \frac{M_j^h(\chi_{1,1}(x+iy)\overline{\xi}(x-iy), \chi_{2,1}(x+iy)\overline{\xi}(x-iy), \xi(x+iy)\overline{\xi}(x-iy))}{(u_\xi^2 + v_\xi^2)^{\alpha_j}}$$

$$= \frac{M_j^h(u_1 + iv_1, u_2 + iv_2, u_\xi^2 + v_\xi^2)}{(u_\xi^2 + v_\xi^2)^{\alpha_j}} = \frac{A_j(x,y) + iB_j(x,y)}{(u_\xi^2 + v_\xi^2)^{\alpha_j}},$$

where $A_j, B_j \in \mathbb{R}[x,y]$ denote the real and imaginary parts, respectively, of the polynomial $M_j^h(u_1 + iv_1, u_2 + iv_2, u_\xi^2 + v_\xi^2)$. Therefore,

$$M(\mathcal{P}(x+iy)) = \frac{M_1(\mathcal{P}(x+iy))}{M_2(\mathcal{P}(x+iy))} = \frac{A_1(x,y) + iB_1(x,y)}{A_2(x,y) + iB_2(x,y)}(u_\xi^2 + v_\xi^2)^{\alpha_2 - \alpha_1}.$$

On the other hand, if $L_1(z) = az + b$, and $L_2(z) = cz + d$, then

$$M(\mathcal{P}(x+iy)) = \varphi(x+iy) = \frac{L_1(x+iy)}{L_2(x+iy)} = \frac{u(x,y)+iv(x,y)}{r(x,y)},$$

where $r = L_2(x+iy) \cdot \overline{L_2}(x-iy)$ and u, v are its real and imaginary parts (see Lemma 7.16). Moreover, by Lemma 7.18, u and v are nonconstant. Thus, one has that

$$\frac{A_1(x,y)+iB_1(x,y)}{A_2(x,y)+iB_2(x,y)}(u_\xi^2+v_\xi^2)^{\alpha_2} = \frac{u(x,y)+iv(x,y)}{r(x,y)}(u_\xi^2+v_\xi^2)^{\alpha_1}.$$

By Lemma 7.18, one deduces that r is irreducible over $\mathbb{R}$, and u, v define either a real line or a real circle. Normalizing the left-hand side of the equality, and taking the imaginary parts, one deduces that

$$r \cdot (u_\xi^2+v_\xi^2)^{\alpha_2} \cdot (A_2 B_1 - A_1 B_2) = (A_2^2+B_2^2) \cdot (u_\xi^2+v_\xi^2)^{\alpha_1} \cdot v.$$

Now, let $G = \gcd(v_1, v_2)$. Note that $G \neq 0$, since $v_1 \cdot v_2 \neq 0$ (see Lemma 7.16). Furthermore, note that

$$\mathcal{P}(x+iy) = \left(\frac{u_1+iv_1}{u_\xi^2+v_\xi^2}, \frac{u_2+iv_2}{u_\xi^2+v_\xi^2} \right).$$

Thus, since $\operatorname{card}(\mathcal{C} \cap \mathbb{R}^2) = \infty$, there exist infinitely many $(x_0, y_0) \in \mathbb{R}^2$ such that $v_1(x_0, y_0) = v_2(x_0, y_0) = 0$. Therefore, G is a nonconstant polynomial. In these conditions, we have to prove that G is either a real line or a real circle. We first note that $M_j^h(x, y, w)$ is real and homogeneous. Since B_j is the imaginary part of $M_j^h(u_1+iv_1, u_2+iv_2, u_\xi^2+v_\xi^2)$, we have that G divides B_j, for $j = 1, 2$. Thus, since (see above)

$$r \cdot (u_\xi^2+v_\xi^2)^{\alpha_2} \cdot (A_2 B_1 - A_1 B_2) = (A_2^2+B_2^2) \cdot (u_\xi^2+v_\xi^2)^{\alpha_1} \cdot v,$$

we deduce that G divides $(A_2^2+B_2^2) \cdot (u_\xi^2+v_\xi^2)^{\alpha_1} \cdot v$. On the other hand, taking into account Lemma 7.17, one has that $\gcd(v_1, v_2, u_\xi^2+v_\xi^2) = 1$. Hence, G divides $(A_2^2+B_2^2) \cdot v$. Let us see that $\gcd(G, A_2^2+B_2^2) = 1$. Indeed, if $\gcd(G, A_2^2+B_2^2) \neq 1$, then we deduce that there exists an irreducible real common factor, H, of G and $A_2^2+B_2^2$. Since H divides G, and G divides B_2, it follows that H divides B_2, and then H divides also A_2. Now, we consider the complex polynomial

$$N(t) = \gcd(M_2^h(\chi_{1,1}(t), \chi_{2,1}(t), \xi(t)), \xi(t)^{\alpha_2}).$$

Let N_1, N_2 be the components of $N(x+iy)$. We observe that (see above)

$$\frac{M_2^h(\chi_{1,1}(x+iy), \chi_{2,1}(x+iy), \xi(x+iy))}{\xi(x+iy)^{\alpha_2}} = \frac{A_2(x,y)+iB_2(x,y)}{(u_\xi^2+v_\xi^2)^{\alpha_2}}.$$

Therefore, by Lemma 7.16, $\gcd(A_2, B_2) = N_1^2 + N_2^2$. Thus, H divides $N_1^2 + N_2^2$. Moreover, since $N(t)$ divides $\xi(t)^{\alpha_2}$, then $N(x+iy)\overline{N}(x-iy)$ divides $\xi(x + iy)^{\alpha_2} \cdot \overline{\xi}(x - iy)^{\alpha_2}$. Thus H divides $(u_\xi^2 + v_\xi^2)^{\alpha_2}$. However, since H is irreducible, one deduces that H divides $u_\xi^2 + v_\xi^2$, and then, H divides $\gcd(v_1, v_2, u_\xi^2 + v_\xi^2)$. Thus, by Lemma 7.17, one concludes that $H = 1$ which implies that $\gcd(G, A_2^2 + B_2^2) = 1$. Therefore, since v is irreducible, then G and v are associated, and therefore G defines a real line or a real circle (note that v defines a real line or a real circle).   □

In Theorem 7.20, which reproduces Theorem 3.2 in [ReS97a], we see how to reparametrize over $\mathbb{R}$ real curves given by complex rational parametrizations.

Theorem 7.20. *Let C and $\mathcal{P}(t)$ be as above. Let C be real and let*

(i) $u_1(x, y), v_1(x, y)$ be the real and imaginary parts of $\chi_{1,1}(x+iy) \cdot \overline{\xi}(x-iy)$,
(ii) $u_2(x, y), v_2(x, y)$ be the real and imaginary parts of $\chi_{2,1}(x+iy) \cdot \overline{\xi}(x-iy)$,
(iii) u_ξ, v_ξ be the components of the analytic polynomial generated by $\xi(z)$,
(iv) $\mathcal{M}(t) = (m_1(t), m_2(t))$ be a real proper parametrization of the curve defined by $\gcd(v_1, v_2)$.

Then

$$\mathcal{P}(m_1(t) + i m_2(t)) = \left(\frac{u_1(\mathcal{M}(t))}{u_\xi(\mathcal{M}(t))^2 + v_\xi(\mathcal{M}(t))^2}, \frac{u_2(\mathcal{M}(t))}{u_\xi(\mathcal{M}(t))^2 + v_\xi(\mathcal{M}(t))^2} \right)$$

is a real proper rational parametrization of C.

Proof. Since G is real, by Theorem 7.19, $G = \gcd(v_1, v_2)$ defines either a real line or a real circle. Let us first see that $m_1(t) + im_2(t)$ is an invertible rational function. If G is a real line, then there exists a polynomial parametrization $\mathcal{L}(t) = (L_1(t), L_2(t))$ of G, where $\deg(L_i) \leq 1$. Moreover, since $\mathcal{M}(t)$ is proper, there exists a linear rational function $\varphi(t)$ such that $\mathcal{L}(\varphi(t)) = \mathcal{M}(t)$. Furthermore, $L_1(t) + iL_2(t)$ is a polynomial of degree 1 (note that if the polynomial were constant this would imply that G would be a nonreal line), say $\lambda + \mu t$, with $\mu \neq 0$. Therefore, $m_1(t) + im_2(t) = L_1(\varphi(t)) + iL_2(\varphi(t)) = \lambda + \mu\varphi$, which is linear. Now, assume that G is a real circle of the form $(x-a)^2 + (y-b)^2 = c^2$, with $a, b, c \in \mathbb{R}$. Then,

$$\mathcal{N}(t) = \left(a + c\frac{t^2 - 1}{t^2 + 1}, b + c\frac{2t}{t^2 + 1} \right),$$

is a real proper parametrization of G. Therefore, since $\mathcal{M}(t)$ is another proper parametrization of the same circle, there exists an invertible rational function $\phi \in \mathbb{C}(t)$ such that $\mathcal{M}(t) = \mathcal{N}(\phi(t))$. Thus,

$$m_1(t) + i\,m_2(t) = (a + i\,b) + c\frac{(\phi(t) + i)^2}{\phi(t)^2 + 1} = (a + i\,b) + c\frac{\phi(t) + i}{\phi(t) - i},$$

which is an invertible rational function. In both situations, one deduces that $m_1(t) + im_2(t)$ is invertible, and hence $\mathcal{P}(m_1(t) + im_2(t))$ is again a proper rational parametrization of $\mathcal{C}$. Furthermore, by Lemma 7.17, one has that $\gcd(v_1, v_2, u_\xi^2 + v_\xi^2) = 1$, and hence $(u_\xi^2 + v_\xi^2)(\mathcal{M}(t)) \neq 0$. Finally, taking into account that $\mathcal{M}(t)$ parametrizes $G = \gcd(v_1, v_2)$, one has that (see proof of Theorem 7.19)

$$
\begin{aligned}
&\mathcal{P}(m_1(t) + im_2(t)) \\
&= \left(\frac{u_1(\mathcal{M}(t)) + iv_1(\mathcal{M}(t))}{u_\xi(\mathcal{M}(t))^2 + v_\xi(\mathcal{M}(t))^2}, \frac{u_2(\mathcal{M}(t)) + iv_2(\mathcal{M}(t))}{u_\xi(\mathcal{M}(t))^2 + v_\xi(\mathcal{M}(t))^2} \right) \\
&= \left(\frac{u_1(\mathcal{M}(t))}{u_\xi(\mathcal{M}(t))^2 + v_\xi(\mathcal{M}(t))^2}, \frac{u_2(\mathcal{M}(t))}{u_\xi(\mathcal{M}(t))^2 + v_\xi(\mathcal{M}(t))^2} \right) \in \mathbb{R}(t)^2,
\end{aligned}
$$

which implies that $\mathcal{P}(m_1(t) + im_2(t))$ is a real proper rational parametrization of $\mathcal{C}$. $\qquad\square$

The combination of all these ideas leads to the algorithm REAL-REPARAMETRIZATION.

Algorithm REAL-REPARAMETRIZATION

Given $\mathcal{P}(t) = \left(\frac{\chi_{11}(t)}{\xi(t)}, \frac{\chi_{21}(t)}{\xi(t)} \right) \in \mathbb{C}(t)^2$, with $\gcd(\chi_{11}, \chi_{21}, \xi) = 1$, a proper rational parametrization defining an affine plane curve $\mathcal{C}$ over $\mathbb{C}$, the algorithm decides whether $\mathcal{C}$ can be parametrized over the reals and, in the affirmative case, it computes a proper real parametrization of $\mathcal{C}$.

1. If χ_{11}/ξ is constant (similarly if χ_{21}/ξ is constant), then if it is real return $(\chi_{11}/\xi, t)$ else return that $\mathcal{C}$ is not real.
2. Compute $u_1 + iv_1 := \chi_{11}(x + iy)\overline{\xi}(x - iy)$, $u_2 + iv_2 := \chi_{21}(x + iy)\overline{\xi}(x - iy)$ and $w := \xi(x + iy)\overline{\xi}(x - iy)$, where $u_1, v_1, u_2, v_2, w \in \mathbb{R}[x, y]$.
3. Determine $G(x, y) = \gcd(v_1, v_2) \in \mathbb{R}[x, y]$.
4. If $\deg(G(x, y)) \notin \{1, 2\}$, then return that $\mathcal{C}$ is not real.
5. If $\deg(G(x, y)) = 1$, then
   5.1. Compute a real proper rational parametrization $\mathcal{M}(t) = (m_1(t), m_2(t)) \in \mathbb{R}(t)^2$ of the real line that $G(x, y)$ defines.
   5.2. Return $\mathcal{Q}(t) = \mathcal{P}(m_1(t) + im_2(t)) = \left(\frac{u_1(\mathcal{M}(t))}{w(\mathcal{M}(t))}, \frac{u_2(\mathcal{M}(t))}{w(\mathcal{M}(t))} \right) \in \mathbb{R}(t)^2$.
6. If $\deg(G(x, y)) = 2$, then
   6.1. If $G(x, y)$ does not define a real circle, then return that $\mathcal{C}$ is not real
   6.2. If $G(x, y)$ defines a real circle $\mathcal{D}$ then
      6.2.1. Compute a real proper rational parametrization $\mathcal{M}(t) \in \mathbb{R}(t)^2$ of $\mathcal{D}$.
      6.2.2. Return $\mathcal{Q}(t) = \mathcal{P}(m_1(t) + im_2(t)) = \left(\frac{u_1(\mathcal{M}(t))}{w(\mathcal{M}(t))}, \frac{u_2(\mathcal{M}(t))}{w(\mathcal{M}(t))} \right) \in \mathbb{R}(t)^2$.

Example 7.21. Let C be the affine plane curve defined by the proper complex parametrization $\mathcal{P}(t) =$

$$\left(\frac{(t+2)^3}{-t^3 - 5t^2 + t + 3 - 2it^2 - 7it + it^3}, \; -\frac{it^3 + 4it^2 + 4it - t^2 - 4t - 4}{-t^3 - 5t^2 + t + 3 - 2it^2 - 7it + it^3} \right).$$

We compute the polynomials u_1, u_2, v_1, v_2, w:

$u_1 := 24 + 44x + 56y - 8y^5 - 11x^5 - 42x^2y + 32xy - 10x^2 - 19y^4 - 39xy^2 - 60x^2y^2 - 59x^3 - 41x^4 + 34y^2 - 10y^3 - x^6 - 3x^2y^4 - 16x^2y^3 - 38x^3y - y^6 - 38xy^3 - 8x^4y - 3x^4y^2 - 22x^3y^2 - 11xy^4$

$v_1 := -4x^5 + 7x^4 - y^6 + 17y^4 + y^5 - x^6 + 68y^2 + 55y^3 + 58x^3 + 100x^2 + 26xy^3 - 4xy^4 + 90xy^2 - 3x^2y^4 + 116xy + x^4y + 26x^3y - 8x^3y^2 + 28y + 24x^2y^2 - 3x^4y^2 + 2x^2y^3 + 99x^2y + 56x$

$u_2 := 12 + 16x + 40y - 2y^5 - 3x^5 + 22x^2y + 40xy + 15x^2 + 4y^4 + 29xy^2 + 6x^2y^2 + 13x^3 + 2x^4 + 49y^2 + 26y^3 - x^6 - 3x^2y^4 - 4x^2y^3 + 2x^3y - y^6 + 2xy^3 - 2x^4y - 3x^4y^2 - 6x^3y^2 - 3xy^4$

$v_2 := 24x^3 + 20x^2 + 28y^2 + 8x^5 + 21x^4 + y^6 + 23y^4 + 7y^5 + 30xy^3 + 8xy^4 + x^6 + 16x + 8y + 30x^3y + 16x^3y^2 + 3x^2y^4 + 14x^2y^3 + 3x^4y^2 + 7x^4y + 52xy^2 + 53x^2y + 44x^2y^2 + 46xy + 37y^3$

$w := 9 + 6x + 42y + 14y^5 + 6x^5 + 56x^2y + 24xy + 20x^2 + 45y^4 + 36xy^2 + 58x^2y^2 + 12x^3 + 13x^4 + 80y^2 + 80y^3 + 2x^6 + 6x^2y^4 + 28x^2y^3 + 24x^3y + 2y^6 + 24xy^3 + 14x^4y + 6x^4y^2 + 12x^3y^2 + 6xy^4$

We have that

$$G(x, y) = \gcd(v_1, v_2) = 2x + x^2 + y + y^2,$$

which defines the real circle $(x+1)^2 + (y+\frac{1}{2})^2 - \frac{5}{4} = 0$. So C can be parametrized over the reals. We consider the real proper rational parametrization

$$\mathcal{M}(t) = (m_1(t), m_2(t)) = \left(-\frac{t+2}{1+t^2}, -\frac{t(2+t)}{1+t^2} \right).$$

Finally, Algorithm REAL-REPARAMETRIZATION returns

$$\mathcal{Q}(t) = \left(\frac{u_1(\mathcal{M}(t))}{w(\mathcal{M}(t))}, \frac{u_2(\mathcal{M}(t))}{w(\mathcal{M}(t))} \right) = \left(\frac{t^3}{t-1}, \frac{-t^2}{t-1} \right).$$

Example 7.22. Let C be the affine plane curve defined by the proper complex parametrization

$$\mathcal{P}(t) = \left(\frac{\chi_{11}(t)}{\xi(t)}, \frac{\chi_{21}(t)}{\xi(t)} \right) =$$

$$\left(-\frac{7 + 32it - 64t^2 - 96it^3 + 80t^4 + 32it^5}{-7 - 10it + 12t^2 + 8it^3}, \frac{-7 - 26it + 28t^2 + 16it^3}{-7 - 10it + 12t^2 + 8it^3} \right).$$

We compute the polynomials v_1, v_2:

$v_1 := 832x^3y + 64x^5 + 154x + 256x^7 - 360x^3 + 2104xy^2 - 728xy + 3840x^3y^4 + 2304x^5y^2 - 3200xy^5 - 3840x^3y^3 - 640x^5y - 1536x^3y^5 - 1536x^5y^3 - 512x^7y + 1792xy^6 - 512xy^7 + 640x^3y^2 - 3648xy^3 + 4160xy^4$

$v_2 := -(-7 + 26y + 28x^2 - 28y^2 - 48x^2y + 16y^3)(24xy + 8x^3 - 24xy^2 - 10x) - (56xy + 16x^3 - 48xy^2 - 26x)(7 + 12y^2 - 10y - 12x^2 - 8y^3 + 24x^2y)$

We have that $G(x,y) = \gcd(v_1, v_2) = 2x$, so $\mathcal{C}$ can be parametrized over the reals. In Step 5 of the algorithm we consider a real proper rational parametrization $\mathcal{M}(t) = (m_1(t), m_2(t)) \in \mathbb{R}(t)^2$ of the real line defined by $G(x,y)$. Let $\mathcal{M}(t) = (0, t)$. Finally, Algorithm REAL-REPARAMETRIZATION returns

$$\mathcal{Q}(t) = \mathcal{P}(m_1(t) + im_2(t))$$

$$= \left(\frac{-7 - 80t^4 + 32t - 64t^2 + 96t^3 + 32t^5}{-7 - 12t^2 + 10t + 8t^3}, \frac{-7 - 28t^2 + 26t + 16t^3}{-7 - 12t^2 + 10t + 8t^3} \right).$$

Example 7.23. Let $\mathcal{C}$ be the affine plane curve defined by the proper complex parametrization $\mathcal{P}(t) =$

$$\left(-\frac{1}{2} \frac{-12 + i - 24it + 6it^2 + 24it^3 + it^4 + 12t^4}{-2it^3 + 2it + t^4 - 1}, \frac{1}{2} \frac{i(1 + 2it + t^2)^2}{-2it^3 + 2it + t^4 - 1} \right).$$

The gcd computed in Step 3 of the algorithm is $G(x,y) = x^2 + y^2 + 1$. Although $\deg(G(x,y)) = 2$, it does not define a real circle. Thus, $\mathcal{C}$ cannot be parametrized over the reals. In fact, $\mathcal{C}$ is the curve defined implicitly by the polynomial $f(x,y) = -48x^2y + 144y^4 + 146x^2y^2 + 2x^4 + 5329y^2 - 1752y^3$, and applying for instance Theorem 7.7 one can check that $\mathcal{C}$ is not a real curve. Indeed, it only has the two real points $(0, 0)$ and $(0, \frac{73}{12})$ and they are singular (compare Theorem 7.4).

7.3 Normal Parametrization

In Sect. 6.3 we have studied the normality of parametrizations over an algebraically closed field. In this section we analyze the normality problem for real parametrizations; i.e., we study the surjectivity of the rational map induced by the real parametrization being restricted to the field of real numbers. As in Sect. 6.3 we follow the ideas in [Sen02]; see also [BaR95]. A generalization to the case of real space curves, based on valuation rings, can be found in [AnR06].

Throughout this section, we consider a real affine plane curve $\mathcal{C}$, and we assume w.l.o.g. that we are given an affine rational parametrization, not necessarily proper, $\mathcal{P}(t)$ of $\mathcal{C}$ over $\mathbb{R}$ that is expressed, in reduced form, as

$$\mathcal{P}(t) = \left(\frac{\chi_{11}(t)}{\chi_{12}(t)}, \frac{\chi_{21}(t)}{\chi_{22}(t)} \right) \in \mathbb{R}(t)^2.$$

In addition, for every rational parametrization $\mathcal{Q}(t)$ we consider the polynomials $H_1^{\mathcal{Q}}(t, x)$ and $H_2^{\mathcal{Q}}(t, y)$ introduced in Chap. 4. Furthermore, we consider the rational map

$$\varphi_{\mathcal{P}}|_{\mathbb{R}} : \mathbb{R} \longrightarrow \mathcal{C} \cap \mathbb{R}^2$$
$$t \longmapsto \mathcal{P}(t).$$

As in previous sections, we denote the fiber of a point $P \in \mathcal{C}$ as

$$\mathcal{F}_{\mathcal{P}}(P) = \{t_0 \in \mathbb{C} \,|\, \mathcal{P}(t_0) = P\}.$$

Note that this is the fiber of the map $\varphi_{\mathcal{P}}$ and not of $\varphi_{\mathcal{P}}|_{\mathbb{R}}$.

Definition 7.24. *A rational affine real parametrization $\mathcal{P}(t)$ of an affine curve $\mathcal{C}$ is $\mathbb{R}$-normal, or normal over $\mathbb{R}$, if $\varphi_{\mathcal{P}}|_{\mathbb{R}}(\mathbb{R}) = \mathcal{C} \cap \mathbb{R}^2$, or equivalently iff for all $P \in \mathcal{C} \cap \mathbb{R}^2$ there exists $t_0 \in \mathbb{R}$ such that $\mathcal{P}(t_0) = P$. Furthermore, if there exists an $\mathbb{R}$-normal parametrization of $\mathcal{C}$ we say that $\mathcal{C}$ can be $\mathbb{R}$-normally parametrized.*

In Sect. 6.3, when dealing with the normality problem over an algebraically closed field, the picture was very clear; namely, every rational parametrization reaches all points on the curve with the possible exception of one point (the critical point), and every rational curve can always be parametrized by means of a proper normal parametrization (see Theorem 6.26). Now, in this new setting, the situation is different: not all rational real curves can be $\mathbb{R}$-normally parametrized and, for those that can, we will only ensure (in general) the existence of $\mathbb{R}$-normal parametrizations of tracing index two (see Sect. 4.3). The difficulty behind this complication is that some points on the curve, although reachable by the real parametrization, are only generated by nonreal complex values of the parameter. For instance, let us consider the proper polynomial parametrization $\mathcal{P}(t) = (t^4, t^3 + t)$ of the curve $\mathcal{C}$ defined by $f(x, y) = 4y^2x - y^4 + x - 2x^2 + x^3$. By Corollary 6.21 to Lemma 6.19, we know that $\mathcal{P}(t)$ is normal. However, $\mathcal{P}(t)$ is not $\mathbb{R}$-normal since the point $(1, 0) \in \mathcal{C}$ is only reachable via $\mathcal{P}(t)$ by the parameter values $\pm i$. Based on the existence of this phenomenon we will introduce the notions of degenerated and nondegenerated parametrization. In addition, we will distinguish between two different types of degeneration: weak and strong. In this situation, the picture will be as follows: nondegenerate and weakly degenerate parametrizations can be reparametrized into $\mathbb{R}$-normal parametrizations with tracing index at most 2. However, if there exists a proper strong degenerated parametrization, then the corresponding curve cannot be $\mathbb{R}$-normally parametrized.

Definition 7.25. *Let* $\mathcal{P}(t) \in \mathbb{R}(t)^2$ *be a not necessarily proper parametriza-tion of an affine curve* $\mathcal{C}$. *Then,*

(1) we define the set of degenerations *of* $\mathcal{P}(t)$, *denoted by* $\mathcal{D}^\mathcal{P}$, *as*

$$\mathcal{D}^\mathcal{P} = \{P \in \mathcal{C} \cap \mathbb{R}^2 \,|\, \mathcal{F}_\mathcal{P}(P) \neq \emptyset \text{ and } \mathcal{F}_\mathcal{P}(P) \subset \mathbb{C} \setminus \mathbb{R}\} \,;$$

(2) we say that $\mathcal{P}(t)$ *is* degenerate *if* $\mathcal{D}^\mathcal{P} \neq \emptyset$, *otherwise we say that* $\mathcal{P}(t)$ *is* nondegenerate;
(3) we say that $\mathcal{P}(t)$ *is* weakly degenerate *if it is degenerated, the critical point of* $\mathcal{P}(t)$ *exists (see Definition 6.24), and* $\mathcal{D}^\mathcal{P}$ *contains only the critical point;*
(4) we say that $\mathcal{P}(t)$ *is* strongly degenerate *if it is degenerate but not weakly degenerate.*

Remarks. We observe that

(1) a degenerate parametrization is not $\mathbb{R}$-normal;
(2) the condition on the existence of the critical point, required in Defini-tion 7.25 (3), is equivalent to $\deg(\chi_{i\,1}) \leq \deg(\chi_{i\,2})$ for $i = 1, 2$ (see Defi-nition 6.24);
(3) degenerate polynomial parametrizations are strongly degenerate;
(4) if $\mathcal{P}(t)$ is nondegenerate, then every point $P \in \mathcal{C} \cap \mathbb{R}^2$ different from the critical point is reachable by at least one real parameter value. More-over, if the critical point exits, then either it is not reachable via $\mathcal{P}(t)$ (i.e., its fibre is empty) or it is reachable by at least one real parameter value;
(5) every real proper parametrization of a line is nondegenerate.

Now, let us see how to compute $\mathcal{D}^\mathcal{P}$, when $\mathcal{P}(t)$ is assumed to be proper. For further details see Propositions 1 and 2 in [Sen02]. We consider a real function $S(x, y)$ representing the inverse of $\mathcal{P}(t)$ (note that $\mathcal{P}(t)$ is real and proper). We observe that for those $P \in \mathcal{C} \cap \mathbb{R}^2$, such that $S(P)$ and $\mathcal{P}(S(P))$ are defined, the fiber $\mathcal{F}_\mathcal{P}(P)$ is not empty and contains at least a real value, namely $S(P)$. Therefore, these points are not in $\mathcal{D}^\mathcal{P}$. This implies that $\mathcal{D}^\mathcal{P}$ is either empty or zero-dimensional. Moreover, the missing points satisfy that either the denominator of S or some of the denominators of $\mathcal{P}(S(x, y))$ vanishes at them. Therefore, considering the intersection of the curve $\mathcal{C}$ defined by $\mathcal{P}(t)$ and each of curves defined by the denominators of S and $\mathcal{P}(S(x, y))$ we have a description of the possible elements in $\mathcal{D}^\mathcal{P}$. Finally, taking into account that parametrizations are given in reduced form, it is clear that for every point $P = (a, b) \in \mathcal{C}$ the fiber $\mathcal{F}_\mathcal{P}(P)$ is nonempty iff $\gcd(H_1^\mathcal{P}(t, a), H_2^\mathcal{P}(t, b))$ has positive degree. Moreover, if $P \in \mathcal{C} \cap \mathbb{R}^2$, $\mathcal{F}_\mathcal{P}(P) \cap \mathbb{R}$ is the set of real roots of $\gcd(H_1^\mathcal{P}(t, a), H_2^\mathcal{P}(t, b))$. Therefore, $P \in \mathcal{D}^\mathcal{P}$ iff $\gcd(H_1^\mathcal{P}(t, a), H_2^\mathcal{P}(t, b))$ has positive degree and no real roots. From this reasoning, we derive the following algorithm.

Algorithm DEGENERATIONS

Given a rational proper parametrization $\mathcal{P}(t) = \left(\frac{\chi_{11}(t)}{\chi_{12}(t)}, \frac{\chi_{21}(t)}{\chi_{22}(t)} \right) \in \mathbb{L}(t)^2$, where $\mathbb{L}$ is a computable subfield of $\mathbb{R}$, of an affine rational curve $\mathcal{C}$, the algorithm computes $\mathcal{D}^{\mathcal{P}}$.

1. Apply Algorithm INVERSE (see Sect. 4.4) to compute the inverse of $\mathcal{P}(t)$. Let $S(x,y) = A(x,y)/B(x,y)$ be a representative of the inverse over $\mathbb{R}$, such that $\gcd(A, B) = 1$.
2. For $i = 1, 2$ compute the denominator $q_i^*(x,y)$ of the reduced expression of $\frac{\chi_{i1}(S(x,y))}{\chi_{i2}(S(x,y))}$.
3. Compute the set $\mathcal{B}$ of affine intersection points of $\mathcal{C}$ and each of the curves defined by $B(x,y), q_1^*(x,y), q_2^*(x,y)$.
4. $\mathcal{D} := \emptyset$. For every $(a, b) \in \mathcal{B} \cap \mathbb{R}^2$ check whether $\gcd(H_1^{\mathcal{P}}(t, a), H_2^{\mathcal{P}}(t, b))$ has positive degree and whether it has a root in $\mathbb{R}$. If not then $\mathcal{D} = \mathcal{D} \cup \{(a, b)\}$.
5. Return $\mathcal{D}$.

Example 7.26. Let us consider the parametrization mentioned in the introduction to the section. That is, $\mathcal{P}(t) = (t^4, t^3 + t)$. In Step 1, applying Algorithm INVERSE (see Sect. 4.4), we get that $S(x,y) = A(x,y)/B(x,y)$, where $A = -x + x^2 + y^2$ and $B = y(1 + x)$. In Step 2 we obtain $q_1^* = (-x + x^2 + y^2)^4$ and $q_2^* = (-x + x^2 + y^2)(x^2 - 2x^3 + x^4 + 3y^2x^2 + y^4 + y^2)$. The set $\mathcal{B}$ in Step 3 is $\mathcal{B} = \{(0,0), (1,0), (-1, \pm\sqrt{2}i)\}$. Moreover, $\gcd(H_1^{\mathcal{P}}(t, 0), H_2^{\mathcal{P}}(t, 0)) = t$ and $\gcd(H_1^{\mathcal{P}}(t, 1), H_2^{\mathcal{P}}(t, 0)) = t^2 + 1$ (see Step 4). Thus, $\mathcal{D}^{\mathcal{P}} = \{(1,0)\}$. So, since the critical point of $\mathcal{P}(t)$ does not exists, $\mathcal{P}(t)$ is strongly degenerate.

Example 7.27. We consider the rational proper parametrization

$$\mathcal{P}(t) = \left(\frac{1}{t^2}, \frac{t+1}{t^3} \right) \in \mathbb{R}(t)^2.$$

In Step 1, applying Algorithm INVERSE, we get that $S(x,y) = A(x,y)/B(x,y)$, where $A = x$ and $B = y - x$. In Step 2 we obtain $q_1^* = (x - y)^2$ and $q_2^* = (x - y)^2 y$. The set $\mathcal{B}$ in Step 3 is $\mathcal{B} = \{(0,0), (1,0)\}$. Moreover, $\gcd(H_1^{\mathcal{P}}(t, 0), H_2^{\mathcal{P}}(t, 0)) = 1$ and $\gcd(H_1^{\mathcal{P}}(t, 1), H_2^{\mathcal{P}}(t, 0)) = t + 1$ (see Step 4). Thus, $\mathcal{D}^{\mathcal{P}} = \{(0,0)\}$. So, since $(0,0)$ is the critical point of $\mathcal{P}(t)$, the parametrization $\mathcal{P}(t)$ is weakly degenerate.

Before dealing with the problem of computing $\mathbb{R}$-normal parametrizations, it is natural to have an algorithmic criterion to check whether a given real parametrization is already $\mathbb{R}$-normal or not. As remarked after Definition 7.25, if the parametrization is degenerate then it is not $\mathbb{R}$-normal. This can be decided by Algorithm DEGENERATIONS. So we only need to know how to proceed when the parametrization is nondegenerate. Because of remark (4) to Definition 7.25 we have Theorem 7.28.

Theorem 7.28. *Let $\mathcal{P}(t) \in \mathbb{R}(t)^2$ be a nondegenerate parametrization. Then $\mathcal{P}(t)$ is $\mathbb{R}$-normal if and only if $\mathcal{P}(t)$ is normal.*

A combination of Algorithms DEGENERATIONS and NORMALITY-TEST (see Sect. 6.3) leads to an algorithm for deciding the $\mathbb{R}$-normality of proper real parametrizations. Theorem 7.29 shows how to reparametrize nondegenerate parametrizations into $\mathbb{R}$-normal parametrizations (see also [BaR95]).

Theorem 7.29. *Let $\mathcal{P}(t)$ be a nondegenerate real proper parametrization of $\mathcal{C}$. Then the following hold:*

(1) $\mathcal{C}$ can be $\mathbb{R}$-normally parametrized as

$$\mathcal{P}\left(\frac{t}{t^2 - 1}\right).$$

(2) If $\chi_{12}(t)\chi_{22}(t)$ has a real root α, then $\mathcal{C}$ can be $\mathbb{R}$-normally and properly parametrized as

$$\mathcal{P}\left(\frac{\alpha t + 1}{t}\right).$$

Proof. Let us prove (1). Let $\mathcal{Q}(t)$ be the reparametrization of $\mathcal{P}(t)$ in Statement (1), expressed in reduced form. First observe that if $P = \mathcal{P}(t_0)$ with $t_0 \in \mathbb{R}$, then

$$P = \mathcal{Q}(0) \text{ if } t_0 = 0, \quad \text{and} \quad P = \mathcal{Q}(\tfrac{1 \pm \sqrt{1 + 4t_0^2}}{2t_0}) \text{ if } t_0 \neq 0.$$

Therefore, $\varphi_\mathcal{P}|_\mathbb{R}(\mathbb{R}) \subset \varphi_\mathcal{Q}|_\mathbb{R}(\mathbb{R}) \subset \mathcal{C} \cap \mathbb{R}^2$. Now, if $\mathcal{P}(t)$ is $\mathbb{R}$-normal then $\varphi_\mathcal{P}|_\mathbb{R}(\mathbb{R}) = \mathcal{C} \cap \mathbb{R}^2$, and hence $\mathcal{Q}(t)$ is $\mathbb{R}$-normal too. On the other hand, if $\mathcal{P}(t)$ is not $\mathbb{R}$-normal, then by Theorem 7.28, $\mathcal{P}(t)$ is not normal. Therefore, by Theorem 6.22, the critical point of $\mathcal{P}(t)$ exists. Let us call it $P_\mathcal{P}$. Moreover, by remark (4) to Definition 7.25, one has that $\varphi_\mathcal{P}|_\mathbb{R}(\mathbb{R}) = [\mathcal{C} \cap \mathbb{R}^2] \setminus \{P_\mathcal{P}\}$. However, since $\deg(\chi_{i1}) \leq \deg(\chi_{i2})$ for $i = 1, 2$, $\mathcal{Q}(1)$ is defined and $\mathcal{Q}(1) = P_\mathcal{P}$. Therefore $\mathcal{Q}(t)$ is $\mathbb{R}$-normal.

In order to prove (2), let $R(t) = \frac{\alpha t + 1}{t}$ and let $\mathcal{Q}(t)$ be the reduced expression of $\mathcal{P}(R(t))$. First observe that the properness of $\mathcal{Q}(t)$ follows from Lemma 4.17. Now, by the reasoning in the proof of Theorem 6.26, we know that $\mathcal{Q}(t)$ is normal. Thus, by Theorem 7.28, we only need to prove that $\mathcal{Q}(t)$ is nondegenerate. For this purpose, let us assume that $\mathcal{D}^\mathcal{Q} \neq \emptyset$, and consider $P \in \mathcal{D}^\mathcal{Q}$. Then $P \in \mathcal{C} \cap \mathbb{R}^2$, and $\emptyset \neq \mathcal{F}_\mathcal{Q}(P) \subset \mathbb{C} \setminus \mathbb{R}$. So, there exists $t_0 \in \mathbb{C} \setminus \mathbb{R}$, in particular $t_0 \neq 0$, such that $\mathcal{Q}(t_0) = P$. Therefore, $R(t_0) \in \mathcal{F}_\mathcal{P}(P)$, and hence $\mathcal{F}_\mathcal{P}(P) \neq \emptyset$. Now, since $\mathcal{P}(t)$ is nondegenerate and $\mathcal{F}_\mathcal{P}(P) \neq \emptyset$, there exists $t_1 \in \mathbb{R}$ such that $\mathcal{P}(t_1) = P$. $\mathcal{P}(t)$ is defined at t_1, so $t_1 \neq \alpha$, and we have $1/(t_1 - \alpha) \in \mathcal{F}_\mathcal{Q}(P) \cap \mathbb{R}$. This, however, is a contradiction. $\qquad\square$

Now we analyze the case of weakly degenerate parametrizations.

Theorem 7.30. *Let $\mathcal{P}(t)$ be a weakly degenerate real proper parametrization of $\mathcal{C}$. Then the following hold:*

(1) $\mathcal{C}$ can be $\mathbb{R}$-normally parametrized as

$$\mathcal{P}\left(\frac{t^2 + at - 1}{t^2 + t - 1}\right),$$

where $a \in \mathbb{R} \setminus \{1\}$ is such that $\mathcal{F}_{\mathcal{P}}(\mathcal{P}(a)) = \{a\}$.

(2) If $\chi_{12}(t)\chi_{22}(t)$ has a real root α, then $\mathcal{C}$ can be $\mathbb{R}$-normally and properly parametrized as

$$\mathcal{P}\left(\frac{\alpha t + 1}{t}\right).$$

Proof. In order to prove (1), we observe that because of Theorem 7.29 (1), it is enough to prove that

$$\mathcal{Q}(t) := \mathcal{P}\left(\frac{at + 1}{t + 1}\right)$$

is nondegenerate; as always $\mathcal{Q}(t)$ is considered in reduced form. Since $\mathcal{P}(t)$ is weakly degenerate, the critical point $P_{\mathcal{P}}$ of $\mathcal{P}(t)$ exists and therefore $\deg(\chi_{i1}) \leq \deg(\chi_{i2})$ for $i = 1, 2$. Moreover, $\chi_{12}(a)\chi_{22}(a) \neq 0$ because $\mathcal{P}(a)$ is defined by hypothesis. Therefore, the degree of each numerator of $\mathcal{Q}(t)$ is less or equal to the degree of the corresponding denominator. Thus, the critical point $P_{\mathcal{Q}}$ of $\mathcal{Q}(t)$ exists, and $P_{\mathcal{Q}} = \mathcal{P}(a)$. Moreover, since $a \neq 1$, $\mathcal{Q}(-1)$ is defined and $\mathcal{Q}(-1) = P_{\mathcal{P}}$. Now, we prove that $\mathcal{D}^{\mathcal{Q}} = \emptyset$. Clearly $P_{\mathcal{P}} \notin \mathcal{D}^{\mathcal{Q}}$. Let us see that $P_{\mathcal{Q}} \notin \mathcal{D}^{\mathcal{Q}}$. If there exists $t_0 \in \mathbb{C} \setminus \mathbb{R}$, in particular $t_0 \neq -1$, such that $\mathcal{Q}(t_0) = P_{\mathcal{Q}}$, then $\mathcal{P}((at_0 + 1)/(t_0 + 1)) = P_{\mathcal{Q}} = \mathcal{P}(a)$. But $\mathcal{F}_{\mathcal{P}}(P_{\mathcal{Q}}) = \mathcal{F}_{\mathcal{P}}(\mathcal{P}(a)) = \{a\}$. So, $(at_0 + 1)/(t_0 + 1) = a$, and this is impossible because $a \neq 1$. Thus, $P_{\mathcal{Q}} \notin \mathcal{D}^{\mathcal{Q}}$. Now, let $P \in [\mathcal{C} \cap \mathbb{R}^2] \setminus \{P_{\mathcal{P}}, P_{\mathcal{Q}}\}$. Since $\mathcal{P}(t)$ is weakly degenerate and $P \neq P_{\mathcal{P}}$, there exists $t_0 \in \mathbb{R}$ such that $\mathcal{P}(t_0) = P$; recall from Sect. 6.3 that every point different from the critical point is reachable. Moreover, $t_0 \neq a$ since $P \neq P_{\mathcal{Q}}$. Therefore, $(t_0 - 1)/(a - t_0) \in \mathcal{F}_{\mathcal{Q}}(P) \cap \mathbb{R}$. Thus, $P \notin \mathcal{D}^{\mathcal{Q}}$. So, $\mathcal{Q}(t)$ is nondegenerate.

The proof of (2) is essentially analogous to the proof of (2) in Theorem 7.29. Properness is clear. Let us see that all real points on $\mathcal{C}$ are reachable by real values via $\mathcal{Q}(t) := \mathcal{P}(\frac{\alpha t + 1}{t})$; where $\mathcal{Q}(t)$ is considered in reduced form. Indeed, first we observe that $\mathcal{Q}(0)$ is the critical point $P_{\mathcal{P}}$ of $\mathcal{P}(t)$. Now, take $P \in [\mathcal{C} \cap \mathbb{R}^2] \setminus \{P_{\mathcal{P}}\}$. Then, since $P \neq P_{\mathcal{P}}$ and since $\mathcal{D}^{\mathcal{P}} = \{P_{\mathcal{P}}\}$, there exists $t_0 \in \mathbb{R}$ such that $\mathcal{P}(t_0) = P$. Note that $t_0 \neq \alpha$ because $\mathcal{P}(t_0)$ is defined. Then $\mathcal{Q}(1/(t_0 - \alpha)) = P$ and $1/(t_0 - \alpha) \in \mathbb{R}$. $\square$

Remarks. Note that, since in Theorem 7.30 the parametrization $\mathcal{P}(t)$ is assumed to be proper, there exist infinitely many real values a satisfying the condition imposed in Statement (1). Furthermore, in order to check whether a particular a is valid one only needs to check whether a is the only root of $\gcd(H_1^P(t, \chi_{11}(a)/\chi_{12}(a)), H_2^P(t, \chi_{21}(a)/\chi_{22}(a)))$.

Finally, we deal with the strongly degenerate case. We will see that in this situation, the corresponding curve cannot be normally parametrized over $\mathbb{R}$. Geometrically, the reason is that the existence of strong degeneration is related to the existence of isolated singularities (compare Exercise 7.16). Intuitively speaking, observe that if one has an isolated singular, there is no real place centered at it, and hence it is impossible to approach the point via a real parametrization.

We start with some technical lemmas (see Exercise 7.17).

Lemma 7.31. *Let $\mathcal{P}(t) \in \mathbb{R}(t)^2$ be a rational parametrization of an affine curve $\mathcal{C}$. Let $R(t) = \frac{M(t)}{N(t)} \in \mathbb{C}(t) \setminus \mathbb{C}$, with $\gcd(M, N) = 1$, and let $\mathcal{Q}(t) = \mathcal{P}(R(t))$. Then,*

$$H_1^{\mathcal{Q}}(t, x) = N^{\deg\left(\frac{\chi_{11}}{\chi_{12}}\right)} \cdot H_1^P(R(t), x), \quad H_2^{\mathcal{Q}}(t, y) = N^{\deg\left(\frac{\chi_{21}}{\chi_{22}}\right)} \cdot H_2^P(R(t), y).$$

Lemma 7.32. *Let $\mathcal{P}(t) \in \mathbb{R}(t)^2$ be a proper parametrization of $\mathcal{C}$. If there exists $(a_1, a_2) \in \mathcal{D}^P$ such that for some $i \in \{1, 2\}$*

$$\deg(H_i^P(t, a_i)) = \deg(\chi_{i1}/\chi_{i2}),$$

then $\mathcal{C}$ cannot be $\mathbb{R}$-normally parametrized.

Proof. Let us assume that the hypothesis on the degrees holds for $i = 1$; similarly, for $i = 2$. Let $\ell_i = \deg(\chi_{i1}/\chi_{i2})$ and $k_i = \deg(H_i^P(t, a_i))$ for $i = 1, 2$. So, $k_1 = \ell_1$, and $\ell_2 \geq k_2$; note that since $\mathcal{P}(t)$ is degenerate, by the remark to Definition 7.25 the rational functions cannot be zero. The idea is to prove that every real parametrization of $\mathcal{C}$ is degenerate. By hypothesis $\mathcal{D}^P \neq \emptyset$, and hence $\mathcal{P}(t)$ is degenerate. Now, let $\mathcal{Q}(t)$ be another real parametrization of $\mathcal{C}$. Since $\mathcal{P}(t)$ is proper, because of Lemma 4.17, $\mathcal{Q}(t) = \mathcal{P}(R(t))$ where $R(t)$ is a nonconstant rational function. Moreover, because of Sect. 6.2, $R(t) = \mathcal{P}^{-1}(\mathcal{Q}(t))$. Thus, since $\mathcal{P}(t)$ and $\mathcal{Q}(t)$ are real, then $R(t)$ is also real. Let $R(t) = \frac{M(t)}{N(t)}$ be in reduced form. We prove that $(a_1, a_2) \in \mathcal{D}^{\mathcal{Q}}$. This will imply that $\mathcal{Q}(t)$ is degenerate. For this purpose, we have to show that $T(t) := \gcd(H_1^{\mathcal{Q}}(t, a_1), H_2^{\mathcal{Q}}(t, a_2)) \neq 1$, and that $T(t)$ does not have real roots (see paragraph before Algorithm DEGENERATIONS). First we carry out some preparatory steps. By Lemma 7.31,

$$H_1^{\mathcal{Q}}(t, a_1) = \prod_{j=1}^{k_1}(\alpha_j M(t) - \beta_j N(t)), \quad H_2^{\mathcal{Q}}(t, a_2) = N^{\ell_2 - k_2} \prod_{j=1}^{k_2}(\alpha_j' M(t) - \beta_j' N(t)),$$

where β_j/α_j and β'_j/α'_j are the roots in $\mathbb{C}$ of $H_1^{\mathcal{P}}(t, a_1)$ and $H_2^{\mathcal{P}}(t, a_2)$, respectively. Now, since $\alpha_j \neq 0$ and $\gcd(M, N) = 1$, we have $\gcd(N, \alpha_j M - \beta_j N) = 1$. Therefore,

$$T(t) = \gcd\left(\prod_{j=1}^{k_1}(\alpha_j M(t) - \beta_j N(t)), \prod_{j=1}^{k_2}(\alpha'_j M(t) - \beta'_j N(t))\right).$$

Let $K(t) := \gcd(H_1^{\mathcal{P}}(t, a_1), H_2^{\mathcal{P}}(t, a_2))$. Since $(a_1, a_2) \in \mathcal{D}^{\mathcal{P}}$, then $K(t)$ is not constant, and it does not have real roots. Therefore, there exists a common complex, nonreal, root (say β_{i_0}/α_{i_0}) of $H_1^{\mathcal{P}}(t, a_1)$, $H_2^{\mathcal{P}}(t, a_2)$. Thus, $\alpha_{i_0} M(t) - \beta_{i_0} N(t)$ is a common factor of $H_1^{\mathcal{Q}}(t, a_1)$ and $H_2^{\mathcal{Q}}(t, a_2)$. This factor cannot be constant. Indeed, if $\alpha_{i_0} M(t) - \beta_{i_0} N(t) = \lambda \in \mathbb{C}$, taking conjugates, one gets that

$$\begin{pmatrix} \alpha_{i_0} & \beta_{i_0} \\ \overline{\alpha_{i_0}} & \overline{\beta_{i_0}} \end{pmatrix} \cdot \begin{pmatrix} M \\ N \end{pmatrix} = \begin{pmatrix} \lambda \\ \overline{\lambda} \end{pmatrix}.$$

Now, observe that the determinant of the 2×2 matrix cannot be zero, since otherwise one would get that $\beta_{i_0}/\alpha_{i_0} = \overline{\beta_{i_0}}/\overline{\alpha_{i_0}}$ and this implies that the root is real. However, inverting the matrix, this implies that both M and N are constant, which is a contradiction. So, $T(t) \neq 1$. Now, let us assume that there exists a real root ρ of $T(t)$. Then, there exists some i_0 and j_0 such that $\alpha_{i_0} M(\rho) - \beta_{i_0} N(\rho) = \alpha'_{j_0} M(\rho) - \beta'_{j_0} N(\rho) = 0$. Since $\gcd(N, \alpha_j M - \beta_j N) = 1$ for every j, we deduce that $N(\rho) \neq 0$. Thus $R(\rho)$, which is real because $\rho \in \mathbb{R}$ and $R \in \mathbb{R}(t)$, is a real root of $K(t)$. This, however, is a contradiction. $\qquad\square$

Now, finally, we are ready to state the corresponding theorem on strong degenerations.

Theorem 7.33. *Let $\mathcal{P}(t) \in \mathbb{R}(t)^2$ be a strongly degenerate proper parametrization of $\mathcal{C}$. Then, the curve $\mathcal{C}$ cannot be $\mathbb{R}$-normally parametrized.*

Proof. If there exists a point $(a_1, a_2) \in \mathcal{C}$ such that for $i = 1, 2$ we have $\deg(H_i(t, a_i)) \neq \deg(\chi_{i1}/\chi_{i2})$, then (a_1, a_2) is the critical point $P_{\mathcal{P}}$ of $\mathcal{P}(t)$. Now, since $\mathcal{P}(t)$ is strongly degenerate, either $P_{\mathcal{P}}$ exists and $\mathrm{card}(\mathcal{D}^{\mathcal{P}}) > 1$ or $P_{\mathcal{P}}$ does not exists and $\mathcal{D}^{\mathcal{P}} \neq \emptyset$. In any case, the theorem follows from Lemma 7.32. $\qquad\square$

We finish this section by summarizing these results in the following algorithm.

Algorithm REAL-NORMAL-PARAMETRIZATION

Given a proper rational parametrization $\mathcal{P}(t) = \left(\dfrac{\chi_{11}(t)}{\chi_{12}(t)}, \dfrac{\chi_{21}(t)}{\chi_{22}(t)} \right) \in$ $\mathbb{L}(t)^2$, in reduced form, where $\mathbb{L}$ is a computable subfield of $\mathbb{R}$, of an affine rational real curve $\mathcal{C}$, the algorithm decides whether $\mathcal{C}$ can be $\mathbb{R}$-normally parametrized, and in the affirmative case computes an $\mathbb{R}$-normal parametrization of $\mathcal{C}$.

1. Apply algorithm DEGENERATIONS to compute $\mathcal{D}^{\mathcal{P}}$.
2. If $[\text{card}(\mathcal{D}^{\mathcal{P}}) > 1]$ or $[\mathcal{D}^{\mathcal{P}} \neq \emptyset$ and the critical point of $\mathcal{P}(t)$ does not exist] then return "$\mathcal{C}$ can not be $\mathbb{R}$-normally parametrized".
3. If $\mathcal{D}^{\mathcal{P}} = \emptyset$, apply Algorithm NORMALITY-TEST (see Sect. 6.3), and if it outputs that $\mathcal{P}(t)$ is normal then return $\mathcal{P}(t)$.
4. Check whether $\chi_{12}\chi_{22}$ has a real root, say α. If yes, return

$$\mathcal{P}\left(\frac{\alpha t + 1}{t} \right).$$

5. If $\mathcal{D}^{\mathcal{P}} = \emptyset$ return

$$\mathcal{P}\left(\frac{t}{t^2 - 1} \right).$$

6. Compute $\alpha \in \mathbb{R} \setminus \{1\}$ such that $\text{card}(\mathcal{F}_{\mathcal{P}}(\mathcal{P}(\alpha))) = 1$, and return

$$\mathcal{P}\left(\frac{t^2 + \alpha t - 1}{t^2 + t - 1} \right).$$

Remarks. For performing Step 6, see the remark to Theorem 7.30. Also, note that if the algorithm continues through Steps 3 and 4, then the output is proper, otherwise either the curve cannot be $\mathbb{R}$-normally parametrized or the output has tracing index 2.

We finish this section with some examples to illustrate the Algorithm REAL-NORMAL-PARAMETRIZATION.

Example 7.34. We consider the proper parametrization

$$\mathcal{P}(t) = \left(\frac{t+1}{t^3 - 1}, \frac{t+1}{t^3} \right),$$

and we apply the Algorithm REAL-NORMAL-PARAMETRIZATION. In the first step, we apply Algorithm DEGENERATIONS, and we get that $\mathcal{D}^{\mathcal{P}} = \{(0,0)\}$. Since, the origin is the critical point of $\mathcal{P}(t)$, we have that $\mathcal{P}(t)$ is weakly degenerate. In Step 4, taking $\alpha = 0$, we return the $\mathbb{R}$-normal and proper parametrization

$$\mathcal{P}\left(\frac{1}{t} \right) = \left(\frac{(t+1)\left(2 + t^2\right) t}{t^2 + 1}, \frac{(t+1)t}{t^2 + 1} \right).$$

Example 7.35. We consider the proper parametrization

$$\mathcal{P}(t) = \left(t^3 + 1,\, t^3 + 2t + 1\right).$$

Applying Algorithm DEGENERATIONS, we get that $\mathcal{D}^{\mathcal{P}} = \emptyset$. Thus $\mathcal{P}(t)$ is nondegenerate. Moreover, applying Algorithm NORMALITY-TEST, we deduce that $\mathcal{P}(t)$ is normal. Thus, $\mathcal{P}(t)$ is $\mathbb{R}$-normal.

Example 7.36. We consider the proper parametrization

$$\mathcal{P}(t) = \left(\frac{t^3 + 3t + t^2 + 1}{(t^2+1)\,(t^2+3)},\, \frac{1}{(t^2+3)}\right).$$

In the first step, we apply Algorithm DEGENERATIONS, and we get that $\mathcal{D}^{\mathcal{P}}_{\mathbb{R}} = \{(0,0)\}$. Since the origin is the critical point of $\mathcal{P}(t)$, we get that $\mathcal{P}(t)$ is weakly degenerate. Now, since the denominators of $\mathcal{P}(t)$ do not have real roots, we execute Step 6. Then, we compute $\alpha \in \mathbb{R}\setminus\{1\}$ such that $\mathrm{card}(\mathcal{F}_{\mathcal{P}}(\mathcal{P}(\alpha))) = 1$. We take $\alpha = 2$. $\mathcal{P}(2) = (19/35, 1/7)$ and $\gcd(H^{\mathcal{P}}_1(t, 19/35), H^{\mathcal{P}}_2(t, 1/7)) = t - 2$. Therefore, $\mathcal{F}_{\mathcal{P}}(\mathcal{P}(2)) = \{2\}$. Thus, we return the $\mathbb{R}$-normal parametrization (of tracing index 2):

$$\mathcal{P}\left(\frac{t^2 + 2t - 1}{t^2 + t - 1}\right) = \left(\frac{p_1(t)}{q_1(t)},\, \frac{p_2(t)}{q_2(t)}\right),$$

$$p_1(t) = (t^2 + t - 1)(-6 + 26t - 33t^3 - 20t^2 + 20t^4 + 6t^6 + 26t^5),$$

$$q_1(t) = (2t^4 + 6t^3 + t^2 - 6t + 2)(4t^4 + 10t^3 - t^2 - 10t + 4),$$

$$p_2(t) = (t^2 + t - 1)^2, \quad q_2(t) = 4t^4 + 10t^3 - t^2 - 10t + 4.$$

Exercises

7.1. Let $f(x, y) \in \mathbb{R}[x, y]$ be irreducible over $\mathbb{R}$, such that f has infinitely many roots in $\mathbb{R}^2$. Show that f is irreducible over $\mathbb{C}$.

7.2. Let $\mathcal{C}$ be the plane curve defined over $\mathbb{C}$ by the polynomial

$$f(x, y) = 3y^3 - 3xy^2 - 2xy^3 + x^2y^3 + x^3.$$

(i) Apply Algorithm REALITY-TEST to decide whether $\mathcal{C}$ is real and, in the affirmative case, compute one real simple point on $\mathcal{C}$.
(ii) Determine, if possible, a real rational parametrization of $\mathcal{C}$.
(iii) Obtain an algebraically optimal real parametrization of $\mathcal{C}$.

7.3. Let $\mathcal{C}$ be the plane curve defined over $\mathbb{C}$ by the polynomial

$$f(x, y) = (x^2 + 4y + y^2)^2 - 16(x^2 + y^2).$$

Decide if the curve $\mathcal{C}$ is real, and in the affirmative case compute a real rational parametrization.

7.4. Let C be the plane curve defined over $\mathbb{C}$ by the polynomial

$$f(x,y) = -13870y^3x - 542y^3 + 8074x^2y^2 + 686xy^2 + 8y^2 + 40x^3$$
$$+ 2x^2 - 8xy - 2080x^3y - 288x^2y + 8902y^4 + 200x^4.$$

Determine, if possible, a real rational parametrization of C.

7.5. Let C be the plane curve defined over $\mathbb{C}$ by the polynomial

$$f(x,y) = -12x^2 + 8x^2y + 2x^2y^2 + 48y + 12y^2 + x^4 + 8y^3 + y^4 + 4 - 4x^2y\sqrt{2}$$
$$-4y^3\sqrt{2} - 8y\sqrt{2} - 24y^2\sqrt{2} - 16\sqrt{2} - 8x^2\sqrt{2}.$$

Determine, if possible, a real rational parametrization of C.

7.6. Prove Propositions 7.12 and 7.15.

7.7. Let $p(x,y)$ be an analytic polynomial of components u, v and generator f. Prove the following:

(i) If either u or v are constant, then p is constant.
(ii) If $p \in \mathbb{R}[x,y]$, then $p(x,y)$ is constant.
(iii) p is irreducible over $\mathbb{C}$ if and only if f is irreducible over $\mathbb{C}$.
(iv) The factors of p are analytic polynomials.

7.8. (i) Let $p(x,y) = x^3 + xy^2 - x + x^2yi - xi$. Prove that $p(x,y)$ is not analytic and that $\gcd(u,v) \neq 1$, where u and v are the real and imaginary parts of $p(x,y)$, respectively.
(ii) Let $p(x,y) = x^2 + y^2 - 1 - i + xyi$. Prove that $p(x,y)$ is not analytic and that $\gcd(u,v) = 1$, where u and v are the real and imaginary parts of $p(x,y)$, respectively.

7.9. Using the notation of Lemma 7.13 prove that

(i) $\deg(C_u) = \deg(C_v) = \deg(f(z))$.
(ii) C_u and C_v do not have intersection points at infinity.
(iii) The affine intersection points of C_u and C_v are

$$\bigcup_{a\in\mathbb{C},\, f(a)=0} \{(a - it_0, t_0) \mid t_0 \text{ is a root of } \gcd(u(a - it, t), v(a - it, t))\}.$$

(iv) The real affine intersection points of C_u and C_v are

$$\{(x_0, y_0) \in \mathbb{R}^2 \mid f(x_0 + iy_0) = 0\}.$$

7.10. Show that $r(x,y) = \dfrac{1}{x + iy}$ is analytic, but the numerator of $r(x,y)$, when written in normalized form, is not an analytic polynomial.

7.11. Decide whether the curve defined by the parametrization

$$\mathcal{P}(t) = \left(\frac{-6it^2 + i + it^4}{-4t - 4t^3 - 2 + 2t^4}, \frac{i(-4t - 4t^3 + 2 - 2t^4)}{-4t - 4t^3 - 2 + 2t^4} \right)$$

is real, and if so compute a real parametrization.

7.12. Determine the set of degenerations of the parametrization

$$\mathcal{P}(t) = \left(\frac{-t^3 + t + 1}{t}, \frac{t - 1}{t^3 + 2t + 1} \right) \in \mathbb{R}(t)^2.$$

7.13. Give a nonpolynomial proper parametrization that is strongly degenerate.

7.14. By applying the algorithm DEGENERATIONS, decide whether the parametrization

$$\mathcal{P}(t) = \left(\frac{(1 - 10t + 26t^2)t^2}{1 - 20t + 150t^2 - 500t^3 + 626t^4}, \frac{-(1 - 10t + 26t^2)t}{-1 + 15t - 77t^2 + 134t^3} \right)$$

is degenerate or not.

7.15. Given the proper parametrizations

$$\mathcal{P}_1(t) = \left(\frac{(2t - 3)^2(-11 + 8t)}{-2249 + 4794t - 3408t^2 + 808t^3}, \frac{(2t - 3)^3}{-1547 + 3336t - 2400t^2 + 576t^3} \right),$$

and

$$\mathcal{P}_2(t) = \left(\frac{-(2t + 7)(25 + 8t)}{2(337 + 214t + 34t^2)}, \frac{(2t + 7)^2}{4(193 + 121t + 19t^2)} \right),$$

decide whether they are $\mathbb{R}$-normal. In the negative case decide whether they can be $\mathbb{R}$-normally parametrized, and compute a reparametrization of them.

7.16. Prove that a real rational curve is not $\mathbb{R}$-normal if and only if it has isolated singularities.

7.17. Prove Lemma 7.31.

A

The System CASA

In this appendix, we briefly describe the mathematical software package CASA and we illustrate, by means of some examples, how some of the algorithms in the book can be carried out with this package. CASA has been developed by the computer algebra research group at the Research Institute for Symbolic Computation (RISC) of Johannes Kepler University in Linz, Austria. It can be freely downloaded at

http://www.risc.uni-linz.ac.at/software.

CASA or Computer Algebra Software for constructive Algebraic geometry is based on Maple, and it is designed for symbolic manipulation in algebraic geometry, mainly in projective algebraic geometry, over an algebraically closed field of characteristic zero. Essentially, the package provides some basic functions for the algebraic manipulation of geometric objects from the practical point of view. Many of the algorithms in CASA are now integrated into major computer algebra software such as Maple. A reference to CASA can be found in the example worksheet for the algebraic curves package of Maple 10.

The main data structure of CASA is that of an algebraic sets. In the following, we describe some of the functions of the package. For more details we refer to [GKW91], and [HHW03]. In CASA an algebraic set can be given implicitly, parametrically, by places or by projection. The system provides functions for computing with algebraic sets, as well as conversion algorithms between different representations. The functions of CASA can roughly be grouped as follows:

1. computations with ideals
2. determination of algebraic sets in different representations
3. conversion algorithms between representations
4. determination of linear systems of curves
5. intersection, union and difference of algebraic sets
6. computation of tangent spaces
7. computation of the dimension of algebraic sets
8. decomposition of an algebraic set on irreducible components

9. projection of algebraic sets onto hypersurfaces
10. computation of singularities and genus
11. algorithms of parametrization and implicitation
12. determination of Puiseux series
13. multivariate resultants and Dixon resultants
14. Gröbner bases for ideals and modules
15. Gröbner walk
16. computation of rational points
17. plotting of curves and surfaces
18. manipulation of offset curves
19. algorithms in algebraic coding theory

Let us illustrate the usability of CASA by means of some examples. First we read the package within the Maple session.

```
>  with(casa);
```

Welcome to CASA 2.5 for Maple V.5

Copyright (C) 1990-2000 by Research Institute for Symbolic Computation (RISC-Linz), the University of Linz, A-4040 Linz, Austria.

For help type '?casa' or '?casa,<topic>'.

[BCH2, BCHDecode, CyclicEncode, DivBasisL, GWalk, GoppaDecode, GoppaEncode, GoppaPrepareDu, GoppaPrepareSV, GoppaPrepareSa, GoppaPrimary, Groebnerbasis, InPolynomial, NormalPolynomial, OutPolynomial, PolynomialRoots, RPHcurve, SakataDecode, SubsPolynomial, _casaAlgebraicSet, adjointCurve, algset, casaAttributes, casaVariable, computeRadical, conic, decompose, delete, dimension, equalBaseSpaces, equalProjectivePoints, finiteCurve, finiteField, generators, genus, homogeneousForm, homogeneousPolynomial, homogenize, implDifference, implEmpty, implEqual, implIdealQuo, implIntersect, implOffset, implSubSet, implUnion, implUnionLCM, imult, independentVariables, init, isProjective, leadingForm, makeDivisor, mapOutPolynomial, mapSubsPolynomial, mgbasis, mgbasisx, mkAlgSet, mkImplAlgSet, mkParaAlgSet, mkPlacAlgSet, mkProjAlgSet, mnormalf, msolveGB, msolveSP, mvresultant, neighbGraph, neighborhoodTree, numberOfTerms, pacPlot, paraOffset, parameterList, passGenCurve, planecurve, plotAlgSet, pointInAlgSet, projPoint, properParametrization, properties, rationalPoint, realroot_a, realroot_sb, setPuiseuxExpansion, setRandomParameters, singLocus, singularities, ssiPlot, subresultantChain, tangSpace, toAffine, toImpl, toPara, toPlac, toProj, toProjective, tsolve, variableDifferentFrom, variableList]

We start with a complex affine plane curve $\mathcal{C}$ of degree 15 defined by the polynomial:

$$f := -15\,x^5 y^4 - 21\,y^2 x^8 - 7\,y^2 x^7 - 3\,y^3 x^6 - 30\,x^2 y^8 + 2\,y^4 x^9 + 3\,y^6 x^5 + 3\,y^5 x - 5\,yx^8 + y^5 x^8 - 11\,xy^{10} + 11\,x^6 y^8 + y^2 x^3 - 68\,y^5 x^6 - y^8 x^7 + 7\,y^4 x^2 + 15\,x^{10} y^2 - 2\,y^6 x^4 + 2\,y^9 x^3 - 2\,x^7 y - 22\,x^8 y^4 - 11\,x^5 y^5 + 5\,y^3 x^3 - x^5 y^2 - 106\,y^3 x^8 + 5\,x^2 y^7 - 165\,y^4 x^7 + 3\,x^{10} y + 167\,y^7 x^6 + x^{10} + 33\,y^5 x^2 + 13\,y^6 x + 12\,x^{10} y^3 + 21\,y^4 x^3 + 76\,y^8 x^5 - x^8 + 165\,y^6 x^2 - 3\,x^5 y^3 + 33\,y^6 x^6 - 3\,y^8 + 15\,y^7 x^5 - 5\,x^7 y^5 + 106\,y^5 x^3 + 22\,x^3 y^6 - y^7 - y^7 x^3 + 22\,x^7 y^7 + 65\,y^3 x^9 - 167\,y^9 x - x^6 y^2 - 15\,y^9 - 2\,y^7 x^8 - y^3 x^{11} - 33\,y^8 x + 69\,y^7 x + y^{10} x^2 - 13\,y^4 x^6 + 30\,x^7 y^6 - 33\,y^3 x^7 - 22\,x^2 y^9 - 65\,y^5 x^4 - 76\,y^{10} + 2\,y^3 x^2 + y^4 x$$

We declare $\mathcal{C}$ to be the curve defined by f; i.e., we "make an implicitly defined algebraic set" from f and the variables x, y. For a detailed description of this command call help(casa, mkImplAlgSet).

```
> C:=mkImplAlgSet([f],[x,y]);
```

$$C := \text{Implicit\_Algebraic\_Set}([y^{10}\,x^2 - y^7 - 3\,y^8 - y^7 x^3 + 30\,x^7 y^6$$
$$- y^3 x^{11} - x^6 y^2 - 33\,y^3 x^7 - 22\,x^2 y^9 - 3\,y^3 x^6 - 65\,y^5 x^4 - 7\,y^2 x^7$$
$$+ 2\,y^3 x^2 - 13\,y^4 x^6 + 13\,y^6 x - 22\,x^8 y^4 - 2\,y^6 x^4 - 5\,y x^8 + y^4 x + 69\,y^7 x$$
$$- 15\,x^5 y^4 - 21\,y^2 x^8 - 76\,y^{10} - 33\,y^8 x - 2\,y^7 x^8 + 65\,y^3 x^9 - 167\,y^9 x$$
$$+ 22\,x^7 y^7 + 22\,x^3 y^6 + 106\,y^5 x^3 - 5\,x^7 y^5 + 15\,y^7 x^5 + 33\,y^6 x^6 - 3\,x^5 y^3$$
$$+ 76\,y^8 x^5 + 21\,y^4 x^3 + 33\,y^5 x^2 + 12\,x^{10} y^3 + 3\,x^{10} y + 167\,y^7 x^6 - 15\,y^9$$
$$- x^8 + y^5 x^8 + x^{10} + y^2 x^3 - y^8 x^7 - 165\,y^4 x^7 + 5\,x^2 y^7 + 165\,y^6 x^2$$
$$+ 5\,y^3 x^3 - 106\,y^3 x^8 - 11\,x y^{10} + 11\,x^6 y^8 - 68\,y^5 x^6 + 7\,y^4 x^2 + 15\,x^{10} y^2$$
$$+ 2\,y^9 x^3 - 2\,x^7 y - 11\,x^5 y^5 - x^5 y^2 - 30\,x^2 y^8 + 2\,y^4 x^9 + 3\,y^6 x^5$$
$$+ 3\,y^5 x], [x,\ y])$$

Let us compute the components of $\mathcal{C}$ (see Sect. 2.1).

```
> Components:=decompose(C);
```

Components := Implicit_Algebraic_Set($[y^2 - x^5]$, $[x, y]$),
Implicit_Algebraic_Set($[y^5 + 2\,y^4 x - y^2 x - 2\,y x^2 - x^3 + x^4]$, $[x, y]$),
Implicit_Algebraic_Set($[y^3 x^2 - 11\,y^3 x - 76\,y^3 - 15\,y^2$
$-3\,y - 1 - 15\,y^2 x - 3\,y x - x]$, $[x, y]$)

So we see that $\mathcal{C}$ has three irreducible components, each of them of degree 5. Let us call them $\mathcal{C}_1, \mathcal{C}_2$ and $\mathcal{C}_3$, respectively. We give names to them in Maple

```
> for i from 1 to 3 do C[i]:=Components[i] od:
```

Now, we proceed to compute the genus of each component of $\mathcal{C}$ (see Sect. 3.3).

```
>  for i from 1 to 3 do g[i]:=genus(C[i]) od;
```

$$g_1 := 3$$
$$g_2 := 0$$
$$g_3 := 0$$

We can ask for information on algebraic sets:

```
>  properties(C[1]);
```

"The algebraic set is known to have the following properties:"

"It has genus 3"

"A neighborhood graph is: "

$$[[3, 0, 0, 1, \{y = y - x, x = x + y\}, [\,]]]$$

Implicit_Algebraic_Set($[y^5 + 2\,y^4\,x - y^2\,x - 2\,y\,x^2 - x^3 + x^4]$, $[x, y]$)

We determine the singularities of the curves. The singularities are given projectively with their multiplicities (see Sect. 2.1).

```
>  singularities(C[1]);
```

$$\text{table}([3 = [[0, 0, 1]]])$$

```
>  singularities(C[2]);
```

$$\text{table}([2 = [[\frac{40}{21}\,\text{RootOf}(5\,\_Z^2 + 4\,\_Z + 89) - \frac{5}{21},$$
$$-\frac{5}{21} - \frac{2}{21}\,\text{RootOf}(5\,\_Z^2 + 4\,\_Z + 89), 1],$$
$$[0, 1, 0]], 3 = [[1, 0, 0]]])$$

```
>  singularities(C[3]);
```

$$\text{table}([2 = [[0, 0, 1]], 3 = [[0, 1, 0]]])$$

$\mathcal{C}_1$ has one triple point. $\mathcal{C}_2$ has three double points, two of them in an algebraic extension of degree 2 (each of them depending on one root of the polynomial $5x^2 + 4x + 89$), and one triple point. $\mathcal{C}_3$ has one double point and one triple point. Thus, genus($\mathcal{C}_1$) = 3, and genus($\mathcal{C}_2$) = 0. However, in the case of $\mathcal{C}_3$ there must be infinitely near singularities. Let us compute the neighborhood graph for $\mathcal{C}_3$ (see Sect. 3.2).

```
>  neighbGraph(C[3]);
```

$[[2, 0, 0, 1, \{x = x + y, y = y - x\}, [[2, 1, 1, 0, \{\}, []]]],$
$[3, -1, 1, 0, \{x = x + y, y = y - x\}, [[2, 2, 1, 0, \{x = x + 2y, y = y - 2x\}, []]]]$

In the blow-up, every singularity has an infinitely near double point. Note that we also get information on the necessary linear changes for the blowing up (compare Sect. 3.2).

Since the curves C_2 and C_3 have genus zero, we may determine rational parametrizations of these curves (see Sect. 4.8). In addition, these parametrizations can be obtained over the original ground field if possible (with the optional argument: OPTIMAL), or otherwise over an algebraic extension field of degree 2 (see Sect. 5.3).

```
>  P[2]:=toPara(C[2],t,["optimal"]);
```

$$P_2 := \text{Parametric\_Algebraic\_Set}([-t^3 + 3t^2 - 3t, -\frac{t-1}{-4+t^2-t}], [t])$$

```
>  P[3]:=toPara(C[3],t);
```

$$P_3 := \text{Parametric\_Algebraic\_Set}([\frac{1}{t^2}, -\frac{1}{t^5}], [t])$$

Since the curve C_1 has positive genus, we deduce that it is not rational. However, we may compute a local parametrization with center, for instance, at $(0, 0)$ (see Sect. 2.5).

```
>  properties(toPlac(C[1]));
```

"The algebraic set is known to have the following properties:"

"There are at least 3 terms to show in the Puiseux expansion"

$$\text{Places\_Algebraic\_Set}([[x^2, -x^2 - x^3 - \frac{1}{2}x^5 - \frac{3}{2}x^6 - \frac{7}{8}x^7 + x^8 - \frac{3}{16}x^9$$
$$+ O(x^{10})], [x^3, x + \frac{1}{3}x^5 - \frac{2}{3}x^7 - \frac{1}{3}x^8 + x^9 + O(x^{10})]], [x])$$

Now, we can implicitize the rational parametrizations given by P_2 and P_3 (see Sect. 4.5).

```
>  toImpl(P[3]);toImpl(P[2]);
```

$$\text{Implicit\_Algebraic\_Set}([x^5 - y^2], [x, y])$$

$\text{Implicit\_Algebraic\_Set}([y^3 x^2 - 11 y^3 x - 76 y^3 - 15 y^2 - 3y - 1 - 15 y^2 x$
$-3 y x - x], [x, y])$

We may determine a reparametrization of the curve C_2 by performing a substitution of the parameter in the algebraic set P_2 in parametric representation.

```
>   RP[2]:=mkAlgSet(P[2],[t=t^2-2]);
```

$$RP_2 := \text{Parametric\_Algebraic\_Set}([-t^6 + 9\,t^4 - 27\,t^2 + 26, -\frac{t^2 - 3}{t^4 - 5\,t^2 + 2}], [t])$$

Observe that now, the parametrization RP_2 of the curve C_2 is not proper. However, from the parametrization RP_2, we may determine a new proper parametrization of C_2 (see Sect. 6.1).

```
>   properParametrization(RP[2]) ;
```

$$[\frac{-26 + 51\,t - 33\,t^2 + 7\,t^3}{-1 + t^3 - 3\,t^2 + 3\,t}, -\frac{2\,t^2 - 5\,t + 3}{2\,t^2 - t - 2}]$$

CASA can also plot the following types of algebraic sets: planar curves represented implicitly, in parametric form or by places, space curves represented implicitly, in parametric form, in projected form or by places, surfaces represented implicitly or in parametric form. Let us plot the plane curves C_1, C_2, and C_3.

```
>   plotAlgSet(C[1],x=-3..3,y=-3..3,numpoints=200,thickness=5,
color=blue);
```

Time for isolating critical points : , .047

Time for finding intermediate points : , 1.156

Time for others : , .0

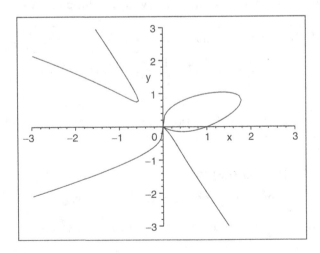

```
>  plotAlgSet(C[2],x=-2..2,y=-2..2,numpoints=200,thickness=5,
color=red);
```

Time for isolating critical points : , .016
Time for finding intermediate points : , 2.703
Time for others : , .032

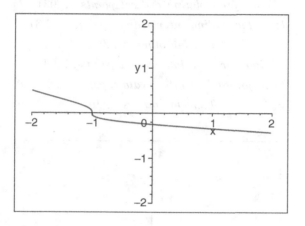

```
> plotAlgSet(C[3],x=-4..4,y=-4..4,numpoints=200,thickness=5);
```

Time for isolating critical points : , .031
Time for finding intermediate points : , .532
Time for others : , .015

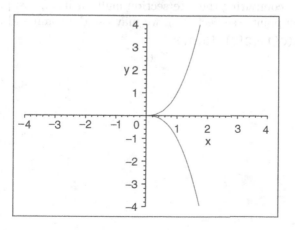

Let us also plot the whole composite curve C.

```
>   plotAlgSet(C,x=-2..2,y=-2..2,numpoints=200, thickness=5,
color=pink)
```

Time for isolating critical points : , .012

Time for finding intermediate points : , 0.516

Time for others : , .0

Time for isolating critical points : , .031

Time for finding intermediate points : , 2.375

Time for others : , .031

Time for isolating critical points : , .047

Time for finding intermediate points : , 1.500

Time for others : , .0

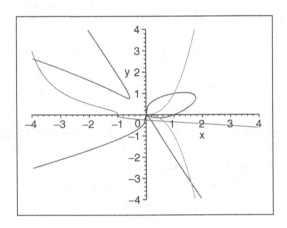

Finally, for computing the intersection multiplicity of two plane curves at an intersection point (see Sect. 2.3), we may use the command IMULT:

```
>   imult(C[1],C[3],[0,0]);
```

B

Algebraic Preliminaries

For a thorough introduction to algebra we refer the reader to any of a great number of classical textbooks, e.g., [BiM79], [Lan84], [ZaS58] or [VaW70]. We denote the set of natural numbers (including 0) by $\mathbb{N}$, the integers by $\mathbb{Z}$, the rational numbers by $\mathbb{Q}$, the real numbers by $\mathbb{R}$, and the complex numbers by $\mathbb{C}$.

B.1 Basic Ring and Field Theory

A *semigroup* $(S, \circ)$ is a set S together with an associative binary operation $\circ$ on S. A semigroup is *commutative* iff the operation $\circ$ is commutative. A *monoid* $(S, \circ, e)$ is a semigroup with an identity element e; that is $e \circ x = x \circ e = x$, for all $x \in S$. A monoid is *commutative* iff the operation $\circ$ is commutative. For a monoid S, we denote by S^* the set $S \setminus \{e\}$. A *group* $(G, \circ, \diamond, e)$ is a monoid $(G, \circ, e)$ together with a unitary inverse operation $\diamond$; that is, $x \circ (\diamond x) = (\diamond x) \circ x = e$, for all $x \in G$. G is *commutative* or *abelian group* iff the operation $\circ$ is commutative.

A *ring* $(R, +, \cdot, 0)$ is an abelian group $(R, +, 0)$, and a semigroup $(R, \cdot)$ satisfying the laws of distributivity $x \cdot (y + z) = x \cdot y + x \cdot z$, and $(x + y) \cdot z = x \cdot z + y \cdot z$. A *commutative ring* is one in which the operation $\cdot$ is commutative. A *ring with identity* is a ring R together with an element 1 ($\neq 0$), such that $(R, \cdot, 1)$ is a monoid. If R is a ring (with identity or not), by R^* we denote the set $R \setminus \{0\}$. Unless stated otherwise, we will always use the symbols $+, -, \cdot, 0, 1$ for the operations of a ring. We call these operations *addition, subtraction, multiplication, zero,* and *one*. The *subtraction* operation is defined as $x - y := x + (-y)$, for $x, y \in R$.

The *characteristic* of a commutative ring with identity R, char(R), is the least positive integer m such that $\underbrace{1 + \cdots + 1}_{m \text{ times}} = 0$. char$(R) = 0$ if no such m exists.

Let $(R, +, \cdot, 0)$ and $(\widetilde{R}, \widetilde{+}, \widetilde{\cdot}, \widetilde{0})$ be rings. A *ring homomorphism* h is a function from R to $\widetilde{R}$ satisfying the conditions

$$h(r + s) = h(r) \widetilde{+} h(s), \quad h(r \cdot s) = h(r) \widetilde{\cdot} h(s).$$

Furthermore, if R and $\widetilde{R}$ are rings with identities 1 and $\widetilde{1}$, respectively, and h is not the zero-homomorphism, then $h(1) = \widetilde{1}$. Moreover, if h is one-to-one and onto then h is called an *isomorphism from R to $\widetilde{R}$*. In this case, we say that R and $\widetilde{R}$ are *isomorphic*, and we write $R \cong \widetilde{R}$.

A nonzero element a of R is a *zero divisor* iff for some nonzero $b \in R$, we have that $a \cdot b = 0$. An *integral domain* or a *domain* D is a commutative ring with identity having no zero divisors.

A *field* $(K, +, \cdot, 0, 1)$ is a commutative ring with identity $(K, +, \cdot, 0, 1)$, and simultaneously a group $(K^\star, \cdot, 1)$. If all the operations on K are computable, we call K a *computable field*.

Let D be an integral domain. The *quotient field* $Q(D)$ of D is defined as

$$Q(D) = \{\frac{a}{b} \mid a, b \in D,\ b \neq 0\}/\sim,$$

where $\frac{a}{b} \sim \frac{c}{d}$ if and only if $ad = bc$. The operations $+, -, \cdot, {}^{-1}$, can be defined on representatives of the elements of $Q(D)$ as:

$$\frac{a}{b} + \frac{c}{d} = \frac{ad + bc}{bd}, \quad \frac{a}{b} \cdot \frac{c}{d} = \frac{ac}{bd}, \quad -\frac{a}{b} = \frac{-a}{b}, \quad \left(\frac{a}{b}\right)^{-1} = \frac{b}{a}.$$

$Q(D)$ is the smallest field containing D.

Let $(R, +, \cdot, 0, 1)$ be a commutative ring with identity. A nonempty subset I of R is an *ideal* in R iff $a + b \in I$, and $a \cdot c \in I$ for all $a, b \in I$, and $c \in R$. Moreover we say that I is a *proper ideal* iff $\{0\} \neq I \neq R$. I is a *maximal ideal* if it is not contained in a bigger proper ideal. I is a *prime ideal* iff $a \cdot b \in I$ implies that $a \in I$ or $b \in I$. I is a *primary ideal* if $a \cdot b \in I$ implies that $a \in I$ or $b^n \in I$ for some $n \in \mathbb{N}$. I is a *radical ideal* iff $a^n \in I$ for some $n \in \mathbb{N}$ implies that $a \in I$. Moreover, the *radical* of the ideal I is the ideal $\{a \mid a^n \in I$ for some $n \in \mathbb{N}\}$, and we denote it by $\sqrt{I}$ or radical(I). A set $B \subseteq R$ *generates the ideal I* or B is a *generating set* or a *basis for I* iff

$$I = \{\sum_{i=1}^{n} r_i b_i \mid n \in \mathbb{N}^\star,\ r_i \in R,\ b_i \in B\}.$$

In this case, we say that the ideal I is *generated* by B, and we denote this by $I = \langle B \rangle$. Furthermore, we say that the ideal I is *finitely generated* if it has a finite generating set. If the cardinality of the generating set is 1, we say that I is a *principal ideal*.

An ideal I in R generates a *congruence relation*, $\equiv_I$ on R by $a \equiv_I b$ or $a \equiv b \mod I$ iff $a - b \in I$. In this case we say that *a is congruent to b modulo I*. Observe that the factor ring R/I consisting of the congruence classes w.r.t. $\equiv_I$ inherits the operations of R in a natural way. If R is a commutative ring with identity 1 and I is a prime ideal of R, then R/I is an integral domain. If I is maximal, then R/I is a field.

In the following considerations, we take nonzero elements of a commutative ring R with identity 1. Invertible elements of R are called *units*. If $a = b \cdot u$ for a unit u, then a and b are called *associated*. b *divides* a iff $a = b \cdot c$ for some $c \in R$. If c divides $a - b$ we say that a is *congruent* to b modulo c, and we write this as $a \equiv_c b$ or $a \equiv b \mod c$ or $a \equiv b \mod < c >$. The congruence modulo c is an equivalence relation.

An element a of R is *irreducible* iff every b dividing a is either a unit or associated to a. An element a of R is *prime* iff a is not a unit, and whenever a divides a product $b \cdot c$, then a divides either b or c; i.e. if $< a >$ is a proper prime ideal. In general prime and irreducible elements can be different; for instance, 6 has two different factorizations into irreducibles in $\mathbb{Z}[\sqrt{-5}]$, and none of these factors is prime.

A *principal ideal domain* D is a domain in which every ideal is principal. An integral domain D is a *unique factorization domain* iff every nonunit of D is a finite product of irreducible factors and every such factorization is unique up to reordering and unit factors. In a unique factorization domain prime and irreducible elements are the same. An element a is *squarefree* iff every nonunit factor of a occurs with multiplicity exactly 1 in a.

Let D be an integral domain, and let $a, b \in D$ such that at least one of them is not zero. We say that $d \in D$ is a *greatest common divisor (gcd)* of a and b iff (i.) d divides both a and b, and (ii.) if c is a common divisor of a and b then, c divides d. In a unique factorization domain the gcd always exists, and it is determined up to associates. Moreover, if R is a principal ideal domain, the greatest common divisor $d \in R$, can be written as a linear combination $d = s \cdot a + t \cdot b$, for some $s, t \in R$. This equation is called the *Bézout equality*, and s, t are the *Bézout cofactors*. If $\gcd(a, b) = 1$, we say that a and b are *relatively prime*.

In $\mathbb{Z}$ we have the well known *Euclidean Algorithm* for computing a gcd. In general, an integral domain D in which we can execute the Euclidean algorithm, i.e., we have division with quotient and remainder such that the remainder is less than the divisor, is called a *Euclidean Domain*. More precisely, a Euclidean domain D is an integral domain together with a degree function $\deg : D^* \to \mathbb{N}$, such that

1. $\deg(a \cdot b) \geq \deg(a)$ for all $a, b \in D^*$,
2. (division property) for all $a, b \in D$, $b \neq 0$, there exists a *quotient q* and a *remainder r* in D such that $a = q \cdot b + r$ and $r = 0$ or $\deg(r) < \deg(b)$.

Every Euclidean domain is a principal ideal domain and every principal ideal domain is a unique factorization domain. However, the reverse implications do not hold in general.

In a Euclidean Domain, the Euclidean algorithm can be adapted such that the Bézout cofactors are also computed. Usually, this extension is called the extended Euclidean algorithm.

B.2 Polynomials and Power Series

Let R be a ring. A *(univariate) polynomial over R* is a mapping $p : \mathbb{N} \longrightarrow R$, $n \longmapsto p_n$, such that $p_n = 0$ nearly everywhere, i.e., for all but finitely many values of n. If $n_1 < n_2 < \cdots < n_r$ are the nonnegative integers for which p yields a nonzero result, then we usually write

$$p = p(x) = \sum_{i=1}^{r} p_{n_i} x^{n_i}.$$

p_j is the *coefficient* of x^j in the polynomial p, and we denote it by $\mathrm{coeff}(p, j)$. If p is the zero mapping, we say that p is the zero polynomial. The set of all polynomials over R together with the usual addition and multiplication of polynomials form a ring over R, which is denoted by $R[x]$. The *degree* of a nonzero polynomial p, $\mathrm{degree}(p)$, is the maximal $n \in \mathbb{N}$ such that $p_n \neq 0$. We say that the degree of the zero polynomial is -1. The *leading term* of a nonzero polynomial p is $x^{\mathrm{degree}(p)}$, denoted by $\mathrm{lt}(p)$. The *leading coefficient* of a nonzero polynomial p is the coefficient of $\mathrm{lt}(p)$, denoted by $\mathrm{lc}(p)$. For the zero polynomial the leading coefficient and the leading term are undefined. A polynomial p is *monic* iff $\mathrm{lc}(p) = 1$.

If R is an integral domain, then also the ring of polynomials $R[x]$ over R is an integral domain. Furthermore, if R is a unique factorization domain, then also the ring of polynomials $R[x]$ over R is a unique factorization domain.

An *n-variate polynomial over R* is a mapping $p : \mathbb{N}^n \longrightarrow R$, $(i_1, \ldots, i_n) \longmapsto p_{i_1, \ldots, i_n}$, such that $p_{i_1, \ldots, i_n} = 0$ nearly everywhere. We usually write

$$p = p(x_1, \ldots, x_n) = \sum_{i_1, \ldots, i_n} p_{i_1, \ldots, i_n} x_1^{i_1} \cdots x_n^{i_n}.$$

The set of all n-variate polynomials over R together with the usual addition and multiplication of polynomials form a ring over R, which is denoted by $R[x_1, \ldots, x_n]$. The n-variate polynomial ring can be viewed as built up successively from R by adjoining one polynomial variable at a time. In fact, $R[x_1, \ldots, x_n]$ is isomorphic to $(R[x_1, \ldots, x_{n-1}])[x_n]$. The *total degree* of an n-variate polynomial p is defined as the maximal $\sum_{j=1}^{n} i_j$ such that $p_{i_1, \ldots, i_n} \neq 0$. We denote this total degree by $\mathrm{degree}(p)$. In addition, we write $\mathrm{coeff}(p, x_n, j)$ for the coefficient of x_n^j in the polynomial p, where p is considered in $(R[x_1, \ldots, x_{n-1}])[x_n]$. The *degree in the variable x_n* of

$$p = p(x_1, \ldots, x_n) = \sum_{i=0}^{m} p_i(x_1, \ldots, x_{n-1}) x_n^i \in (R[x_1, \ldots, x_{n-1}])[x_n]^{*}$$

is m if $p_m \neq 0$, and we denote this by $\deg_{x_n}(p)$. By reordering the set of variables we get $\deg_{x_i}(p)$ for all $1 \leq i \leq n$. In a similar way, we get $\mathrm{lt}_{x_i}(p)$ and $\mathrm{lc}_{x_i}(p)$. If all the terms occurring (with nonzero coefficient) in the polynomial p have the same (total) degree, then p is call a *form* or a *homogeneous polynomial*.

An n-variate polynomial $p(x_1, \ldots, x_n)$ of total degree d can be written as

$$p(x_1, \ldots, x_n) = p_d(x_1, \ldots, x_n) + p_{d-1}(x_1, \ldots, x_n) + \cdots + p_0(x_1, \ldots, x_n),$$

where $p_i(x_1, \ldots, x_n)$ are forms of degree i, respectively (i.e., all the terms occurring in p_i are of the same degree and it is i). The homogenization $p^*(x_1, \ldots, x_n, x_{n+1})$ of the polynomial $p(x_1, \ldots, x_n)$ is given as

$$p^*(x_1, \ldots, x_n, x_{n+1}) =$$

$$p_d(x_1, \ldots, x_n) + p_{d-1}(x_1, \ldots, x_n)x_{n+1} + \ldots + p_0(x_1, \ldots, x_n)x_{n+1}^d.$$

The polynomial $p^*(x_1, \ldots, x_n, x_{n+1})$ is homogeneous. For all $\alpha \in K^*$, where K is a field, we have that $p^*(a_1, \ldots, a_n, a_{n+1}) = 0$ if and only if it holds that $p^*(\alpha\, a_1, \ldots, \alpha\, a_n, \alpha\, a_{n+1}) = 0$. Furthermore $p(a_1, \ldots, a_n) = 0$ if and only if $p^*(a_1, \ldots, a_n, 1) = 0$. In addition, $p^*(a_1, a_2, \ldots, a_n, 0) = p_d(a_1, \ldots, a_n) = 0$ means that there is a zero $(a_1, a_2, \ldots, a_n, 0)$ of p^* at infinity in the direction $(a_1, \ldots, a_n)$. By adding these points at infinity to affine space we get the corresponding projective space.

For homogeneous polynomials, we have the following well known *Euler's Formula*: let $F(x_1, \ldots, x_r)$ be an homogeneous polynomial of degree d. Then,

$$\sum_{i=1}^{r} x_i \cdot \frac{\partial F}{\partial x_i} = d \cdot F.$$

Let K, L be fields such that $K \subset L$. Let $\alpha \in L$ such that $f(\alpha) = 0$ for some irreducible $f \in K[x]$. Then, α is called *algebraic* over K of degree $\deg(f)$. If α is not algebraic over K, the we say that α is *transcendental* over K. The polynomial f is determined up to a constant and it is called the *minimal polynomial* of α over K. In addition, by $K(\alpha)$ we denote the smallest field containing K and α. $K(\alpha)$ is called a *(simple) algebraic extension field* of K. For representing the elements in the algebraic extension field $K(\alpha)$ of K, we use the isomorphism $K(\alpha) \cong K[x]/\langle f(x) \rangle$, where $\langle f(x) \rangle$ denotes the ideal generated by $f(x)$ in $K[x]$. Every polynomial can be reduced modulo $f(x)$ to some $r(x)$, with $\deg(r) < \deg(f)$. On the other hand, two different polynomials $r(x), s(x)$ with $\deg(r), \deg(s) < \deg(f)$ cannot be congruent modulo $f(x)$, since otherwise $r - s$, a nonzero polynomial of degree less than $\deg(f)$, would be a multiple of f. Thus, every element $a \in K(\alpha)$ has a unique representation

$$a = \underbrace{a_{m-1}x^{m-1} + \ldots + a_1 x + a_0}_{a(x)} + \langle f(x) \rangle, \quad a_i \in K.$$

We call $a(x)$ the *normal representation* of a. Observe that from this unique normal representation we can immediately deduce that $K(\alpha)$ is a vector space over K of dimension $\deg(f)$, and $\{1, \alpha, \ldots, \alpha^{m-1}\}$ is a basis of this vector space.

The field $\overline{K}$ is called the *algebraic closure* of K if $\overline{K}$ is algebraic over K and every polynomial $f(x) \in K$ has a root over $\overline{K}$, so that $\overline{K}$ can be said to contain all the elements that are algebraic over K. We say that K is algebraically closed if $K = \overline{K}$.

Let K be a field, and let $\overline{K}$ be the algebraic closure of K. A polynomial $f(x_1, \ldots, x_n) \in K[x_1, \ldots, x_n]$ is *irreducible* over K if and only if every $g(x_1, \ldots, x_n) \in K[x_1, \ldots, x_n]$ dividing f is either a unit or an associate of f. Moreover, if $f(x_1, \ldots, x_n)$ has no nontrivial factor in $\overline{K}[x_1, \ldots, x_n]$, then $f(x_1, \ldots, x_n)$ is called *absolutely irreducible*. A factorization over $\overline{K}$ is called an *absolute factorization*.

Let I be an ideal. A univariate polynomial $p(x)$ over I is *primitive* if and only if there is no prime in I which divides all the coefficients in $p(x)$. Every polynomial $q(x) \in I[x]$ can be decomposed uniquely, up to multiplication by units, as

$$q(x) = \mathrm{cont}(q) \cdot \mathrm{pp}(q),$$

where $\mathrm{cont}(q) \in I$, and $\mathrm{pp}(q)$ is the primitive polynomial in $I[x]$. We call $\mathrm{cont}(q)$ the *content* of $q(x)$, and $\mathrm{pp}(q)$ the *primitive part* of $q(x)$.

If K is a field, then $K[x]$ is a Euclidean domain, so $h = \gcd(f, g)$, for $f, g \in K[x]$ can be computed by means of the Euclidean Algorithm.

A polynomial $p(x_1, \ldots, x_n) \in K[x_1, \ldots, x_n]$ is *squarefree* if and only if every nontrivial factor $q(x_1, \ldots, x_n)$ of p (i.e, q not associated to p and not a constant) occurs with multiplicity exactly 1 in p. There is a simple criterion for deciding squarefreeness (see for instance [Win96], pp.101). More precisely, let $q(x)$ be a nonzero polynomial in $K[x]$, where $\mathrm{char}(K)$ is either zero or prime. Then, $q(x)$ is squarefree if and only if $\gcd(q(x), q'(x)) = 1$ ($q'(x)$ is the derivative of $q(x)$). The problem of squarefree factorization of $q(x) \in K[x]$ consists of determining the squarefree pairwise relatively prime polynomials $q_1(x), \ldots, q_s(x)$ such that

$$q(x) = \prod_{i=1}^{s} q_i(x)^{e_i},$$

where $e_i \in \mathbb{N}$. The representation of $q(x)$ as above is called the *squarefree factorization of $q(x)$*. For a thorough introduction to factorization of polynomials we refer the reader to Chapter 5 in [Win96].

Let K be a field. A *power series* $A(x)$ over K is a mapping $A : \mathbb{N} \longrightarrow K$. Usually we write a power series as $A(x) = \sum_{i=0}^{\infty} a_i x^i$, where a_i is the image of i under the mapping A.

The set of all power series over K form a commutative ring with 1 and we denote this ring by $K[[x]]$. The *order* of the power series A is the smallest i such that $a_i \neq 0$.

Taylor's Theorem *Let D be a unique factorization domain of characteristic zero, and let $f(x_1, \ldots, x_n) \in D[x_1, \ldots, x_n]$. Let $p = (a_1, \ldots, a_n)$, $a_i \in D$, and $h = (h_1, \ldots, h_n) = (x_1 - a_1, \ldots, x_n - a_n)$. Then,*

$$f(x_1, \ldots, x_n) = f(p) + \sum_{i_1=1}^{n} \frac{\partial f(p)}{\partial x_{i_1}} h_{i_1} + \frac{1}{2!} \sum_{i_1, i_2=1}^{n} \frac{\partial^2 f(p)}{\partial x_{i_1} \partial x_{i_2}} h_{i_1} h_{i_2} + \cdots +$$

$$\frac{1}{k!} \sum_{i_1, \ldots, i_k=1}^{n} \frac{\partial^k f(p)}{\partial x_{i_1} \ldots \partial x_{i_k}} h_{i_1} \cdots h_{i_k} + \cdots.$$

We call this expression the *Taylor expansion of the polynomial f at p.*

Taylor's Theorem for univariate analytic functions: *Let $f(z)$ be a complex analytic function in an open disk centered at z_0 and radius r. Then, for $|z - z_0| < r$, the power series $\sum_{n=0}^{\infty} \frac{f^{(n)}(z_0)}{n!} (z - z_0)^n$ converges to $f(z)$.*

Implicit Mapping Theorem (see [Gun90]): *Let F be an holomorphic mapping from an open neighborhood of a point $A \in \mathbb{C}^n$ into $\mathbb{C}^m$ for some $m \leq n$, such that $F(A) = 0$ and rank($J_F''(A)$) $= m$, where $J_F(A) = (J_F'(A), J_F''(A))$ is the Jacobian matrix of F at A, and $J_F'(A)$ is an $m \times (n - m)$ matrix, and $J_F''(A)$ is an $m \times m$ matrix. Then, for some open polydisc $U(A, R) = U(A', R') \times U(A'', R'') \subset \mathbb{C}^{n-m} \times \mathbb{C}^m = \mathbb{C}^n$, there exists an holomorphic mapping $G : U(A', R') \to U(A'', R'')$ such that $G(A') = A''$, and $F(Z) = 0$ for some point $Z = (Z', Z'') \in U(A, R)$, precisely when $Z'' = G(Z')$.*

B.3 Polynomial Ideals and Elimination Theory

Let R be a commutative ring with identity 1 and $R[x_1, \ldots, x_n]$ the polynomial ring in n indeterminates over R. A commutative ring with identity R is called a *Noetherian ring* if and only if the *basis condition* holds in R, i.e., every ideal in R is finitely generated.

A commutative ring with identity R is Noetherian if and only if there are no infinitely ascending chains of ideals in R. I.e., if $I_1 \subseteq I_2 \subseteq \ldots \subseteq R$, then there is an index k such that $I_k = I_{k+1} = \cdots$.

Hilbert's Basis Theorem: *If R is a Noetherian ring then also the ring of polynomials $R[x]$ is Noetherian.*

Hilbert's basis theorem implies that the multivariate polynomial ring $K[x_1, \ldots, x_n]$ over a field K is Noetherian. So every ideal $I \in K[x_1, \ldots, x_n]$ has a finite basis, and if we are able to effectively compute with finite bases then we are dealing with all the ideals in $K[x_1, \ldots, x_n]$.

B.3.1 Gröbner Bases

The method of Gröbner bases was introduced by Buchberger in [Buc65], where he also developed an algorithm for computing it. Gröbner bases are very special and useful bases for polynomial ideals. The Buchberger algorithm for constructing Gröbner bases is at the same time a generalization of the Euclidean Algorithm and of Gauss' triangularization algorithm for linear systems. Intuitively speaking, Gröbner bases can be motivated from different points of view. The first one is based on the theory of polynomials ideals, and the second one focuses on the application to the solution of systems of algebraic equations. In both cases, the difficulty of the problem comes from the need to generalize the Euclidean division to the non-Euclidean domain $K[x_1, \ldots, x_n]$, where K is a field.

Ideal theoretically the goal is to decide the "main problem in ideal theory", namely the question whether a polynomial $f \in K[x_1, \ldots, x_n]$ is contained in a given ideal I of the ring $R = K[x_1, \ldots, x_n]$. Observe that if R is univariate, then R is a Euclidean domain. Therefore, I is a principal ideal, and it can be expressed as $I = \langle g(x) \rangle$. So $f \in I$ if and only if f is a multiple of g if and only if the remainder of f on division by g is 0. In the multivariate case a Gröbner basis for the ideal I admits a generalization of the division algorithm such that $f \in I$ if and only if the remainder of f on division by the Gröbner basis is 0.

The Buchberger algorithm for computing a Gröbner basis for an ideal I can also be considered as a generalization of Gaussian elimination to the multivariate case. Given a system of algebraic, i.e., polynomial, equations

$$f_i(x_1, \ldots, x_n) = 0, \quad i = 1, \ldots, m,$$

where $f_i \in K[x_1, \ldots, x_n]$, we observe that the solutions of these equations over the algebraic closure of K remain unchanged when we replace the f_i by g_j, where $\{g_j | 1 \leq j \leq k\}$ is another basis for the ideal generated by $\{f_i | 1 \leq i \leq m\}$. If we determine a Gröbner basis w.r.t a lexicographic ordering of the terms, then we get a triangular basis comparable to a triangular system of linear equations. This property is expressed in the following theorem.

Theorem (elimination property of Gröbner bases): *Let G be a Gröbner basis for the ideal I in $K[x_1, \ldots, x_n]$ w.r.t. the lexicographic ordering $x_1 < x_2 < \cdots < x_n$. Then for every $i \in \{1, \ldots, n\}$ the ideal $I \cap K[x_1, \ldots, x_i]$ is generated by $G \cap K[x_1, \ldots, x_i]$.*

Any Gröbner basis allows to decide the solvability of the corresponding system of algebraic equations, simply by checking whether the Gröbner basis contains a constant. We can also read off whether the system has finitely or infinitely many solutions.

For a thorough introduction to Gröbner bases we refer the reader to [AdL94], [BeW93], [BCL83], [CLO97] or [Win96].

B.3.2 Resultants

Let D be a unique factorization domain, and let $f(x), g(x) \in D[x]$ be the polynomials

$$f(x) = a_n x^n + \cdots + a_0, \quad a_n \neq 0, \qquad g(x) = b_m x^m + \cdots + b_0, \quad b_m \neq 0.$$

The *resultant*, $\mathrm{res}_x(f, g)$, of the univariate polynomials $f(x), g(x)$ over D is the determinant of the Sylvester matrix of f and g, consisting of shifted lines of coefficients of f and g. More precisely, $\mathrm{res}_x(f, g) = \det(\mathrm{Syl}_x(f, g))$, where

$$
\mathrm{Syl}_x(f, g) = \begin{pmatrix}
a_n & \cdots & a_0 & & & & \\
 & a_n & \cdots & a_0 & & & \\
 & & \ddots & \ddots & \ddots & & \\
 & & & a_n & \cdots & a_0 & \\
b_m & \cdots & b_0 & & & & \\
 & b_m & \cdots & b_0 & & & \\
 & & \ddots & \ddots & \ddots & & \\
 & & & b_m & \cdots & b_0 &
\end{pmatrix}.
$$

$\mathrm{Syl}_x(f, g)$ contains m rows of coefficients of f, and n rows of coefficients of g.

Resultants have important properties (see [BrK86], [CLO98], [VaW70]). Some of the more important ones are the following:

1. $\mathrm{res}_x(f, g) = 0$ if and only if f and g have a common root.
2. $\mathrm{res}_x(f, g) = (-1)^{mn}\mathrm{res}_x(g, f)$.
3. Let α_i, $i = 1, \ldots, n$, be the roots of f, and let β_i, $i = 1, \ldots, m$, be the roots of g. Then

$$
\mathrm{res}_x(f, g) = \mathrm{lc}(f)^m \mathrm{lc}(g)^n \prod_{i=1}^{n}\prod_{j=1}^{m}(\alpha_i - \beta_j), \text{ and } \mathrm{res}_x(f, g) = \mathrm{lc}(f)^m \prod_{j=1}^{n} g(\alpha_j).
$$

4. There exist polynomials $a(x)$ and $b(x)$ over D such that $af + bg = \mathrm{res}_x(f, g)$.

The notion of resultant of two univariate polynomials can be generalized to multivariate polynomials. Let K be an algebraically closed field, and let $f(x_1, \ldots, x_n), g(x_1, \ldots, x_n) \in K[x_1, \ldots, x_n]$. We view the polynomials f and g as elements of $K[x_1, \ldots, x_{n-1}][x_n]$ where the coefficients are in $K[x_1, \ldots, x_{n-1}]$, and the main variable is x_n. So we may determine $R(x_1, \ldots, x_{n-1}) := \mathrm{res}_{x_n}(f, g) \in K[x_1, \ldots, x_{n-1}]$, and we have the following important property (see [Mis93]): if $(a_1, \ldots, a_n) \in K^n$ is a common root of f and g, then $R(a_1, \ldots, a_{n-1}) = 0$. Conversely, if $R(a_1, \ldots, a_{n-1}) = 0$, then one of the following holds:

1. $\mathrm{lc}_{x_n}(f)(a_1, \ldots, a_{n-1}) = \mathrm{lc}_{x_n}(g)(a_1, \ldots, a_{n-1}) = 0$,
2. $f(a_1, \ldots, a_{n-1}, x_n) = 0$ or $g(a_1, \ldots, a_{n-1}, x_n) = 0$,
3. for some $a_n \in K$, $(a_1, \ldots, a_n)$ is a common root of f and g

For a thorough introduction to resultants we refer the reader to [BrK86], [CLO98] or [Mis93].

B.4 Algebraic Sets

For any field K, the *n-dimensional affine space* over K is defined as

$$\mathbb{A}^n(K) := K^n = \{(a_1, \ldots, a_n) \mid a_i \in K\} .$$

$\mathbb{A}^2(K)$ is the affine plane over K. The *n-dimensional projective space* over K is defined as

$$\mathbb{P}^n(K) := \{(a_1 : \cdots : a_{n+1}) \mid (a_1, \ldots, a_{n+1}) \in K^{n+1} \setminus \{(0, \ldots, 0)\}\} ,$$

where $(a_1 : \cdots : a_{n+1}) = \{(\alpha a_1, \ldots, \alpha a_{n+1}) \mid \alpha \in K^\star\}$. So a point in $\mathbb{P}^n(K)$ has many representations as an $(n+1)$-tuple, since $(a_1 : \cdots : a_{n+1})$ and $(\alpha a_1 : \cdots : \alpha a_{n+1})$, for any $\alpha \in K^\star$, denote the same projective point P. $(a_1 : \ldots : a_{n+1})$ are *homogeneous coordinates* for P. $\mathbb{P}^2(K)$ is the projective plane over K.

For any ideal I in $K[x_1, \ldots, x_n]$ we denote by $V(I)$ the set of all the points in $\mathbb{A}^n(\overline{K})$, the n-dimensional affine space over the algebraic closure of K, which are common zeros of all the polynomials in I. Such sets $V(I)$ are called *algebraic sets*. In commutative algebra and algebraic geometry there is a 1–1 correspondence between radical polynomial ideals and algebraic sets, the zeros of such ideals over the algebraic closure of the field of coefficients. *Hilbert's Nullstellensatz* relates the radical of an ideal I to the set of common roots $V(I)$ of the polynomials contained in I. More precisely, the radical of I consists of exactly those polynomials in $K[x_1, \ldots, x_n]$ which vanish on all the common roots of I.

On the other hand, for any subset V of $\mathbb{A}^n(\overline{K})$ we denote by $I(V)$ the ideal of all the polynomials vanishing at V. Then for radical ideals I and algebraic sets $V(\cdot)$ and $I(\cdot)$ are inverses of each other, i.e., $V(I(V)) = V$ and $I(V(I)) = I$. This correspondence extends to operations on ideals and algebraic sets (see, for instance [CLO97], Chap. 4).

An algebraic set V is called *irreducible* if it cannot be expressed as a union of two algebraic sets, both of them different from V. In fact, V is irreducible if and only if $I(V)$ is a prime ideal. Irreducible algebraic sets are called *algebraic varieties*. In general, an algebraic set can be written uniquely as the finite union of algebraic varieties. These notions can be extended analogously to projective space.

The intersection of two algebraic sets is an algebraic set defined by the union of the two ideals. In fact, the intersection of an arbitrary number of algebraic sets is again an algebraic set. However, in general, only finite unions of algebraic sets are algebraic. The empty set is the algebraic set generated by any nonzero constant polynomial, for example 1, and the whole affine space is the algebraic set generated by the zero polynomial. Consequently, the algebraic sets are the closed sets in a topology, called the *Zariski topology*. In the Zariski topology, any two nonempty open sets have a nonempty intersection.

References

[Abh66] Abhyankar, S.S.: Resolution of Singularities of Embedded Algebraic Surfaces. Academic Press (1966)

[AbB87a] Abhyankar, S.S., Bajaj, C.L.: Automatic Parametrization of Rational Curves and Surfaces I: Conics and Conicoids. Computer Aided Geometric Design; **19**, no. 1: 11–14 (1987)

[AbB87b] Abhyankar, S.S., Bajaj, C.L.: Automatic Parametrization of Rational Curves and Surfaces II: Cubics and Cubicoids. Computer Aided Geometric Design; **19**, no. 9: 499–502 (1987)

[AbB88] Abhyankar, S.S., Bajaj, C.L.: Automatic Parametrization of Rational Curves and Surfaces III: Algebraic Plane Curves. Computer Aided Geometric Design; **5**, 390–321 (1988)

[AbB89] Abhyankar, S.S., Bajaj, C.L.: Automatic Rational Parametrization of Curves and Surfaces IV: Algebraic Space Curves. Transactions on Graphics; **8**, no. 4, 325–334 (1989)

[AdL94] Adams, W.W., Loustaunau, P.: An Introduction to Gröbner Bases. AMS, Providence, RI, Graduate studies in Mathematics; **3** (1994)

[ASS07] Alcázar, J.G., Schicho, J., Sendra, J.R.: A Delineability-based Method for Computing Critical Sets of Algebraic Surfaces. Journal of Symbolic Computation; **42**, no. 6, 678–691 (2007)

[AlS07] Alcázar, J.G., Sendra, J.R.: Local Shape of Offsets to Algebraic Curves. Journal of Symbolic Computation; **42**, no. 3, 338–351 (2007)

[AGR95] Alonso, C., Gutierrez, J., Recio, T.: Reconsidering Algorithms for Real Parametric Curves. Journal of Applicable Algebra in Engineering, Communication and Computing; **6**, 345-352 (1995)

[AnR06] Andradas, C., Recio, T.: Plotting missing points and branches of real parametric curves. Applicable Algebra in Engineering, Communication and Computing; **18(1-2)**, 107–126 (2007)

[ARS97] Andradas, C., Recio, T., Sendra, J.R.: A Relatively Optimal Rational Space Curves Reparametrization Algorithm through Canonical Divisors. Proceedings of the 1997 International Symposium on Symbolic and Algebraic Computation, Kchlin W. (ed.); 349–356. ACM Press, New York (1997)

[ARS99] Andradas, C., Recio, T., Sendra, J.R.: Base Field Restriction Techniques for Parametric Curves. Proceedings of the 1999 International Symposium on Symbolic and Algebraic Computation, Dooley S. (ed.); 17–22. ACM Press, New York (1999)

[ARS04] Andradas, C., Recio, T., Sendra, J. R.: La Variedad de Weil para Variedades Unirracionales. Book in honor of Prof. Outerelo. Editorial de la Universidad Complutense de Madrid; 33–51 (2004)

[ACFG05] Aroca, J.M., Cano, J., Feng, R., Gao, X.S.: Algebraic General Solutions of Algebraic Ordinary Differential Equations. Proceedings of the 2005 International Symposium on Symbolic and Algebraic Computation. Gutierrez J. (ed.); 29–36. ACM Press, New York (2005)

[ASS97] Arrondo, E., Sendra, J., Sendra, J.R.: Parametric Generalized Offsets to Hypersurfaces. Journal of Symbolic Computation; **23**, 267–285 (1997)

[ASS99] Arrondo, E., Sendra, J., Sendra, J.R.: Genus Formula for Generalized Offset Curves. Journal of Pure and Aplied Algebra; **136**, Issue 3, 199–209 (1999)

[BaN82] Back, J., Newman, D.J.: Complex Analysis Undergraduate Text in Mathematics. Springer-Verlag. New York (1982)

[Baj94] Bajaj, C. (ed.): Algebraic Geometry and Its Applications. Springer-Verlag, Berlin Heidelberg New York (1994)

[BaR95] Bajaj, C.L., Royappa, A.V.: Finite Representation of Real Parametric Curves and Surfaces. International Journal on Computational Geometry and Applications; **5**, no. 3, 313–326 (1995)

[BLM97] Bajaj, C., Lee, H., Merkert, R., Pascucci, V.: NURBS based B-rep Models from Macromolecules and their Properties. In Proceedings of Fourth Symposium on Solid Modeling and Applications. Atlanta,Georgia, 1997. C. Hoffmann and W. Bronsvort(ed.); 217–228. ACM Press (1997)

[BPR03] Basu, S., Pollack, R., Roy, M-F.: Algorithms in Real Algebraic Geometry. Springer-Verlag Heidelberg; Series: Algorithms and Computation in Mathematics (2003)

[BeW93] Becker, T., Weispfenning, V.: Gröbner Bases - A Computational Approach to Commutative Algebra. Springer-Verlag, Berlin, Graduate texts in Mathematics (1993)

[BiH96] Bilu, Y., Hanrot, G.: Solving the Thue Equations of High Degree. Journal of Number Theory; **60(2)**, 373–392 (1996)

[BiM79] Birkhoff, G., MacLane, S.: Algebra. 2nd Edition. Macmillan, New York (1979)

[BSS99] Blake, I., Seroussi, G., Smart, N.: Elliptic Curves in Cryptography. Cambridge Univ. Press. (1999)

[BrK86] Brieskorn, E., Knörrer, H.: Plane Algebraic Curves. Birkhäuser, Basel (1986)

[Buc65] Buchberger, B.: Ein Algorithmus Zum Baseselemente de Restklassenringen nach einen nulldimensionen Polynomideal. Ph. D. Thesis Math Ins., Univ of Innsbruk, Austria (1965)

[BCL83] Buchberger, B., Collins, G.E., Loos, R.: Computer Algebra, Symbolic and Algebraic Computation. 2nd ed., Springer-Verlag, Wien New York (1983)

[Buc01] Buchmann, J.A.: Introduction to Cryptography. Springer-Verlag (2001)

[BCD03] Busé, L., Cox, D., D'Andrea, C.: Implicitization of surfaces in $\mathbb{P}^3$ in the presence of base points. Journal of Algebra and Applications; **2**, 189–214 (2003)

[BuD06] Busé, L., D'Andrea, C.: A Matrix-Based Approach to Properness and Inversion Problems for Rational Surfaces. Applicable Algebra in Engineering, Communication and Computing; **17**, no. 6, 393–407 (2006)

[CaM91] Canny, J. F., Manocha, D.: Rational Curves with Polynomial Parametrizations. Computer Aided Design; **23**, no. 9: 645–652 (1991)

[Coh00] Cohen, H.: A Course in Computational Number Theory. GTM 138, Springer Verlag (2000)

[CLO97] Cox, D.A., Little, J., O'Shea, D.: Ideals, Varieties, and Algorithms. Springer-Verlag, New York (1997)

[CLO98] Cox, D.A., Little, J., O'Shea, D.: Using Algebraic Geometry. Graduate Texts in Mathematics, 185. Springer–Verlag (1998)

[CoS97b] Cox, D.A., Sturmfels, B. (eds.): Applications of Computational Algebraic Geometry. Proceedings of Symposia in Applied Mathematics; **53**, AMS, Providence (1997)

[CrR03] Cremona, J.C., Rusin, D.: Efficient Solutions of Rational Conics. Math. of Computation; **72**, 1417–1441 (2003)

[ChG92a] Chionh, E.-W., Goldman, R.N.: Using multivariate resultants to find the implicit equation of a rational surface. The Visual Computer; **8**, 171–180 (1992)

[ChG92b] Chionh, E. W., Goldman, R. N.: Degree, Multiplicity and Inversion Formulas for Rational Surfaces Using u-Resultants. Computer Aided Geometric Design; **9**, no. 2, 93–109 (1992)

[CGS06] Chionh, E. W., Gao, X.S., Shen, L.Y.: Inherently improper surface parametric supports. Computer Aided Geometric Design; **23**, no. 8: 629–639 (2006)

[ChG91] Chou, S. C., Gao, X. S.: On the Normal Parametrization of Curves and Surfaces. International Journal on Computational Geometry and Applications; **1**, no. 2: 125–136 (1991)

[Duv87] Duval, D.: Diverses questions relatives au calcul formel avec des nombres algébriques (Some Questions Concerning Algebraic Numbers in Symbolic Computation). PhD Thesis, Institut Fourier, Grenoble, France (1987)

[Duv89] Duval, D.: Rational Puiseux Expansion. Compositio Mathematica; **70**, 119–154 (1989)

[Far93] Farin, G.: Curves and Surfaces for Computer Aided Geometric Design. A Practical Guide (Second Edition), Academic Press (1993)

[FHK02] Farin, G., Hoschek, J., Kim, M.–S.: Handbook of Computer Aided Geometric Design. North–Holland (2002)

[FaN90a] Farouki, R. T., Neff, C.A.: Analytic Properties of Plane Offset Curves. Computer Aided Geometric Design; **7**, 83–99 (1990)

[FaN90b] Farouki, R. T., Neff, C.A.: Algebraic Properties of Plane Offset Curves. Computer Aided Geometric Design; **7**, 100–127 (1990)

[FaS90] Farouki, R., Sakkalis, T.: Singular Points on Algebra Curves. Journal of Symbolic Computation; **9/4**, 405–421 (1990)

[FeG04] Feng, R., Gao, X.-S.: Rational General Solutions of Algebraic Ordinary Differential Equations. Proceedings of the 2004 International Symposium on Symbolic and Algebraic Computation, Gutierrez J. (ed.); 155–162. ACM Press, New York (2004)

[For92] Forsman, K.: On rational state space realizations. In M. Fliess, editor. Proceeding NOLCOS'92: 197–202. Bordeaux. IFAC (1992)

[Ful89] Fulton, W.: Algebraic Curves – An Introduction to Algebraic Geometry. Addison-Wesley, Redwood City CA (1989)

[GKW91] Gebauer, R., Kalkbrener, M., Wall, B., Winkler, F.: CASA: A Computer Algebra Package for Constructive Algebraic Geometry. Proceedings of the 1991 International Symposium on Symbolic and Algebraic Computation, Watt S.M (ed.); 403–410. ACM Press (1991)

[GCL92] Geddes, K.O., Czapor, S. R., Labahn, G.: Algorithms for Computer Algebra. Kluwer Academic Publishers, Boston (1992)

[GSA84] Goldman, R. N., Sederberg, T. W., Anderson, D. C.: Vector Elimination: A Technique for the Implicitization, Inversion, and a Intersection of Planar Parametric Rational Polynomial Curves. Computer Aided Design; **1**, 337–356 (1984)

[Gon97] González-Vega, L.: Implicitization of Parametric Curves and Surfaces. Journal of Symbolic Computation; **23**, 137–152 (1997)

[Gop77] Goppa, V.D.: Codes associated with divisors. Problems of Information Transmission; **12**, n.1, 22–27 (1977)

[Gop81] Goppa, V.D.: Codes on algebraic curves. Soviet Math. Dokl. (translation); 207–214 (1981)

[GMS02] Götz, R.M., Maymeskul, V., Saff, E.B.: Asymptotic Distribution of Nodes for Near-Optimal Polynomial Interpolation on Certain Curves in $\mathbb{R}^2$. Constr. Approx; **18**, 255–283 (2002)

[GuR65] Gunning, R. C., Rossi, H.: Analytic Functions of Several Complex Variables. Prentice–Hall, Inc., N.J (1965)

[Gun90] Gunning, R. C.: Introduction to Holomorphic Functions of Several Variables. CRC Press (1990)

[GuR92] Gutierrez, J., Recio, T.: Rational Function Decomposition and Gröbner Basis in the Parameterization of a Plane Curve. LATIN 92, LNCS 583; 231–246. Springer Verlag (1992)

[GRS02] Gutierrez, J., Rubio, R., Schicho, J.: Polynomial Parametrization of Curves whithout Affine Singularities. Computer Aided Geometric Design; **19**, 223–234 (2002)

[GRY02] Gutierrez, J., Rubio, R., Yu, J-T.: D-Resultant for Rational Functions. Proceedings of the American Mathematical Society; **130 (8)**, 2237–2246 (2002)

[Har95] Harris, J.: Algebraic Geometry. A First Course. Springer-Verlag (1995)

[Har01] Hartmann, E.: Parametric G^n-Blending of Curves and Surfaces. Visual Computer; **17**, 1–13 (2001)

[HHW03] Hemmecke, R., Hillgarter, E., Winkler, F.: The CASA system, in Handbook of Computer Algebra: Foundations, Applications, Systems. J. Grabmeier, E. Kaltofen, V. Weispfenning (eds.) Springer-Verlag (2003)

[HiH90] Hilbert, D., Hurwitz, A.: Über die Diophantischen Gleichungen vom Geschlecht Null. Acta math; **14**, 217–224 (1890)

[HiW98] Hillgarter, E., Winkler, F.: Points on Algebraic Curves and the Parametrization Problem. Automated Deduction in Geometry. Lecture Notes in Artif. Intell. 1360: 185–203. D. Wang (ed.). Springer Verlag Berlin Heidelberg (1998)

[Hof93] Hoffmann, C. M.: Geometric and Solid Modeling. Morgan Kaufmann Publ., Inc (1993)

[HSW97] Hoffmann, C.M., Sendra, J.R., Winkler, F. (eds.): Parametric Algebraic Curves and Applications. Special Issue on Parametric Curves and Applications of the Journal of Symbolic Computation; **23/2&3** (1997)

[HoL93] Hoschek, J., Lasser, D.: Fundamentals of Computer Aided Geometric Design. A.K. Peters, Ltd. Natick, MA, USA (1993)

[IrR82] Ireland, K., Rosen, M.: A classical introduction to modern number theory. Springer Verlag (1982)

[Jac74] Jacobson, N.: Basic Algebra I. Freeman, San Francisco (1974)

[Jac80] Jacobson, N.: Basic Algebra II. Freeman, San Francisco (1980)

[Joh98] Johnson, J.R.: Algorithms for Real Root Isolation. Quantifier Elimination and Cylindrical Algebraic Decomposition. Text and Monographs in Symbolic Computation. Springer Verlag: 269–289 (1998)

[Kiy93] Kiyosi, I. (ed.): Encyclopedic Dictionary of Mathematics. Vol. 1. Mathematical Society of Japan (1993)

[Kob98] Koblitz, N.: Algebraic Aspects of Cryptography. Springer-Verlag Berlin (1998)

[Kob02] Koblitz, N.: Good and bad uses of elliptic curves in cryptography. Moscow Math. J; **2**, n.4: 693–715 (2002)

[Kot04] Kotsireas I. S.: Panorama of Methods for Exact Implicitization of Algebraic Curves and Surfaces. Geometric Computation. Falai Chen and Dongming Wang (eds.). Lecture Notes Series on Computing; **11**, Chapter 4, World Scientific Publishing Co., Singapore (2004)

[Krä81] Krätzel, E.: Zahlentheorie. VEB Dt. Verlag der Wissenschaften (1981)

[Lan84] Lang, S.: Algebra. 2nd Edition. Addison-Wesley, Reading M.A. (1984)

[LiV00] Li, H., Van Oystaeyen, F.: A Primer of Algebraic Geometry. Marcel Dekker, New York – Basel (2000)

[LiN94] Lidl, R., Niederreiter, H.: Introduction to Finite Fields and Their Applications. Cambridge University Press, Cambridge, UK (1994)

[Lü95] Lü, W.: Offset-Rational Parametric Plane Curves. Computer Aided Geometric Design; **12**, 601–617 (1995)

[MPL96] Mignotte, M., Pethö, A., Lemmermeyer, F.: On the Family of Thue Equations $x^3 - (n-1)x^2y - (n+2)xy^2 - y^3 = k$. Acta Arithm., LXXVI.3; 245–296 (1996)

[Mir99] Miranda, R.: Linear Systems of Plane Curves. Notices of AMS; **46**, no. 2: 192–201 (1999)

[Mis93] Mishra, B.: Algorithmic Algebra. Springer Verlag (1993)

[Noe83] Noether, M.: Rationale Ausführung der Operationen in der Theorie der Algebraischen Funktionen. Math. Ann; **23**, 311–358 (1883)

[Orz81] Orzech, G., Orzech, M.: Plane Algebraic Curves. An Introduction Via Valuations. Marcel Dekker, New York (1981)

[PeD06] Pérez–Díaz, S.: On the Problem of Proper Reparametrization for Rational Curves and Surfaces. Computer Aided Geometric Design; **23/4**, 307–323 (2006)

[PDS01] Pérez–Díaz, S., Sendra, J.R.: Parametric G^1–Blending of Several Surfaces. Computer Algebra in Scientific Computing CASC' 01. Lectures Notes in Computer Science. Springer Verlag, XII; 445–461 (2001)

[PDSS02] Perez-Díaz, S., Schicho, J., Sendra, J.R.: Properness and Inversion of Rational Parametrizations of Surfaces. Applicable Algebra in Engineering Communication and Computing; **13**, 29–51 (2002)

[PDS03] Pérez–Díaz, S., Sendra, J.R.: Computing All Parametric Solutions for Blending Surfaces. Journal of Symbolic Computation; **26/6**, 925–964 (2003)

[PDS04] Perez-Díaz, S., Sendra, J.R.: Computation of the Degree of Rational Surface Parametrizations. Journal of Pure and Applied Algebra; **193(1-3)**, 99–121 (2004)

[PDS05] Pérez–Díaz, S., Sendra, J.R.: Partial Degree Formulae for Rational Algebraic Surfaces. Proceedings of the 2005 International Symposium on Symbolic and Algebraic Computation, Kauers M. (ed.); 301–308. ACM Press, New York (2005)

[PeP98a] Peternell, M., Pottmann, H.: Applications of Laguerre Geometric in CAGD. Computer Aided Geometric Design; **15**, 165–186 (1998)

[PeP98b] Peternell, M., Pottmann, H.: A Laguerre Geometric Approach to Rational Offsets. Computer Aided Geometric Design; **15**, 223–249 (1998)

[Pot95] Pottmann, H.: Rational Curves and Surfaces with Rational Offsets. Computer Aided Geometric Design; **12**, 175–192 (1995)

[PoW97] Pottman, H., Wallner, J.: Rational Blending Surfaces Between Quadrics. Computer Aided Geometric Design; **14**, 407–419 (1997)

[PoV00] Poulakis, D., Voskos, E.: On the Practical Solutions of Genus Zero Diophantine Equations. Journal of Symbolic Computation; **30**, 573–582 (2000)

[PoV02] Poulakis, D., Voskos, E.: Solving Genus Zero Diophantine Equations with at Most Two Infinity Valuations. Journal of Symbolic Computation; **33**, 479–491 (2002)

[Pre98] Pretzel, O.: Codes and Algebraic Curves. Oxford Univ. Press (1998)

[ReS97a] Recio, T., Sendra, J.R.: Real Reparametrizations of Real Curves. Journal of Symbolic Computation; **23**, 241–254 (1997)

[ReS97b] Recio, T., Sendra, J.R.: A Really Elementary Proof of Real Lüroth Theorem. Revista Matematica de la Universidad Complutense de Madrid; **10**, 283–291 (1997)

[RSV04] Recio, T., Sendra, J.R., Villarino, C.: From hypercircles to units. Proceedings of the 2004 International Symposium on Symbolic and Algebraic Computation, Gutierrez J. (ed.); 258–265. ACM Press, New York (2004)

[Ros88] Rose, H.E.: A Course in Number Theory. Oxford Science Publications (1988)

[Rud66] Rudin, W.: Real and Complex Analysis. McGraw-Hill (1966)

[SSeS05] San Segundo, F., Sendra, J.R.: Degree Formulae for Offset Curves. Journal of Pure and Applied Algebra; **195(3)**, 301–335 (2005)

[Sch92] Schicho, J.: On the choice of pencils in the parametrization of curves. Journal of Symbolic Computation; **14**, 557–576 (1992)

[Sch98a] Schicho, J.: Rational Parametrization of Surfaces. Journal of Symbolic Computation; **26**, 1–9 (1998)

[Sch98b] Schicho, J.: Inversion of Birational Maps with Gröbner Bases. Gröbner Bases and Applications; 495–503. B. Buchberger, F. Winkler (eds.). Cambridge Univ. Press (1998)

[ScS07] Schicho, J., Sendra, J.R., (Guest editors): Special Issue of the Applicable Algebra in Engineering, Communication and Computing on Algebraic Curves; **18**, (2007).

[Sed86] Sederberg, T.W.: Improperly parametrized rational curves. Computer Aided Geometric Design; **3**, 67–75 (1986)

[Sed98] Sederberg, T. W.: Applications to Computer Aided Geometric Design. Applications of Computational Algebraic Geometry, Proceedings of Symposia in Applied Mathematics; **53**, 67–89. AMS (1998)

[SGD97] Sederberg, T.W., Goldman, R., Du, H.: Impliciticing Rational Curves by the Method of Moving Algebraic Curves. Journal of Symbolic Computation; **23**, 153–176 (1997)

[Sen02] Sendra, J. R.: Normal Parametrizations of Algebraic Plane Curves. Journal of Symbolic Computation; **33**, 863–885 (2002)

[Sen04] Sendra, J.R.: Rational Curves and Surfaces: Algorithms and Some Applications. Geometric Computation. F. Chen and D. Wang (eds.) Lecture Notes on Computing; **11**, Chapter 3, 63–125, World Scientific Publishing Co., Singapure (2004).

[SeS99] Sendra, J., Sendra, J.R.: Algebraic Analysis of Offsets to Hypersurfaces. Mathematische Zeitschrift; **234**, 697–719 (1999)

[SeS00] Sendra, J., Sendra, J.R.: Rationality Analysis and Direct Parametrization of Generalized Offsets to Quadrics. Applicable Algebra in Engineering, Communication and Computing; **11**, no. 2, 111–139 (2000)

[SeV01] Sendra, J.R., Villarino, C.: Optimal Reparametrization of Polynomial Algebraic Curves. International Journal of Computational Geometry and Applications; **11**, no. 4, 439–453 (2001)

[SeV02] Sendra, J.R., Villarino C.: Algebraically Optimal Reparametrizations of Quasi-Polynomial Algebraic Curves. Journal of Algebra and Its Applications; **1**, no. 1, 51–74 (2002)

[SeW89] Sendra, J.R., Winkler, F.: A Symbolic Algorithm for the Rational Parametrizatioin of Algebraic Plane Curves. Techn. Rep. RISC 89–41, RISC-Linz, J. Kepler Univ. Linz, Austria (1989)

[SeW91] Sendra, J.R., Winkler, F.: Symbolic Parametrization of Curves. Journal of Symbolic Computation; **12**, 607–631 (1991)

[SeW97] Sendra, J.R., Winkler, F.: Parametrization of Algebraic Curves over Optimal Field Extensions. Journal of Symbolic Computation; **23**, 191–207 (1997)

[SeW99] Sendra, J.R., Winkler, F.: Algorithms for Rational Real Algebraic Curves. Fundamenta Informaticae; **39**, no. 1–2, 211–228 (1999)

[SeW01a] Sendra, J.R., Winkler, F.: Tracing Index of Rational Curve Parametrizations. Computer Aided Geometric Design; **18**, 771–795 (2001)

[SeW01b] Sendra, J.R., Winkler, F.: Computation of the Degree of a Rational Map between Curves. Proceedings of the 2001 International Symposium on Symbolic and Algebraic Computation, Mourrain B. (ed.): 317–322. ACM Press, New York (2001)

[Sha94] Shafarevich, I.R.: Basic Algebraic Geometry I and II. Springer-Verlag, Berlin New York (1994)

[Sta00] Stadelmeyer, P.: On the Computational Complexity of Resolving Curve Singularities and Related Problems. Ph.D. thesis, RISC-Linz, J. Kepler Univ. Linz, Austria, Techn. Rep. RISC 00–31 (2000)

[TzW89] Tzanakis, N., de Weger, B.M.M.: On the Practical Solution of the Thue Equation. Journal of Number Theory; **31(2)**, 99–132 (1989)

[VaW70] van der Waerden, B.L.: Algebra I and II. Springer-Verlag, New York (1970)

[VaH94] van Hoeij, M.: Computing Parametrizations of Rational Algebraic Curves. Proceedings of the 1994 International Symposium on Symbolic and Algebraic Computation, von zur Gathen J. (ed.); 187–190. ACM Press, New York (1994)

[VaH97] van Hoeij, M.: Rational parametrization of curves using canonical divisors. Journal of Symbolic Computation; **23**, 209–227 (1997)

[vGG99] von zur Gathen, J., Gerhard, J.: Modern Computer Algebra. Cambridge University Press, New York (1999)

[Wal50] Walker, R.J.: Algebraic Curves. Princeton Univ. Press (1950)

[Win96] Winkler, F.: Polynomial Algorithms in Computer Algebra. Springer-Verlag, Wien New York (1996)

[Zar39] Zariski, O.: The reduction of singularities of algebraic surfaces. Annals of Mathematics; **40**, 639–689 (1939)

[ZaS58] Zariski, O., Samuel, P.: Commutative Algebra I, II. Springer-Verlag, New York Heidelberg Berlin (1958)

[Zip91] Zippel, R.: Rational Function Decomposition. Proceedings of the 1991 International Symposium on Symbolic and Algebraic Computation, Watt S.M (ed.): 1–6. ACM Press, New York (1991)

Index

adjoint curve, 122
affine change of coordinates, 17, 28
affine geometry, 28
affine plane algebraic curve, 16
affine rational parametrization, 89
affine rational parametrization in
    reduced form, 89
algebraic curve, 16, 19
algebraic set, 256
algebraic transform, 71
algebraic variety, 256
algebraically optimal parametrization,
    151
analytic polynomial, 217
analytic rational function, 218
associated equation to Legendre
    equation, 163

Bézout's Theorem, 36
base point, 41
birational isomorphism, 31
branch of a curve, 55

CASA, 239, 255
center of a local parametrization, 53
center of a place, 54
character of a point, 18
components of a bivariate complex
    polynomial, 217
components of a bivariate complex
    rational function, 218
conic, 16
coordinate ring, 24
critical point, 205

cubic, 16
curve $\mathbb{R}$-normally parametrized,
    227
curve normally parametrized, 201
curve parametrizable, 89
curve parametrizable by a linear system
    of curves, 120
curve parametrizable by lines, 118

defining polynomial of a curve, 16, 19
degenerate parametrization, 228
degree of a divisor, 68
degree of a curve, 16
degree of a rational mapping, 32
degree of a rational parametrization,
    96
divisor, 41, 68
domain of definition, 29
dominant mapping, 32
double point, 16

effective divisor, 41, 68
equivalent local parametrizations, 53
Euler's Formula, 251

family of conjugate r-fold points, 79
family of conjugate points, 39, 78
field of definition, 150
field of parametrization, 150
field of rational functions on a variety,
    28
formal Laurent series, 51
formal power series, 51
formal Puiseux series, 52

general position, 47
generating polynomial of a family of
    conjugate points, 79
genus, 69
global parametrization, 59, 90
Gröbner basis, 253
ground field, 78

Hilbert's Basis Theorem, 253
Hilbert's Nullstellensatz, 256
Hilbert–Hurwitz Theorem, 152
homogeneous components of a
    polynomial, 16

Implicit Mapping Theorem, 253
implicitization problem, 108
inversion of a proper parametrization,
    95
inversion problem, 105
irreducible component of a curve, 16
irreducible curve, 16
irreducible local parametrization, 54

Lüroth's Theorem, 94, 188
Legendre equation, 157
Legendre's Theorem, 161
line, 16
linear space of a divisor, 69
linear system of curves, 41, 42
local parametrization, 53
local ring, 29

multiple point, 16
multiple component of a curve, 16
multiplicity of a point, 16
multiplicity of a tangent, 17
multiplicity of intersection, 36, 39

neighborhood graph, 74
neighborhood tree, 74
Noetherian ring, 26
nondegenerate parametrization, 228
nonordinary point, 18
nonsingular point, 16
normal parametrization, 201
normal parametrization over $\mathbb{R}$,
    227

optimal field of parametrization, 151
order of a Puiseux series, 52

order of a formal Laurent series, 51
ordinary point, 18

parametrizing by lines, 114
parametrizing field, 150
parametrizing with adjoints of degree d,
    141
parametrizing with adjoints of degree
    $d - 1$, 139
parametrizing with adjoints of degree
    $d - 2$, 138
pencil of curves, 41
place of a curve, 54
pole of a rational function, 29
polynomial curve, 195
polynomial function, 25
polynomial generator, 217
polynomial mapping, 26
polynomial parametrization, 195
positive divisor, 41, 68
projections, 27
projective change of coordinates, 28
projective geometry, 28
projective local parametrization, 53
projective plane algebraic curve, 19
projective rational parametrization, 90
proper linear subsystem, 170
proper parametrization, 95
proper reparametrization problem, 188
Puiseux's Theorem, 56

quadratic residue, 157
quadratic transform, 71
quadratic transformation, 71

r-fold point, 16
ramification points, 33
rational curve, 89
rational function generator, 218
rational linear subsystem, 170
rational mapping, 30
rational point, 150
real affine plane curve, 209
real projective plane curve, 209
Recio T., 206
reducible local parametrization, 54
regular isomorphism, 26
regular mapping, 26
regular position, 80

regularity of a rational function, 29
regularity of a rational mapping, 30
resultant, 34, 254
Riemann's Theorem, 69

set of degenerations, 228
simple point, 16
singular point, 16
singular locus, 70
singularity, 16
standard Cremona transformation, 71
standard decomposition of the
     neighborhood graph, 85
standard decomposition of the singular
     locus, 83
standard quadratic transformation, 71
strongly degenerate parametrization,
     228

support, 68
system of adjoints, 122

tangent, 17, 20
Taylor's Theorem, 253
tracing index, 101
triple point, 16

value of a rational function at a point,
     29
varieties birationally isomorphic, 31
varieties regularly isomorphic, 26

weakly degenerate parametrization, 228

Zariski topology, 256

Table of Algorithms

- Algorithm GENUS 76
- Algorithm STANDARD-DECOMPOSITION-SINGULARITIES 83
- Algorithm TRACING INDEX 103
- Algorithm INVERSE 107
- Algorithm CONIC-PARAMETRIZATION 115
- Algorithm PARAMETRIZATION-BY-LINES 116
- Algorithm PARAMETRIZATION-BY-ADJOINTS 133
- Algorithm SYMBOLIC-PARAMETRIZATION-BY-DEGREE-d-ADJOINTS .. 142
- Algorithm HILBERT–HURWITZ 153
- Algorithm ASSOCIATED LEGENDRE SOLVE 168
- Algorithm LEGENDRE SOLVE 169
- Algorithm OPTIMAL-PARAMETRIZATION 181
- Algorithm PROPER-REPARAMETRIZATION 193
- Algorithm POLYNOMIAL-REPARAMETRIZATION 199
- Algorithm NORMALITY-TEST 205
- Algorithm NORMAL-PARAMETRIZATION 206
- Algorithm REAL-REPARAMETRIZATION 224
- Algorithm DEGENERATIONS 229
- Algorithm REAL-NORMAL-PARAMETRIZATION 234